PRINCIPLES OF DIFFUSE LIGHT PROPAGATION

Light Propagation in Tissues with
Applications in Biology and Medicine

PRINCIPLES OF DIFFUSE LIGHT PROPAGATION

Light Propagation in Tissues with Applications in Biology and Medicine

Jorge Ripoll Lorenzo

Foundation for Research and Technology — Hellas, Greece

World Scientific

NEW JERSEY · LONDON · SINGAPORE · BEIJING · SHANGHAI · HONG KONG · TAIPEI · CHENNAI

Published by

World Scientific Publishing Co. Pte. Ltd.

5 Toh Tuck Link, Singapore 596224

USA office: 27 Warren Street, Suite 401-402, Hackensack, NJ 07601

UK office: 57 Shelton Street, Covent Garden, London WC2H 9HE

British Library Cataloguing-in-Publication Data
A catalogue record for this book is available from the British Library.

PRINCIPLES OF DIFFUSE LIGHT PROPAGATION
Light Propagation in Tissues with Applications in Biology and Medicine

ISBN-13 978-981-4293-76-1
ISBN-10 981-4293-76-8

Printed in Singapore.

To Alicia

Foreword

Probing disease with light is the oldest and still most wide-spread clinical examination method world-wide. Every single day at every hospital in the world, doctors will 'look' at patients to see discolorations and anatomical changes, they will shine light on them with endoscopes or surgical lamps and they will transilluminate them with pulse oximetry devices to obtain blood saturation measurements or to examine the lung of premature newborns. The familiarity of humans with optical imaging usually hides the fact that a significant part of clinical decision making today is still performed with an 'optical imaging' method. It is often believed that whereas a physical examination may be based on a patient visual inspection, more advanced radiological imaging eventually contributes to accurate diagnostics. However, from skin inspection to gastro-intestinal endoscopies, colposcopies and surgical procedures, optical imaging remains a primary diagnostic and theranostic method. Similarly, within the realm of photo-dynamic therapy, light is also employed therapeutically.

In contrast to human vision or photography typically based on light reflected from surfaces, optical imaging of tissues is a significantly more complex procedure which at times can become highly confusing. This is because tissue is not a perfect light reflector; instead, depending on the wavelength employed, a significant proportion of the incident light will distribute in tissues. This portion of light will encounter cellular organelles and interfaces which will change the original direction of propagation leading eventually to light diffusion. Part of this light will encounter absorbing molecules such as hemoglobin or melanin or will excite natural fluorochromes such as NADH or structural proteins. Some will eventually pass through a detection device such as a lens, or a fiber interface. These natural and seemingly straightforward procedures contribute to a physical system that can be very perplexed. The macroscopic appearance of these processes leads to a non-linear problem that, to the experience of many experimentalists, presents multiple challenges when it comes to quantifying such measurements, generating images with them and more importantly leading to accurate medical interpretation. Addressing these challenges requires a thorough theoretical understanding of light propagation in tissues. Using quantitative models of light propagation optical imaging of tissue can gain in terms of diagnostic and theranostic potential, beyond the one achieved today through simple photographic inspection of tissues.

It is when an inspired theoretical physicist in optical diffusion spends many years with the capricious experimental intricacies of diffusive measurements and tomographic reconstructions that you get a book like the one you read now: an account of light propagation in tissues from the view point of the measurement. This book is an account of a large volume of knowledge that has been developed over decades on scattering theory and bio-photonics, including the authors own contributions. However it does not only serve the purpose of presenting the theoretical abstraction but also prioritize this knowledge and present it from the view point of the experimental relevance. I can only simile the development of the book herein in the context of the phenomena it describes, a tortuous move of the author within complex optical definitions and matching experimental measurements. The result is a rigorous theoretical presentation and analysis of light propagation and diffusion that carries the weathering of experimental application of the concepts described to match real-life measurements. Unequivocally a most comprehensive account of diffusion theory for bio-optical imaging; not only for the theorist but more importantly for the experimentalist. I wish I had this book in 1995!

Vasilis Ntziachristos

Preface

I will never forget the moment when, in the context of a meeting, I was with Simon Arridge and mentioned I was writing a book. He smiled and told me a joke, which on general terms sounded something like: 'two scientists meet at a conference and one says to the other "I'm writing a book" and the other one answers "me neither"'. I have to admit that I did not find it that funny. At first. As the months became a year and my book was not even half-way through, I started to panic and thought how exquisitely accurate this joke was, and have been laughing at it ever since. I guess — hope, rather — that this is what happens to all on their first book, where you realize time is not available in the quantities you expected it to be.

Light propagation in biological tissues, and in particular, applications of light diffusion, is a relatively new field that during the past two decades has been causing a strong impact in the fields of medicine and biology, mainly due to its capabilities to image *in-vivo* in small animals and humans. Due to the fact that most of the on-going research in this area is applied, there are very few basic theoretical works in the area of light diffusion. On the other hand, there are numerous peer-reviewed papers dealing solely with numerical simulations or numerical solutions of complex problems applied to diffuse light imaging. This has caused a gap between the basic theory and application, which I hope that this book can cover at least partially. I believe a basic book oriented towards providing the post-graduate student with a strong foundation of the principles of light propagation in biological media is still lacking, showing how several simple, yet extremely useful problems may be approached using analytical expressions.

The main idea behind this book is to present a rigorous derivation of the equations that govern light propagation in highly scattering media, with a particular emphasis in their application in imaging in biology and medicine. I have included the basis for modeling diffuse light propagation starting from the very beginning and have presented, hopefully with enough detail, all the steps leading to the final expressions. The way I conceived this book was so that it could be read from beginning to end, each chapter building on the formulas and knowledge gained in the previous chapters. Even though I have tried hard for this book to be as enjoyable to read as possible, I understand that it takes great skill to manage this, so beforehand my apologies if some (hopefully not all) areas of the book are too dense. So as to make clear what contributions from each chapter are the most relevant, I have included two ways of highlighting those important parts: 'building blocks' and 'important notes'. Building blocks, as their name clearly implies, are the equations and expressions we will be using to arrive to the diffusion equation and to study diffuse light propagation. Important notes refer to issues which are important and are usually not that well-known.

This book provides the basic framework to work with reflection and transmission of diffuse waves at interfaces and their behavior at non-scattering interfaces, in a manner very similar to that taught to undergraduates when working with Electromagnetic propagation (see Born and Wolf (1999), for example). It explains in detail what approximations are taken when deriving the diffusion approximation and their implication, both from a fundamental physics point of view and with respect to their biological and medical applications. It begins with an overview of the problem (Chapters 1 and 2) and thorough step-by-step derivations until the diffusion equation is reached (Chapters 3 to 5). From then on, it concentrates on the basis of diffuse light propagation (Chapters 6 and 7), diffuse light at interfaces (Chapter 8), the ill-posed nature of diffuse waves (Chapter 9), and briefly presents some approximations to solve the inverse problem with analytical expressions (Chapter 10). Since my main concern was to explain in detail the basics of diffuse light propagation, and the subject of solving the inverse problem is such an extensive one, I have unfortunately dedicated very little space to this issue. I hope, however, that enough material and references are presented so that those willing to approach the inverse problem have a solid starting point.

Most of what is explained in this book is in the context of the new emerging field of optical molecular imaging, to which imaging with diffuse light became an integrating part mainly thanks to Vasilis Ntziachristos, who was the first to publish the use of diffuse light to image activatable fluorescent probes *in-vivo*. This technique, which he termed Fluorescence

Molecular Tomography (FMT), makes use of the principles developed for diffuse optical tomography and applies them to image molecular function by using fluorescent probes. In this sense, the physical governing principles are equivalent to fluorescence diffuse optical tomography (fDOT), however it is the whole concept and application to imaging molecular function which was a breakthrough and clearly distinguishes FMT from fDOT as a molecular imaging modality.

Since this book concentrates on the basics of diffuse light propagation, I have opted for not including too many citations and refer the reader, when possible, to complete and thus more comprehensive volumes such as books or Thesis dissertations. In doing so I understand many peer-reviewed papers that have been very important in this field have been omitted. To those that feel that their contribution is not well represented, my sincere apologies. Additionally, since this book is almost entirely based on formulas, I am certain that some errors might have sneaked through my radar. I would be very grateful if you inform me if one of these unfortunate enemies is detected.

One of the things I would like to single out from this book is the explicit use of units. I remember being grilled in college on the importance of units, and I have come to understand why. All quantities, when presented for the first time in each chapter, are always introduced with the appropriate units, and whenever writing any expression I always check to see if the units are correct. If units are incorrect, there is an error in the formula, plain and simple. The reason I go over such an obvious matter is that lately a vast majority of peer-reviewed papers published (at least in our field) have expressions with different units on each side of the equation. I hope this book at least contributes to establishing what units are relevant in diffuse light propagation.

With regards to the technicalities of this book, all figures were generated with Inkscape, an open source vector graphics editor, http://inkscape.org; Google Sketchup, a freeware 3D modeling program, http://sketchup.google.com; and Octave 3.4.2, an open source high-level interpreted language intended for numerical computations http://www.gnu.org/software/octave/, was used to generate all the data. This book was written in LaTex using TeXworks, a TeX front-end program, http://www.tug.org/texworks/.

Among the many people I am indebted for help in the preparation of this book, I would like to single out Juan Aguirre for thoroughly studying and commenting on the formulas and explanations in the first four chapters of this book; my father, Pedro Ripoll, for reading and editing the whole book — I can now say that at least two people have read all of it — and my mother, Elena Lorenzo, for formatting and editing; Florian Stuker for his useful comments; Eleftherios Economou, Juan José Saenz and Manuel Nieto-Vesperinas for answering questions regarding the physics and equations in the second chapter on basic electromagnetism; and above all Alicia Arranz for not only editing parts of this book but for her endless help and support. Without her this book would have not been possible.

There are also several researchers from whom I have learned much and have been a direct influence in the material that follows, in particular Manuel Nieto-Vesperinas, Eleftherios N. Economou, Simon Arridge, Arjun Yodh, Rémi Carminati, Frank Scheffold, Juan José Saenz and especially Vasilis Ntziachristos, with whom most of the formulas presented in this book were applied experimentally and who has always been the best at finding both interesting and useful problems. Even though they are completely unaware of it, two great Bills were of great help in the long hours until this book's completion: Bill Evans, for providing the perfect background, and Bill Bryson, for providing the best of distractions and also, the most enjoyable of scientific references (if you have not read 'A short history of nearly everything', I can't recommend it enough).

I would also like to gratefully acknowledge the support of the Institute for Electronic Structure and Laser, from the Foundation for Research and Technology-Hellas, where most of this book developed, and Markus Rudin for hosting me at the ETH-Zürich where the final chapters were written.

Jorge Ripoll Lorenzo
November 2011

Contents

Part II: Diffuse Light

PART 1

Light Propagation in Tissues

Chapter 1

Light Absorbers, Emitters and Scatterers: The Origins of Color in Nature

Summary. In this first introductory chapter the basic origins of color in nature will be presented, separating color originating from the emission of light from color generated by light interacting with materials through absorption and scattering. Within these three categories the main concepts and phenomena that will be the basis of this book will be introduced, such as absorption, scattering, and luminescence. Finally, the role of biological light emitters such as fluorescent or bioluminescent systems will be discussed in the context of the novel field of 'Optical Molecular Imaging' which is the framework of application of most of the formulas presented in this book.

1.1 Introduction

Color is all around us, and even though most of the concepts that will be covered by this introductory chapter are common knowledge, I believe it is very useful to go over them in order to understand the real physical meaning of the dense formulas that will follow in the next chapters and their implication in describing the origins of color in nature. One could rightfully argue the need for such complex formulation to describe something we all perceive without the need of understanding surface integrals, energy conservation, or Maxwell's equations, to name but just a few. However, it is this same basic fact of color being so present in our everyday lives that makes the use of light in biology and medicine so attractive: we can inspect, analyze and diagnose without the need of advanced equipment, although with the aid of one of the most complex optical imaging instrument, the human eye. Of course, there are limitations to what can be done through this approach, and here is where the need for the field of Bio-Optics emerges.

To begin with, simple visual inspection is subjective and is very difficult to quantify. Additionally, visible light loses its original direction as it travels through tissue due to scattering and is also significantly absorbed, making inspection of deep lesions in tissue almost impossible for the naked eye.

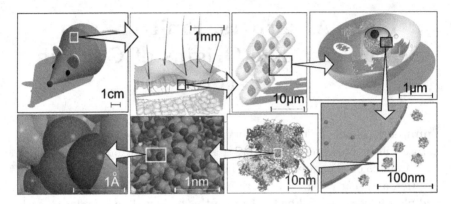

Fig. 1.1   Different scales present in tissue, adapted from Alberts *et al.* (2009). These should be compared with the wavelengths of light typically used in biology, which are in the 400nm-1000nm range. Light incident in tissue will interact at all these levels.

In this chapter we will consider the interaction of light with matter and how this introduces scattering, absorption and light emission. In order to model light propagation in tissue, it is very helpful to have a clear picture of the different sizes light encounters while traversing tissue, and how these scales affect light that propagates within it. In Figure 1.1 we show the different scales from a living animal, to living cells, down to the ribosome and finally to the single atoms that compose the ribosome. For comparison, Table 1.1 and Table 1.2 show the range of visible/near infra-red wavelengths and typical sizes present in biological structures. Electromagnetic radiation will interact with the molecules present in tissue at all levels, starting from the atomic level (if, of course, light managed to reach those molecules). If we can accurately describe how light travels within tissue, then it will also be possible to predict measurements taken at the surface and therefore quantify and produce estimations of the spatial distribution and properties of the lesion or compound we are interested in. This is the main philosophy behind biomedical optics: the use of light to *non-invasively*[1] retrieve information of a compound or lesion present in tissue. It is this philoso-

---

[1]Surgical procedures are divided into non-invasive, minimally invasive, and invasive procedures. Typically, a non-invasive procedure is one that does not require an incision.

phy, similar to remote sensing and transport in many ways, that has driven the need for the formulas that will follow throughout this book. As we will see in following chapters, the exact distribution and composition of all molecules in a tissue is not necessary to obtain most — if not all — of the information needed for current biomedical applications, and a statistical description is in most cases satisfactory.

Table 1.1   Wavelengths and Frequencies in the visible and near-infrared

Color	Frequency	Wavelength
violet	668-789 THz	380-450 nm
blue	630-668 THz	450-475 nm
cyan	606-630 THz	475-495 nm
green	526-606 THz	495-570 nm
yellow	508-526 THz	570-590 nm
orange	484-508 THz	590-620 nm
red	400-484 THz	620-750 nm
near infra-red	214-400 THz	750-1400 nm

Table 1.2   Characteristic Sizes of Biological Structures

Structure	Average Size	With respect to wavelength [a]
Dermis thickness[b]	0.2 - 3 mm	$\sim 1000\lambda$
Epidermis thickness	0.05 - 0.13 mm [c]	$\sim 100\lambda$
Fat cell diameter	80 - 120 μm	$\sim 100\lambda$
Cell diameter (on average)	30 μm	$\sim 40\lambda$
Mitochondria length	1 - 2 μm	$\sim \lambda$
Ribosome width	20 - 30 nm	$\sim \lambda/20$
Microtubule width	25 nm	$\sim \lambda/30$
Actin filament width	5-9 nm	$\sim \lambda/100$

[a] The far-red wavelength of $\lambda = 700$nm was used for comparison.
[b] Hypodermis thickness is not included since it varies immensely throughout the body.
[c] In palms and soles can amount up to 1.5mm.

As a mean of introducing main concepts such as sources, absorption and scattering, this chapter approaches the origins of color in nature by assuming three categories: *light absorbers* (or light 'sinks', if they may be called so), *light emitters* (light sources), and *light scatterers* (where the interaction of light with matter does not involve a loss of radiant energy and occurs at the same frequency). By light emitters we will refer to all the different mechanisms and processes generating visible light in nature, leaving it to the light 'sinks' to absorb part of that visible light. In some cases this absorbed energy will be re-emitted in the form of visible light, as in the case of fluorescence and phosphorescence. It will be the interaction of the radiation emitted by the light sources with the medium it traverses what will yield the colors we see or measure[2]. This interaction will consist on two main processes: radiative (scattering and emission) and non-radiative (absorption). It is scattering that can selectively change the propagation direction, polarization and/or phase of some wavelengths and cause interference patterns, producing for example iridescence in abalone or a soap bubble, dispersive refraction such as in rainbows and halos, the blue color of the sky due to multiple scattering, or diffraction effects such as in glories and opals. On the other hand, it is the absorption present in the medium that selectively robs the incident light from specific spectral components, by exciting molecules and transforming that energy into vibrations (in the form of heat, for example), by re-emitting this energy in the form of light of a different wavelength (as in fluorescence and phosphorescence) or by inducing a chemical reaction (as in photoreduction and photooxidation, the triggering reactions of photosinthesis)[3].

By dividing the origins of color in nature into these three basic categories I am considering 'scattering' as the root cause for all electromagnetic wave interaction with matter as long as no absorption is involved[4]. This division fits the purpose of this book; a more rigorous classification of the causes of color can be found in Nassau (1983). In his book Nassau presents a total of fifteen causes of color, which are included in Table 1.3 for completeness.

Taking a careful look at Nassau's classification we can see that 'scattering' is grouped under the category 'Geometrical and Physical Optics' and refers only to the effect of light changing its original direction upon interaction with a particle or group of particles. It must not be confused with the more general use of 'scattering' used in this book, which includes all the effects present under this category (dispersion, interference and diffraction) — more on this in Chapter 2. Also, we can see that 'absorption' is not present on its own in Nassau's description. This is because many of the

---

[2]In order to be rigorous we should also consider the spectral response of the detector. This response should be accounted for if one wishes to obtain a quantitative measurement.

[3]We could also include here the conversion of light into an electric current through the photovoltaic effect, or the ejection of electrons upon absorption of light of sufficient energy via the photoelectric effect.

[4]This will be the main subject of Chap. 2 and a more detailed description will be given there.

Table 1.3   Examples of the Fifteen Causes of Color, following [Nassau (1983)]

Vibrations and Simple Excitations
  1. Incandescence (flames, lamps, carbon arc, limelight)
  2. Gas Excitations (vapor lamps, lightning, auroras, some lasers)
  3. Vibrations and Rotations (water, ice, iodine, blue gas flame)

Transitions Involving Ligand Field Effects
  4. Transition Metal Compounds (turquoise, many pigments, some fluorescence, lasers and phosphors)
  5. Transition Metal Impurities (ruby, emerald, red iron ore, some fluorescence and lasers)

Transitions between Molecular Orbitals
  6. Organic Compounds (most dyes, most biological colorations, some fluorescence and lasers)
  7. Charge Transfer (blue sapphire, magnetite, lapis lazuli, many pigments)

Transitions Involving Energy Bands
  8. Metals (copper, silver, gold, iron, brass)
  9. Pure Semiconductors (silicon, galena, cinnabar, diamond)
  10. Doped or Activated Semiconductors (blue and yellow diamond, light-emitting diodes, some lasers and phosphors)
  11. Color Centers (amethyst, smoky quartz, some fluorescence and lasers)

Geometrical and Physical Optics
  12. Dispersive Refraction, Polarization, etc. (rainbow, halos, sun dogs, green flash of sun, 'fire' in gemstones)
  13. Scattering (blue sky, red sunset, blue moon, moonstone, Raman scattering, blue eyes and some other biological colors)
  14. Interference (oil slick on water, soap bubbles, coating on camera lenses, some biological colors)
  15. Diffraction (aureole, glory, diffraction gratings, opal, some biological colors, most liquid crystals)

---

effects presented in Table 1.3 can be involved both in the absorption and production of light as for example number 6, organic compounds, which are capable of absorbing light but also of producing it in the form of fluorescence, bioluminescence, or laser light emission. In this sense, we can consider causes 1 through 11 to be involved in the production and/or sequestration of light while causes 12 through 15 are involved solely on the scattering of light.

Since the main interest of this book is light propagation in biological media, it is very helpful to identify which of the fifteen different causes of color presented by Nassau are relevant to our purpose. After a brief introduction on the classical view of interaction of light with matter we will address how the causes of color described by Nassau might fit in the proposed scheme of light absorbers, emitters, and scatterers, placing particular emphasis on the optical properties of tissue. Finally, light absorbers, emitters, and scatterers are all placed into context in the field of Optical 'Molecular Imaging' an approach which makes use of engineered sources (probes) to obtain *in-vivo* information at the molecular level in small animals and humans where tissue is highly scattering.

1.2   The Classical Picture of Light Interaction With Matter

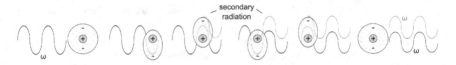

Fig. 1.2   Depiction of the interaction of an electromagnetic wave with matter through the Lorentz model, showing how energy is transferred from the incident radiation to the atom creating a dipole moment which re-radiates what is termed secondary radiation. Depending on its environment this dipole will react differently to the incoming radiation. Note that always both energy and momentum must be conserved.

In order to understand how we might model light propagation within tissue we must first understand the basic interactions of light with matter. It will be the collective interaction of the atoms in molecules and proteins with the incident light that will give rise to macroscopic quantities that can be measured such as the scattering and absorption cross-sections. A classical picture of this interaction is shown in Fig. 1.2 which depicts how electromagnetic radiation can interact with an atom. Considering only classical effects[5], part of the incident radiation will transfer its energy to the atom, creating a dipole moment through charge displacement. In this state of higher energy the classical picture depicts the dipole oscillating with the incident field in a manner which depends on its dipole moment and its environment. This is the *Lorentz model*, first developed by H. A. Lorentz in the early 1900s, and is one of the most useful models in electromagnetic theory. What this model predicts is that the oscillating dipole will generate a secondary radiation: if the surroundings of the dipole permit it to react to the incoming field it will re-radiate at the same frequency and scatter

---

[5]As will be mentioned several times throughout this book, the concept of photon and the quantization of light will be eluded and the classical picture of electromagnetic waves will be used.

light. The energy distribution of this secondary radiation will depend on the dipole moment generated. At the same time, part of the energy transfered to the dipole might be lost by non-radiative coupling of the dipole to the surrounding medium and delivered as vibrations (heat), in which case we would have absorption. Additionally, if the energy delivered to the atom is enough to place it in an excited state, the atom or molecule also has the possibility to re-radiate this energy but this time the frequency and direction will depend on the final dipole transition from the excited to the ground state, resulting in emission. In whatever way this atom gives out the energy that was obtained from the incident radiation two quantities must always be conserved: the total energy and the total momentum. As will be briefly shown in our derivation in the next chapter, rigorously keeping track of either of these is not an easy task when dealing with light interaction with matter.

The amount of each of these mechanisms (scattering, absorption and emission) present in the interaction with the atom depends on its environment. Assuming this environment is homogeneous, it is defined macroscopically through its conductivity (a measure of how well electrons propagate within this medium), magnetic permeability (a measure of how the material reacts to a magnetic field) and electric susceptibility (a measure of how the material is susceptible to polarization under an external electric field)[6]. It is this environment that will determine the dipole moment orientation and if there are any excited states accessible, or if there is non-radiative coupling with the surroundings so that it can give part of its energy as phonons (vibrations). If the environment of the atom is such that it permits re-radiation (i.e. the dipole is capable of aligning itself and in some manner respond to this incident radiation) we will have a re-radiated (scattered) wave with a direct phase relation to the incident field. However, if there are excited states accessible, with this extra energy received it might reach one of these states which will confer the atom or molecule a different vibrational/rotational state. Depending on the lifetime of this state given by the environment of this atom or molecule it will either de-excite through emission of radiation or through coupling with the medium (non-radiative de-excitation). This emission of radiation will also have an associated dipole moment which will re-radiate completely in phase and direction with the

---

[6]It is the relationship between the conductivity σ, the permeability μ and the susceptibility ξ with the electric and magnetic fields that determine what are called the *constitutive relations*, name which reflects the relationship between the electromagnetic fields and the constituents of matter.

incident field and with the same frequency, as in stimulated emission radiation, or with lower frequency as in fluorescence[7] or phosphorescence, in which case there is no (at least apparent) phase or moment relationship with the incoming radiation. In any case, we must remember that the total energy is conserved as long as we account for all energy lost through non-radiative processes, including changes in the momentum which have been transfered to the surrounding medium. It will be the collective response to this incident radiation of all constituents present that will yield the overall absorption, scattering and emission properties that will be presented in this chapter.

1.3   Light Absorbers in Nature

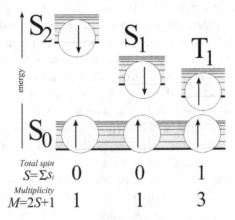

Fig. 1.3   Example symbolizing the distinction between singlet and triplet states, including total spin quantum numbers S and multiplicities M.

Once we have a clearer picture on how the basic constituents of matter interact with electromagnetic radiation, I believe absorption is a good start for our description of the origins of color in nature since light absorption is also present in some of the light producing mechanisms such as fluorescence

---

[7]As an example on the effect that the environment has on the radiative and non-radiative transitions, I recommend reading the works by R. Carminati and co-workers [Pierrat and Carminati (2010); Froufe-Pérez *et al.* (2007); Froufe-Pérez and Carminati (2008)] on this subject, where they show quantitatively how the amount of scattering present in the medium (i.e. the coupling of the dipole with the medium) affects the rate of radiative and non-radiative de-excitation directly.

and lasing. As mentioned in the previous section, in order to understand the mechanisms involved in the absorption of light by a collection of particles we need first to understand the interaction of light with each of the individual molecules that compose this ensemble of particles. Each of these molecules will have different excited states which can be reached by promoting an electron from an orbital in the ground state to an unoccupied orbital in the excited state. This electronic transition takes place through the absorption of the precise amount of energy that the molecule needs to reach that specific unoccupied orbital, and therefore it is discrete in nature. In its relaxed state we may assume that most electrons in the molecule are in their ground state, which would be the configuration expected at room temperature[8].

Depending on the energy of the incident light, when one of the two electrons of opposite spins is promoted from the orbital in the ground state to an orbital of higher energy, the transition involved will only be permitted if the total spin quantum number ($S = \sum s_i$, with $s_i = \pm\frac{1}{2}$) is conserved (see Fig. 1.3). This requires the electron promoted to maintain its spin. Since the multiplicity of a state is given by $M = 2S + 1$, if the total spin quantum number is zero, $S = 0$, we are dealing then with a *singlet state*, and the transition is a singlet-singlet transition. However, there is the possibility for a molecule in its singlet excited state to undergo a conversion to a state where the excited electron has its spin reversed: in this case the total spin quantum number is $S = 1$ and therefore the multiplicity is $M = 3$, being thus termed a *triplet-state*, which will have lower energy according to Hund's rule[9]. Once in a triplet state, the molecule can undergo additional triplet-triplet transitions since these are spin-permitted. The conversion from the singlet to triplet state is termed *intersystem crossing*, which arises from a weak interaction between the wavefunctions of different multiplicities due to spin-orbit coupling. This leads to a small but non-negligible contribution which can be measured effectively. In fact, conversion from the S_1 state to the T_1 state is the basis of phosphorescence, which will be explained in the following section.

---

[8]This is not entirely true, as we will see when describing fluorescence, since a small number of electrons are in excited states causing an overlap between the excitation and emission spectra.

[9]Hund's rule states that every orbital in a subshell is singly occupied with one electron before any one orbital is doubly occupied, and all electrons in singly occupied orbitals have the same spin. This means that a greater total spin state will give higher stability to the atom and therefore a triplet-state will have lower energy than the singlet state it originated from.

Assuming we have the molecule in the ground state, absorption of energy will then trigger a series of events, depending on the energy of the transition. If the transition is permitted (i.e. an $S_1 \rightarrow S_0$ or $S_2 \rightarrow S_0$), there are several paths the excited molecule can take in order to reach the state of minimum energy, the ground state. These events are shown in the Perrin-Jablonksi diagram in Fig. 1.4, where we see that some of the energy can be lost to the system via *vibrational relaxation*, in order to reach the lowest vibrational level of that excited state. The system can also de-excite through a non-radiative transition between two states of the same spin multiplicity, termed *internal conversion*, which is more effective the smaller the energy difference between states. For example, the internal conversion between S_1 and S_0 shown in Fig. 1.4 would be significantly less efficient than that from S_2 and S_1, thus favoring de-excitation through other pathways such as fluorescence or phosphorescence. These other means of bringing the molecule back to the ground state will be described in the next section, since they are clearly part of the light emitting process. Finally, the system can de-excite by relaxing to the lowest vibrational level of an excited state and then, through *intersystem crossing*, undergo a conversion to a triplet state, from example from $S_1 \rightarrow T_1$. Since this triplet state has lower energy, it can easily de-excite through intersystem crossing and vibrational relaxation towards the ground state. This non-radiative transition $T_1 \rightarrow S_0$ is very efficient and is the reason phosphorescence does not occur more often. It is only in those cases where this non-radiative de-excitation is impaired, for example in a rigid medium or at low temperatures, that radiative de-excitation can be observed (remember that the $T_1 \rightarrow S_0$ transition is in principle forbidden, and that it occurs due to spin-orbit coupling). Characteristic times of the non-radiative processes are shown in Table 1.4.

 Note: **On non-radiative processes**

Even though we separate the radiative from the non-radiative contributions by stating that the non-radiative part participates on heating up the medium through vibrations, we must not forget that any charge acceleration or deceleration generates radiation. In this sense, vibrations themselves also produce electromagnetic radiation. It is the contribution of these vibrations to the emission of radiation in the infrared that enables night-vision through infrared cameras, for example. Since both the wavelength and characteristic times of thermal emission are more than an order of magnitude higher than those in the optical range, thermal emission is not included into the radiative contribution as defined in this book.

Table 1.4 Characteristic Times of Non-radiative Processes, from Valeur (2002)

Process	Characteristic Time (s)
Absorption	10^{-15} s
Vibrational Relaxation	$10^{-12} - 10^{-10}$ s
Intersystem Crossing	$10^{-10} - 10^{-8}$ s
Internal Conversion	$10^{-10} - 10^{-8}$ s

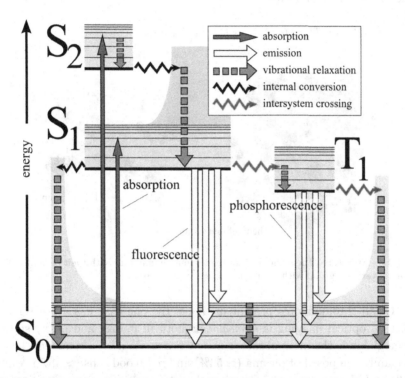

Fig. 1.4 Perrin-Jablonksi diagram illustrating the different pathways for de-exciting a molecule after absorption of radiation and excitation of an electron either to the first (S_1) or second (S_2) excited singlet state. Note that the transitions with different spin multiplicity such as $T_1 \rightarrow S_0$ are not allowed and occur due to spin-orbit coupling.

1.3.1 *Tissue Absorption*

Once we have described the basis of absorption, it is clear that different molecules will absorb different energies depending on the configuration of their excited states. When dealing with an ensemble of different molecules, such as a protein, each molecule will contribute to the absorption bands that this protein will present and the interaction between all components of the protein will also affect the bands present, permitting (in principle) a greater number of different pathways for excitation and de-excitation. Interaction of the protein with its surrounding environment (polarity, pH, pressure, viscosity, temperature, etc.) will also affect the way it is excited and de-excited[10]. A protein in solution, for example, will have greater chances of de-excitation through vibrational relaxation via collisions with the molecules of the solvent.

1.3.1.1 *Blood: The Main Absorber In Tissue*

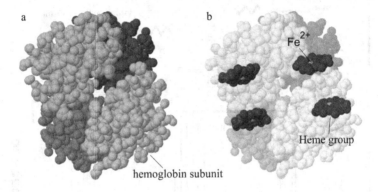

Fig. 1.5   Structure of hemoglobin showing its four subunits (a) with their corresponding heme prosthetic group with a ferrous iron Fe^{+2} atom (b).

Of all chromophores present in tissue, the one that has a greater impact on optical imaging is *hemoglobin*: both in its oxy and deoxy states, hemoglobin is the main absorber in soft tissue. As a brief reminder, blood is mainly composed of plasma ($\simeq 54\%$) and red blood cells ($\simeq 45\%$), white cells and platelets amounting to less than 1% each. The hematocrit represents the fraction of red blood cells present in blood, where we must bear in mind that the composition of red blood cells is primarily hemoglobin (approximately 97% of their dry mass). Plasma is composed mostly of water and plasma proteins, the latter accounting for less than 10% of the volume. Due to this reason plasma does not absorb visible light significantly, and even though plasma carries some oxygen dissolved, hemoglobin is the primary vehicle for transporting oxygen.

---

[10]In terms of what was mentioned in the previous section, the dipolar moments associated with these transitions depend on the enviroment of the dipole.

Hemoglobin is built of four subunits composed of a protein chain associated to a non-protein heme group (see Figure 1.5) which contains the ferrous iron (Fe^{+2}) to which oxygen binds. It is this heme group that gives blood its red color[11], being oxyhemoglobin bright red as opposed to a darker red exhibited by deoxyhemoglobin. Depending on the conditions, the oxygen bound to the hemoglobin might be released into tissue or absorbed from the tissue into the blood. Each hemoglobin molecule can hold up to four oxygen molecules and thus has a limited capacity. The amount of oxygen bound to hemoglobin at any given time is called the *oxygen saturation* (SO_2). Expressed as a percentage, the oxygen saturation is the ratio of the amount of oxygen bound to the hemoglobin (i.e. oxyhemoglobin concentration [HbO_2]) to the oxygen carrying capacity of the hemoglobin, which is given by the total hemoglobin present in blood:

$$SO_2(\%) = \frac{[HbO_2]}{[HbO_2] + [Hb]} \times 100 \qquad (1.1)$$

where [Hb] represents deoxyhemoglobin concentration. How much oxygen is bound to the hemoglobin is partly related to the partial pressure of oxygen in the surrounding environment of the protein. In areas with high partial pressure of oxygen such as in the alveoli in the lungs, oxygen binds effectively to the hemoglobin present. As partial pressure of oxygen decreases in other areas of the body, the oxygen is released since hemoglobin cannot maintain it bound with surrounding low oxygen partial pressures.

Considering that blood amounts to approximately 7% to 8% of our body weight it is clear why light absorption by blood plays a critical role in optical imaging in humans, mice, and vertebrates in general. Typical absorption curves for oxy and deoxyhemoglobin are shown in Figure 1.6, where it should be pointed out that these absorption features are due to the heme group of hemoglobin almost entirely [Itoh *et al.* (2001)]. In this figure the absorption spectrum of hemoglobin is presented in terms of the absorption coefficient, which is the product of the density of the molecule times its absorption cross-section. The absorption coefficient represents the distance light needs to travel to reduce by *e* its intensity, but more details on its physical meaning and implications will be presented in Chapter 2. From studying Figure 1.6 we can easily understand that blood appears red

---

[11]Most mollusks and some arthropods use a different protein to transport oxygen, hemocyanin, the second most popular protein for oxygen transport in nature. Hemocyanin uses copper to bind oxygen, and in this case their blood turns from colorless to blue when oxygenated.

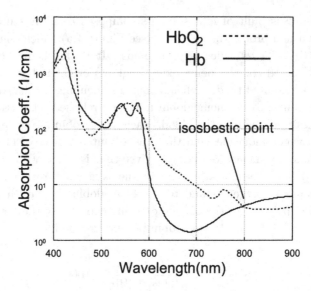

Fig. 1.6   Absorption coefficient of oxy and deoxyhemoglobin. The isosbestic point, i.e. the wavelength at which both oxy and deoxyhemoglobin have the same molar absorptivity, is approximately 800nm. Original data interpolated from Scott Prahl's webpage, http://olmc.ogi.edu/spectra.

due to absorption of the blue and green regions of the spectrum, and we see three orders of magnitude difference between the near infra-red and the blue region. This difference in absorption enables imaging of near infra-red fluorophores deep in tissue, and impairs significantly the use of visible proteins when imaging *in-vivo*. For this reason, new fluorescent proteins and fluorescent probes developed for *in-vivo* imaging applications are in the far-red or in the near infra-red region of the spectrum.

Even though absorption of light by hemoglobin is the main limiting factor for most *in-vivo* imaging applications, its strong absorption when compared to background tissue, combined with the difference in spectra of oxy and deoxyhemoglobin, provide a unique way of directly probing the amount of oxygen present in blood *locally*. This attractive feature was what triggered the development of biomedical optics: by imaging relative changes in blood absorption one can obtain information both on blood volume and oxygen saturation using Eq. (1.1). This is mainly done selecting two wavelengths on each side of the isosbestic point, which is the wavelength at which both oxy and deoxyhemoglobin present the same absorptivity and is approximately 800nm (see Figure 1.6). Measurements of blood volume and oxygen saturation are of great value since they provide information related to angiogenesis, oxygen delivery and vascularization, which can be used as indicators for diagnosis, monitoring and treatment of suspicious lesions.

Once we have established blood as the main absorber of visible light in tissue[12], the next important absorbers we need to consider for *in-vivo* optical imaging applications are water (due to its abundance in tissue), and the effect of skin (since light must traverse it to complete a measurement).

1.3.1.2 *The effect of Water*

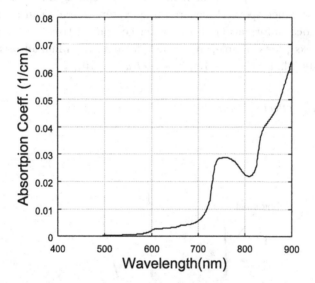

Fig. 1.7   Absorption spectra of water, measured in cm^{-1}. Data interpolated from Scott Prahl's webpage, http://olmc.ogi.edu/spectra. Note the differences in scale when comparing with the blood absorption curves in Figure 1.6.

Even though water does not absorb visible light significantly, since it accounts for approximately 60% of our total body weight its effects on the overall absorption properties of tissue are very important. Additionally, as we move towards the lower energies of the spectrum water shows a very sharp increase in absorption, peaking at approximately 3000nm where its

---

[12]Another absorber commonly used when fitting for tissue optical properties is lipid concentration, which has not been included since its contribution to the overall absorption is lower than that of water. In case it needs to be accounted for, see [van Veen *et al.* (2005)].

absorption coefficient is seven orders of magnitude higher than in the visible[13]. A similar trend occurs in the ultraviolet range of the spectrum where water absorption is almost ten orders of magnitude higher than in the visible range. The absorption of water for the wavelengths of interest for *in-vivo* imaging is shown in Figure 1.7 where we see the sharp increase as we move towards the near infra-red. It is due to this sharp increase that biological tissue is completely opaque to terahertz radiation and the reason the new terahertz body scanners recently installed at some airports can only see through clothing.

Similarly to what was said in the previous subsection regarding the use of the different spectra of hemoglobin depending on its oxygenation state, the distinct spectral features that water presents can also be used to obtain local information on the water content of tissue. This approach has been used, for example, to obtain information on human breast tissue components [Shah *et al.* (2001); Busch *et al.* (2010)].

1.3.1.3   *The effect of Skin*

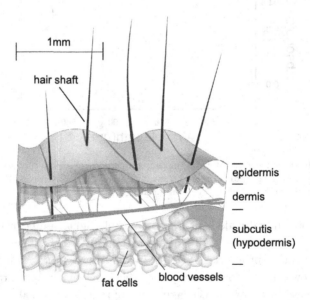

Fig. 1.8   Representation of the main layers of tissue present in skin.

---

[13]Compare this with the three orders of magnitude difference between blue and near-infrared light absorption of hemoglobin.

In any optical imaging approach which attempts to probe subsurface features, light needs to enter the tissue and propagate within it (how it propagates will be the main concern of the following chapters), before exiting and being collected by our detector. In doing so, it must traverse the skin twice if not more, at least in those cases where we are imaging at or deeper than the subcutaneous level. If one wishes to obtain information on the skin, one can measure the wavelength dependence of the reflected intensity which would include both specular and diffuse reflection, i.e. light that has propagated within tissue. In principle, this intensity spectrum contains information on those constituents with a distinguishable absorption spectrum that are present in the three layers of skin (epidermis, dermis, and hypodermis). Apart from the already mentioned absorbers in tissue such as blood and water, the epidermis is composed almost entirely of keratinocytes, that take up *melanosomes*, which are vesicles containing *melanin*.

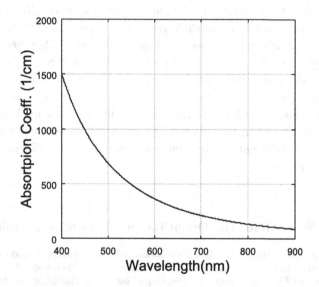

Fig. 1.9   Absorption spectra of melanosomes, measured in cm^{-1}. Data simulated following Steve Jacques' webpage on melanin at `http://olmc.ogi.edu/spectra`, following the formula $\mu_a(cm^{-1}) = 1.7 \times 10^{12}\lambda^{-3.48}$, where λ is the wavelength in nm.

Melanin, the pigment responsible of skin color, controls the amount of ultraviolet radiation from the sun that penetrates the skin. It is therefore a very strong absorber in the 300nm - 400nm, as can be seen from its spectrum, shown in Figure 1.9. Melanin protects the body by absorbing the harmful ultra-violet electromagnetic radiation and then releasing it to its surrounding environment via vibrational relaxation, i.e. heat. The values shown in Figure 1.9 have been obtained using the formula derived by Jacques (1996), $\mu_a(cm^{-1}) = 1.7 \times 10^{12}\lambda^{-3.48}$, where λ is the wavelength in nm. There are two types of melanin, eumelanin and phoemelanin, being skin color determined by the ratio of these two: the more eumelanin present, the darker the tone of the skin. However, the absolute concentration of melanin varies greatly from one subject to another, since the size of the melanosomes varies and so does their overall density.

Finally, an important absorber not to be forgotten is hair. Hair, apart from being a very good scatterer (hence the 'whiteness' of gray hair), is also a great absorber since it is the presence of melanin that gives it color. Pheomelanin colors hair red, while brown or black eumelanin will give either brown or black hair, and the presence of low concentrations of brown eumelanin will result in blond hair. It is for this reason that one of the best animal models for *in-vivo* optical imaging are *nude mice*. Nude mice are genetic mutants that have no thymus gland and thus exhibit an inhibited immune system due to the absence of mature T-cells. They have no hair and are very valuable in research because they do not reject many types of foreign tissues or xenografts. In those cases where nude mice cannot be used (as when we need a complete and working immune system), it is imperative to remove the hair in order to optically image them. Fortunately, hair in most cases can be removed but great care must be taken during the process in order not to cause unwanted skin reactions (for example, to depilatory creams).

 Note: **About the Quantitative Values of Absorption**

Throughout this section you might have noticed that all values presented deal with the main absorbers present in tissue, but the average absorption properties of tissue itself have not been presented. As a matter of fact, even though we do know the absorption spectra of most components present in tissue, their actual *in-vivo* value is for the most part unknown. We might 'assume' some parameters as an indication (i.e. we do know what organs contain more blood), but the truth of the matter is that we do not accurately know the optical properties of whole tissue *a priori*, even if we might know their anatomical distribution. Fitting for the actual *in-vivo* values is still a matter of research, and a quick search through the literature reveals huge discrepancies between the assigned values.

1.4   Light Emitters in Nature

Once we have covered how light is absorbed, we can move on to describing how light is actually generated, focusing mainly on how this takes place in living animals. The history of the understanding of the processes involved in the light production of living animals is a fascinating one and is a great example of truly inter-disciplinary research. Early records of 'living lights' appear in primitive mythology and they are mentioned in Chinese books dating as far back as 1500 BC. Luminescence from dead fish was known to Aristotle, and Pliny mentioned that of damp wood; however, it was not until the 1600's that serious research was devoted to it. The first of the seventeenth-century scientists to pay special attention to luminescence in general was a German priest, Athanasius Kircher (1602-1680). In his works, chapters were devoted to the 'light inherent in animals', the 'marvelous light of certain things that are born in the sea' and to the 'luminescence of stones'. He described in detail the light of fireflies, jellyfish and other luminous animals and attempted to answer questions such as: what is the use of light to a mollusc or a jellyfish? Soon later, in 1668 Robert Boyle using his air pump noted the dependence of the light production from both wood and fish on the amount of air present (even though oxygen was not known at the time). But it is not until the advent of the microscope that Henry Baker in 1742, author of two scientific books on microscopy, hints that this light production is due to living things. Very ingenuous experiments went on to prove the 'cold light' emitted from living organisms, measure their spectra using light emitted by candles as references, and to later disentangle the chemistry and organs involved in producing light. Thanks to the work of Raphael Dubois, Edward. N. Harvey and others during the last decades of 1800's and early 1900's, most of the basics of bioluminescence were established. This paved the road to the discovery of the principles governing the light production of the green fluorescent protein (GFP) which would result in O. Shimomura, M. Chalfie, and R. Tsien sharing the Nobel Prize in 2008. To put the state of research into perspective, I here include an excerpt from *Living Lights* by Charles Frederick Holder, published in 1887:

> *As we have seen, the light emitted by animals, plants, and*
> *minerals, of whatever cause, presents much that is mysteri-*
> *ous; and the problem of animal phosphorescence would seem*
> *no nearer being solved to-day than it was fifty years ago. This*
> *is perhaps due to a lack of study and investigation. A glance*

at the appended bibliography shows that much has been writ-
ten upon the subject; but it is only within the last decade that
serious work in this direction has been done, typified in the su-
perb work of Dubois, and the papers and monographs of the
other scientists mentioned. The naturalists of the 'Albatross',
the government exploring steamer, are to make investigations
regarding the luminosity of the Pacific, during the forthcoming
tour on the western coast. The French Academy of Sciences
offers this year a prize of three thousand francs for the best pa-
per upon animal phosphorescence. From this it would appear
evident that the phenomenon is creating renewed or increasing
interest, and in the following years will be the subject of much
study and investigation; and we may expect in the near future
to have not only its cause explained, but possibly to see a prac-
tical application of its possibilities to the wants of mankind.

Regarding the interest in the matter by the scientific community of the
time, soon after Holder A. S. Packard published in 1896 a paper entitled
The Phosphorescent Organs of Insects where he stated that "*the nature of*
the phosphorescent organs and their physiology has never seriously engaged
the attention of students in this country(referring to the USA)". Towards
the end he commented that "*it is self-evident that a microscopic observa-*
tion of the light of the glow-worm or fire-fly is not possible, but an animal
while giving out its light, or a separated abdomen, may readily be placed
under the microscope and observed under tolerably high powers". During
that time, other researchers like Carlo Emery started using the same light-
producing organs to view their structure under the microscope. This could
very well be the first experiments using luminescent probes as a source of
contrast, even though in these cases bioluminescence was expressed endogenously as opposed to current molecular imaging approaches.

On the other hand, studies on fluorescent and phosphorescent materials had been taking place for several centuries, starting from the famous
Bologna stone near 1602 (more on this soon). However, it is remarkable
that the use of fluorescence in biology came from works trying to reveal
the nature of the green light emitted by the jellyfish *Aequorea victoria*,

which was thought to be bioluminescent instead of fluorescent in origin[14]. It is very encouraging to see that the work started by Kircher, Dubois, Harvey and others in the late 1800's to unravel the mechanism governing bioluminescence culminated in the discovery of a protein and its applications, currently so important in biology, by the Nobel prize winners of 2008. These researchers have once more proven that basic research opens new doors and new applications which, sometimes, are completely unexpected and unimaginable.

In the previous section the basic mechanisms of light absorption where laid out. Now, the main interactions involved in light production that have a direct interest in biology and medicine will be presented. Of the divisions possible, I have chosen to separate light sources we use externally (basically to probe tissue optical properties or excite fluorescence) to light sources that are usually either expressed endogenously or administered externally but which emit light from within the subject. Depending on the nature of light emitted it will be either coherent (like a laser) or incoherent (like a halogen lamp), being this light additionally distinguished by the processes involved in its production. After a brief introduction on coherent and incoherent sources, the two main light producing mechanisms that are currently being used in biology and medicine will be presented: bioluminescence and fluorescence[15].

1.4.1   Coherent and Incoherent Light Sources

The following excerpt from E. N. Harvey's book on *The Nature of Animal Light* [Harvey (1920)] is a good start for understanding light sources:

> Although Dubois and Molisch have both prepared 'bacte-
> rial lamps' and although it has been suggested that this method
> of illumination might be of value in powder magazines where
> any sort of flame is too dangerous, it seems doubtful, to say the
> least, whether luminous bacteria can ever be used for illumi-

---

[14]And it actually somehow is: this jellyfish releases calcium ions, which bind to a bioluminescent protein called *aequorin*. The blue bioluminescence is then very effectively used by the green fluorescent protein to emit green light in the form of fluorescence. This is currently termed BRET (Bioluminescence Resonant Energy Transfer) and had researchers puzzled for quite a while.

[15]Note that there are other ways of producing light that are attracting some interest in biology and medicine. For example, the visible part of the spectrum emitted by the Cherenkov radiation of positron emitting compounds could be another form of light production with biological application.

nation. *Other forms, perhaps, might be utilized, but bacteria produce too weak a light for any practical purposes. The history of Science teaches that it is well never to say that anything is impossible. It is very unlikely that any luminous animal can be utilized for practical illumination, but there is no reason why we cannot learn the method of the firefly. Then we may, perhaps, go one step further and develop a really efficient light along similar lines.*

Following the simplistic picture of light interaction with a dipole shown in Figure 1.2, the nature of coherent and incoherent sources can be understood by assuming that once the atom or molecule receives enough energy to reach an excited state, the emitted light (if this radiative transition is allowed) will be coherent depending on the dipolar moment associated with the de-excitation transition. In other words, if many of the dipoles present in the excited medium are capable of re-emitting with a constant phase relation between them we will have strong coherence of the emitted light. Of course, a single atomic transition emits coherently, it is the *collective behavior* of these transitions which yields the level of coherence. Actually, a truly incoherent source is extremely difficult to obtain, since there is always some level of coherence even though it might be in the order of a few wavelengths. Note that coherence does not necessarily imply directionality, although it is very difficult to have one without the other, as would be the case of the laser. For example, the case presented by Harvey in the excerpt shown above represents a clear example of incoherent source: light produced through chemical reactions can hardly be coherent, since all associated dipole moments will point in all directions in space and will emit depending on the availability of the catalyst, i.e. out of phase. On the other hand, if we have an external field of appropriate energy in a medium which permits population inversion and we have sufficient atoms in the excited state we can collectively measure radiation in-phase from all these dipole transitions through stimulated emission. This is the basis of the laser, and it produces extremely coherent, directional and monochromatic light. A practical way of measuring the coherence of radiation is to check for speckle[16]: if wavefronts are correlated, strong interference effects such

---

[16]As a reminder, speckle are sharp peaks of intensity caused by interference effects of highly coherent light and are typically spread out randomly. Speckle dynamics can be used to measure dynamic quantities such as correlation times, or spatial quantities such as spatial correlation.

as speckle will appear. If they are de-correlated then we will measure an average intensity over which speckle is indistinguishable.

Another example is light from the sun, which is mostly incoherent, but still retains enough coherence to cause some speckle when used in a microscope[17]. Other examples of *relatively* incoherent sources are Light Emitting Diodes (LEDs) and filament lamps. Note, however, that in most microscopic applications fully *incoherent* sources are more suitable. In relation to the matters presented in this book, coherence plays an important role due to its *absence*. This will be introduced and hopefully clearly explained in Chapter 2 when dealing with multiple scattering, but it might be quickly presented this way: due to the fact that when shining light through tissues there is always a great deal of scattering from all different tissue components of different sizes and random orientations (see Figure 1.1) combined with the constant movement present in live tissue, there is a very quick de-correlation of scattered light from the incident light resulting in a series of wavefronts that are very strongly incoherent between them.

Apart from the coherence of a particular wavelength emitted from our source, the other important quantities to be considered are the emission spectrum and the spectral power distribution (power per wavelength, typically measured in *Watts/nm*). For all purposes of this book, our main consideration will be the use of monochromatic sources for excitation (a laser, a light emitting diode (LED) or a lamp with an appropriate filter), but we will need to take into account the spectral properties of the *emitted* light in the case of fluorescent or bioluminescent probes.

Depending on the process involved for excitation, emitted light can be grouped in the different categories presented in Table 1.5, which were obtained from Nassau (1983) and Harvey (1957). From the processes presented in this table, those relevant to biology and medicine that will be referred to in this book are marked with a star.

1.4.2   *Fluorescence*

As was mentioned previously, the first documented (or at least most famous) fluorescence was noted in 1602 and it came from a stone that when heated with charcoal gave out Bologna phosphorus, an impure barium sulphide which 'stored light' and gave it off in the dark. Similarly, John Canton's phosphorous (1768), prepared by heating oyster shells and sulphur,

---

[17]Don't forget either that Young's original experiments where performed with direct sunlight.

Table 1.5   The Various Forms of Light Emission

INCANDESCENCE Thermally produced black body or near-black body radiation [a]

LUMINESCENCE All Incoherent non-thermal light production

   Photoluminescence Luminescence induced by electromagnetic radiation, typically considering only the UV, visible or infrared region of the spectrum

   - FLUORESCENCE ✳ Rapid luminescence, with lifetimes in the $10^{-7} - 10^{-10}$s range
   - PHOSPHORESCENCE Persistent fluorescence, specifically from a triplet state with lifetimes in the $10^{-6} - 1$s range

   Resonance Radiation Immediate re-irradiation of same wavelength
   Cathodeluminescence Fluorescence induced by cathode rays (electrons)
   Anodeluminescence Fluorescence induced by anode rays (beams of positive ions)
   Radioluminescence[b] Fluorescence induced by energetic radiation of high energy particles such as X-rays or γ-rays
   Thermoluminescence Luminescence produced by raising the temperature

   - CANDOLUMINESCENCE Luminescence of incandescent solids emitting light at shorter wavelengths than expected, such as non-blackbody radiation from a flame [a]

   Electroluminescence Luminescence induced by an electric field or current
   Triboluminescence or *piezoluminescence*. Luminescence produced by a mechanical disturbance

   - SONOLUMINESCENCE Luminescence from intense sound waves in solution
   - CRYSTALLOLUMINESCENCE Luminescence produced during crystallization
   - LYOLUMINESCENCE Luminescence obtained from dissolving a substance, typically a crystal

   Chemoluminescence Luminescence accompanying a chemical reaction, whether in a gas, a vapor[a] or in solution

   -BIOLUMINESCENCE ✳ Chemoluminescence produced by living plants or animals

LASING ✳ Any form of coherent luminescence

---

[a] Also *pyroluminescence* in part.
[b] The word *radioluminescence* has often been used as a general term for any luminescence resulting from bombardment by various particles such as electrons, alpha particles (ions), etc., including X-rays and γ-rays. It is also sometimes called *roetgenoluminescence, ionoluminescence, radiography* and *scintillation*.
✳ All forms of light emission marked with a star will be described in this book and will be somehow involved in the modeling of light propagation within diffusive media.

was described by Hulme as the 'light magnet of Canton'. This phenomenon we now know is *phosphorescence* (persistent fluorescence, specifically from a triplet state — see Sec. 1.3). Kircher, with whom we came across in the previous section, was one of the first to thoroughly study what he termed 'the luminescence of stones' in the seventeenth century. However, it was not until 1852 that works by Sir G. Stokes revealed the nature of *fluorescence*, which due to its short lifetime ($10^{-7} - 10^{-10}$s) had been very difficult to detect. At the beginning of the nineteenth century interest in the nature of light rekindled mainly due to the discovery of a number of crystalline minerals (fluorspar, for example) which where found to have, during those times, unexplainable interactions with light. D. Brewster and J. Herschel in the 1840s attempted to explain the color of a beam of light passing through one of these crystals as 'scattering' or what they termed 'internal dispersion'. However, it was Stokes that characterized this light as true emission. After using several terminologies to describe this phenomenon ('true internal dispersion', 'dispersive reflection', amongst others) he finally hit on the term 'fluorescence' which owes its name from fluorspar just like opalescence is used to describe light interaction with opals. During his study Stokes noticed that in producing fluorescence, light is emitted at longer wavelengths. This characteristic of fluorescence emission is what currently is known as Stokes' law.

The processes involved in fluorescence and phosphorescence are shown in Figure 1.10. Going back to what was said in Sec. 1.3 regarding absorption, fluorescence will arise whenever we have a radiative transition from an excited singlet state to a singlet state of lower energy (typically this will always be the ground state S_0). However, if the environment of the molecule favors inter-system crossing we might have a change of spin and end up in a triplet state. Radiative de-excitation from the triplet excited state down to the singlet ground state is termed phosphorescence, and it is the fact that the $T_1 \to S_0$ is a forbidden transition that gives phosphorescence its nature, exhibiting very long lifetimes ($10^{-6} - 1$s). Since the mechanisms that enable inter-system crossing are not typically available in live tissue, phosphorescence will not be considered further in this book.

There are several features to be pointed out from the emission spectra shown in Figure 1.10. Regarding fluorescence, we see there is a shift between the absorption maximum and the fluorescence emission maximum termed 'Stokes' shift'. According to Stokes' law, the fluorescence emission wavelength should always be higher than that of absorption. However, in most cases the absorption spectrum partially overlaps the emission spec-

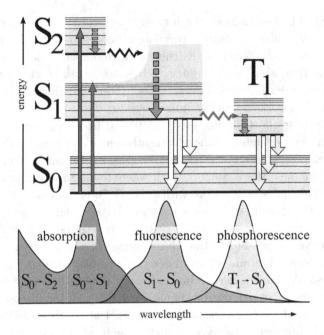

Fig. 1.10   Jablonski diagram for fluorescence and phosphorescence. Example spectra are shown for all possible transitions in order to point out the different shifts present in energies and the spectral overlaps.

trum (see Figure 1.10). This apparent contradiction of energy conservation was first explained by Einstein and it is due to the fact that at room temperature a small fraction of molecules are at vibrational levels higher than the zero level of the ground state, and at very low temperatures this partial overlap should disappear. A realistic example of this can be seen in Figure 1.11 where the absorption and emission spectra of Cy5.5 are shown. This particular example for Cy5.5 shows also clear contribution of higher accessible energy states which account for the 'hump' present in the 625nm range. In general, since the differences between the vibrational levels are similar for the ground and excited states the emission curve from $S_1 \rightarrow S_0$ is practically the $S_0 \rightarrow S_1$ absorption curve reversed and shifted to lower energies (higher wavelengths) by the *Stokes' shift*.

1.4.2.1   *Fluorescence Lifetime*

As we have seen excited molecules can have several pathways for de-excitation, giving out fluorescence emission in the case of a singlet-singlet

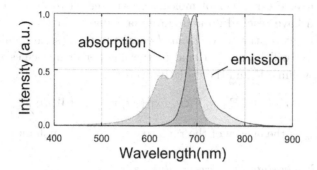

Fig. 1.11 Absorption and emission spectra of the cyanine dye Cy5.5, a commonly used fluorochrome for optical imaging.

transition. From these possible transitions we have seen that an emission spectrum arises. The next important quantity to be understood is how long after excitation does this emission process take place, i.e. what is the *lifetime* of this excited state. Note that the lifetime of the $S_1 \to S_0$ transition will be the same, independent of the specific vibrational mode accessed when reaching the ground state. In order to avoid more rigorous descriptions of the transitions involved in excitation and de-excitation which make use of Boltzmann's Law and Einstein's coefficients, we will introduce a simple definition of fluorescence lifetime as follows. Let us assume we have a collection of molecules that have a concentration $[C]$ in (molecules/cm^3). Let us assume we managed to excite $[C]_0^*$ of these to their excited S_1 state by using a very short pulse, much shorter than the time it took the molecules to absorb this incident energy. Ignoring any possible inter-system conversion, given a rate constant for radiative de-excitation k_r and a rate constant for non-radiative de-exictation k_{nr}, the concentration of molecules still in the excited state $[C]^*$ will change in time as:

$$\frac{\mathrm{d}[C]^*}{\mathrm{d}t} = -(k_r + k_{nr})[C]^*,$$

which can be straightforwardly solved to yield:

$$[C]^*(t) = [C]_0^* \exp(-t/\tau),$$

where we have used the initial condition $[C]^*(t=0) = [C]_0^*$ and introduced the lifetime τ of the excited state as:

$$\tau = \frac{1}{k_r + k_{nr}}.$$

Of course, the number of molecules that ceased to be in the excited state could have reached ground state via radiative or non-radiative processes. Therefore, the total *radiated* energy from these molecules during de-excitation will be proportional to the number of molecules inside the irradiated volume V and the radiative rate k_r:

$$P_{fl}(t) = E_{em} k_r V [C]_0^* \exp(-t/\tau), \qquad \left(Watts\right) \qquad (1.2)$$

where E_{em} is the energy of the radiative transition[18] in Joules.

1.4.2.2 *Steady-state Fluorescence Intensity*

Once we have defined the fluorescence lifetime of an excited state, for the purposes of this book it is convenient to understand how does fluorescence relate to the excitation intensity in the simple case of steady-state illumination. Let us consider the system of molecules mentioned before of concentration $[C]$ inside volume V. If we have a constant intensity U_0 in $Watts/cm^2$ incident on the system, the absorption energy rate due to the of incident radiation will be:

$$P_{abs} = U_0 \sigma_a V [C]_0, \qquad \left(Watts\right) \qquad (1.3)$$

where σ_a is the absorption cross-section for each molecule, and $[C]_0$ is the total number of molecules in the ground state. Given this absorbed energy, the total density of molecules that could have reached the excited state per unit time will therefore be:

$$\frac{d[C]_{0\to*}}{dt} = \frac{P_{abs}/V}{E_{abs}} = \frac{U_0}{E_{abs}} \sigma_a [C]_0,$$

where E_{abs} is the energy required for the transition and P_{abs}/V represents the absorbed power per unit volume. The factor $U_0 \sigma_a / E_{abs}$ can be seen as the effective *absorption rate*:

$$k_{abs} = \frac{U_0}{E_{abs}} \sigma_a, \qquad \left(s^{-1}\right)$$

The variation of concentration of molecules in the excited state $[C]^*$, accounting for those molecules that are de-excited both radiatively and non-radiatively, can be defined as:

$$\frac{d[C]^*}{dt} = \frac{d[C]_{0\to*}}{dt} - (k_r + k_{nr})[C]^*,$$

---

[18]Note that k_r could also be transition-dependent.

which results in:

$$\frac{d[C]^*}{dt} = k_{abs}([C] - [C]^*) - (k_r + k_{nr})[C]^*.$$

Considering the above equation in the steady state ($d[C]^*/dt = 0$), the resulting concentration of excited molecules in the steady state ($[C]^*_{cw}$) is:

$$[C]^*_{cw} = \frac{k_{abs}}{k_r + k_{nr} + k_{abs}}[C], \qquad \left(molecules/cm^3\right) \qquad (1.4)$$

which depends *non-linearly* on the incident average intensity through k_{abs}. For powers typically used in *in-vivo* experiments and typical radiative and non-radiative rates (see Table 1.4) we may assume $k_r + k_{nr} \gg k_{abs}$, in which case we may represent the steady-state emitted fluorescent power, P_{fl}, as:

$$P^{cw}_{fl} = E_{em}k_r V[C]^*_{cw}$$

$$P^{cw}_{fl} \simeq \frac{E_{em}}{E_{abs}}U_0\sigma_a\frac{k_r}{k_r + k_{nr}}V[C], \qquad \left(Watts\right) \qquad (1.5)$$

where we have rewritten k_{abs} in terms of the incident flux. Note that as expected:

$$P^{cw}_{fl} = \frac{1}{\tau}\int_0^\infty P_{fl}(t)dt. \qquad \left(Watts\right)$$

1.4.2.3   *Quantum Yield*

From Eq. (1.5) we may establish a very important quantity for how efficient the excited state of a molecule is in producing fluorescent light, the *quantum yield*. Quantum yield can be defined as the efficiency of conversion of the total absorbed power to radiative energy (i.e. fluorescence emission). Making use of the expression for the radiated power Eq. (1.2), the quantum yield Φ is therefore defined as:

$$\Phi = \frac{P_{fl}}{P_{abs}} = \frac{E_{em}}{E_{abs}}\frac{k_r}{k_r + k_{nr}},$$

Accounting for the fact that we have a two level system and therefore $E_{em} \simeq E_{abs}$ we obtain the commonly used formula for the quantum yield:

$$\Phi = \frac{k_r}{k_r + k_{nr}}. \qquad (1.6)$$

Let us now go back to our definition of the total power emitted, Eq. (1.5). This may be rewritten in terms of the quantum yield as:

$$P_{fl} = U_0\sigma_a\Phi[C]V,$$

or equivalently:

$$P_{fl} = U_0 \mu_a^{fl} \Phi V, \qquad \left(Watts \right) \tag{1.7}$$

where we have made use of the definition for the absorption coefficient previously introduced in Sec. 1.3[19] which is defined in terms of the absorption cross-section as:

$$\mu_a^{fl} = [C]\sigma_a. \qquad \left(cm^{-1} \right) \tag{1.8}$$

In order to avoid confusion one must not forget that the absorption cross-section σ_a, even though usually expressed in units of cm^2 (i.e. barns), is in many occasions presented with alternative units. For example, in fluorophore and fluorescent probe catalogs one usually finds the absorption cross-section or extinction cross-section as *molar absorption coefficient* or *molar absorptivity* in units of $(M^{-1}cm^{-1})$ or equivalently, $(1 \, mol^{-1}cm^{-1})$. Usually, unless otherwise stated, this has been measured in decadic logarithm. In this case, taking great care with the units, one obtains that the relationship between the molar absorptivity ϵ_{abs} and the absorption cross-section σ_a is:

$$\sigma_a = \frac{1000 \log(10)}{N_A} \epsilon_{abs} \left(M^{-1}cm^{-1} \right), \qquad \left(cm^2 \right) \tag{1.9}$$

where N_A is Avogadro's number ($N_A = 6.022 \times 10^{23}$ molecules/mole). Before continuing, let us examine carefully the implications of Eq. (1.7). First of all, Eq. (1.7) is a very clear indicator of what type of problem we are dealing with when working with fluorescence in the sense that, not only several approximations were needed in order to arrive to this simplified expression, but the overall fluorescence intensity depends on several parameters. In particular, if the concentration of the fluorescent protein is not known (as will be the case in most optical tomography applications), we must *assume* we know the absorption cross-section of the protein or fluorophore and its quantum yield at those particular wavelengths. However, the truth of the matter is that all these parameters may vary depending on the environment of the protein or molecule, specially the quantum yield (this very fact is used, for example, to probe pH). This must always be considered, unless working in very controlled conditions, and relates directly to our capability of producing *quantitative* results. Additionally, Eq. (1.7) exhibits a very relevant property in the sense that there is a direct relation between the

---

[19]The relation between the absorption cross-section and the absorption coefficient will be discussed in detail in Chapter 2.

emitted fluorescence and the excitation light. If we know the incident excitation, then we can recover quantitatively the fluorophore concentration by measuring the fluorescence intensity. However, if the main message of this chapter was delivered properly, it is clearly not possible to know in full the optical properties of the tissues that our excitation light traversed, and therefore it is also not possible to predict *exactly* the expression for U_0. Fortunately, as will be seen in later chapters, when dealing with fluorescence tomography both the fluorescence and excitation intensities will traverse approximately the same tissue and therefore their *ratio* will be capable of delivering a good quantitative approximation.

In Table 1.6 a list some of the most commonly used fluorophores and fluorescent proteins is shown, together with their respective molar absorption coefficient, quantum yields and lifetimes (where available). All these flourophores and fluorescent proteins have additional very important characteristics such as molecular weight, photo-stability, and in the case of fluorescent proteins, how many subunits they consist of since this might affect their behavior. In particular, lifetimes of fluorescent proteins are usually not given since they greatly depend on the environment of the protein and hence show a great variation.

Once we have defined the process of fluorescence and the different parameters that it depends on, we should now compare how these parameters are affected by the presence of absorption in tissue. To that end, Figure 1.12 shows measured spectra from four different fluorescent proteins. In this figure the absorption spectra of oxy and deoxyhemoglobin are also included as reference (note that normalized fluorescence intensity is shown in log scale). From this figure and Table 1.6 several very important facts need to be considered when choosing the correct fluorophore or fluorescent protein. Obviously we would like to excite the fluorophore at its maximum absorption (in order to maximize the absorption cross-section σ_a). Additionally, the emission spectra must be considered carefully and as far-red shifted as possible. And, last but not least, the *brightness* of the fluorophore/fluorescent protein must be considered together with the previous factors. Brightness is defined as the product between the absorption cross-section and the quantum yield:

$$\text{Brightness} = \sigma_a \times \Phi,$$

and it is the true representation of how effective the molecule is in converting

Table 1.6   Some Commonly Used Fluorophores and Fluorescent Proteins

Fluorophore	max(λ_{abs})(nm)	max(λ_{em}) (nm)	$\epsilon(\mathrm{M}^{-1}\mathrm{cm}^{-1})$	Φ	τ (ns)
Fluorescein/FITC	495nm	519nm	8×10^4	0.93	4.0ns
Cy3	548nm	562nm	15×10^4	0.04	0.3ns
ATTO590	594nm	624nm	12×10^4	0.80	3.7ns
Cy5	646nm	664nm	25×10^4	0.27	1.0ns
Cy5.5	675nm	694nm	25×10^4	0.23	1.0ns
Cy7	743nm	767nm	11×10^4	0.28	0.6ns
Alexa Fluor 750	749nm	775nm	24×10^4	0.12	0.7ns

Fluo. Protein	max(λ_{abs})(nm)	max(λ_{em})(nm)	$\epsilon(\mathrm{M}^{-1}\mathrm{cm}^{-1})$	Φ	τ (ns)
eGFP	484nm	510nm	2.3×10^4	0.7	-
eYFP	512nm	529nm	4.5×10^4	0.54	-
DsRed	557nm	592nm	2.7×10^4	0.14	-
mCherry	587nm	610nm	7.2×10^4	0.22	-
TurboFP635(katushka)	588nm	635nm	6.5×10^4	0.34	-
mPlum	590nm	649nm	4.1×10^4	0.1	-
iRFP	690nm	713nm	10.5×10^4	0.059	-

$\lambda_{abs}, \lambda_{em}$ = absorption, emission wavelength.
Φ = quantum yield.
τ = fluorescence lifetime.
Note that all values are indicative and depend greatly on the environment of the fluorophore/fluorescent protein. The data in this table were compiled from what is provided by each of the relevant companies.

incident light into fluorescence radiation. As an example, consider the far-red proteins mPlum and Katushka shown in Table 1.6. Based solely on their spectra and considering the absorption spectrum of blood one would expect mPlum to be the best choice. However, due to its low quantum yield, it is expected that using Katushka in this particular case would give better performance[20]. The performance of these new red-shifted fluorescent proteins is studied in detail in [Deliolanis *et al.* (2011)].

1.4.2.4   *Tissue Auto-fluorescence*

The physics of tissue auto-fluorescence is exactly the same fluorescence process that was recently described but with one critical difference: it is (generally speaking) fluorescence that we do not intend to excite, in contrast with

---

[20]In practice this involves a series of several other factors which could affect the results. For example, in the case of proteins it depends on where they are expressed and how efficiently. In this case, it will greatly depend on the protein itself and not so much on its fluorescence properties. Another issue is that each protein will behave differently in different environments giving rise to differences in their brightness.

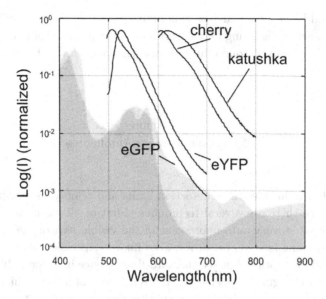

Fig. 1.12 Fluorescence emission spectra of a few of the fluorescent proteins currently available. The absorption coefficients of oxy and deoxyhemogblobin are shown for reference. Note that the fluorescence emission intensity is normalized and presented in log scale.

the fluorophores or fluorescent proteins that are used to probe molecular function and provide the contrast. There are many endogenous fluorescent components in tissue, each one with different absorption/emission properties, quantum yields and lifetimes. Examples are collagen, elastin, NADH, tryptophan and tyrosine, amongst others[21]. Since we do not know exactly what molecules are causing the autofluorescence and, more importantly, we have no clue about their concentration, this autofluorescence will appear as background intensity when we measure our 'true' fluorescent protein or fluorophore. As a practical example on how much of an effect this may have in optical imaging consider the following: suppose we are trying to image a specific fluorophore which is located at very low concentrations deep in tissue, were light propagation is greatly attenuated. If there was absolutely no autofluorescence present and we had managed to develop a perfect emission filter which completely removed the excitation intensity, one could increase the exposure times of our detector in order to recover the scarce radiation emitted by the fluorophore that exits the surface and

---

[21]Collagen and elastin, for example, have such a high quantum yield that are used as a source of contrast in confocal microscopy.

there would be no practical detection limit apart from the sensitivity of the camera. However, ignoring for the time being the presence of noise, in reality it will be the background fluorescence what will establish our detection limits:

$$\text{Sensitivity} \propto \frac{\text{Fluorescence}}{\text{Fluorescence} + \text{Autofluorescence}}.$$

In order to improve these detection limits and remove the effect of autofluorescence there are several techniques. First of all, tissue is — conveniently for us — very auto-fluorescent in the visible part of the spectrum. That is, by using far-red or near infra-red fluorescent molecules we not only avoid blood absorption, but see the auto-fluorescence in tissue reduced several orders of magnitude. Secondly, there are several factors which can be controlled in order to reduce tissue autofluorescence, such as the diet. For example, it is well known that a chlorophyll-free diet greatly reduces tissue autofluorescence[22]. Finally, since the spectra of the fluorescent molecules we intend to measure are, in principle, known *a priori* (and I write 'in principle' since, as mentioned before, this spectrum has some dependence on the environment of the molecule) we can use this information to separate the wheat from the chaff, so to speak. By performing several spectral measurements (for example, measure and/or excite at several wavelengths) we can post-process the data in order to extract the contribution of our 'true' fluorophore to the total measured intensity.

With regards to sensitivity, there is an additional factor which must be considered because it has a similar effect as autofluorescence on the overall sensitivity. This is the effect of excitation light 'leaking' through the filters we use to select the fluorescence emission, which is sometimes termed bleed-through. Since these filters are not perfect, there will always be some amount of light that is not blocked by the filter. It is for this reason that filters must be chosen carefully so that the transmission is maximum for the emission wavelengths and several orders of magnitude lower for the excitation (typically 5 to 7 orders of magnitude, depending on the filter).

---

[22]Chlorophyll is one of those strong fluorescent molecules readily available in nature. It is so strongly fluorescent that its analysis has become one of the most widely used techniques in plant physiology studies.

1.4.3 *Bioluminescence*

As mentioned in the introductory part of this section, bioluminescence is the emission of light from a substrate in a living organism which has undergone a chemical reaction. In particular, light is produced from a chemical reaction which involves a substrate (often but not always referred to as *luciferin*), an enzyme (*luciferase*), oxygen, and ATP. For example:

$$\text{luciferin} + \text{luciferase} + \text{ATP} + O_2 + [...] \longrightarrow [...] + \text{oxyluciferin} + \text{light}.$$

Since the bioluminescent reaction requires ATP, in a carefully controlled medium the amount of light produced from luciferin can be made proportional to the amount of ATP present, to which the bioluminescence reaction is very sensitive. Note that not all bioluminescence reactions require the presence of ATP and/or oxygen, and they might occur in several steps, but the one shown above is the most common. Also, as mentioned before in the context of the discovery of the green fluorescent protein in Section 1.4, some systems like the aequorin use calcium as mediator for the oxidation of the luciferin.

Bioluminescence occurs in a very large number of phyla in nature and it is estimated to have arisen independently many times in the course of evolution (as opposed to fluorescence). Bioluminescence occurs in many organisms such as glowworms and fireflies (such as the *Photinus pyralis*), in earthworms and snails, and in a great deal of ocean creatures such as the jellyfish *Aequoria aequoria* mentioned previously, the sea pansy *Renilla reniformis* and the crustaceans *Cypridina hilgendorfii* and *Cypridina noctiluca*, amongst many others. Most of these bioluminescent reactions involve fluorescent emission from singlet states. In any case the physics of light emission in bioluminescence is equivalent to that of fluorescence emission, the great difference being in the different excitation process.

The luciferins in different species can vary considerably and, even though each one emits light at different energies, most organisms have their emission peaks in the blue/green wavelengths of the spectrum. This is currently being overcome by emulating the bioluminescence resonant energy transfer (BRET) combo shown by *Aequorea victoria*, combining a bioluminescent probe with a far-red emitting protein (see, for example, [Wu *et al.* (2009)]), so it is expected that a large 'palette' of emission wavelengths will possibly be available in the near future. Bioluminescent reporters are very attractive for use in optical imaging due to their lack of photo-toxicity and background auto-fluorescence, although they come with their own specific drawbacks. First of all, the light emitted depends on the luciferin available, which is generally delivered externally. This means that, depending on where the enzyme is, it might have more or less access to the substrate irrespective of its concentration. Secondly, the intensity of the emission is related to changes in local oxygenation and different levels of anesthetic have been seen to have a direct effect on the emitted light. Similarly, the spectrum of emission is known to shift with temperature. Finally, with regards to its application in tomography, since there is no direct relation between the light emitted and an external 'actuator' (an external 'actuator' would be an excitation light source in the case of fluorescence), there is no way of externally modulating the amount of light produced. This considerably complicates the recovery of enzyme localization and concentration[23] *in-vivo*.

1.5   Light Scatterers in Nature

At this point we have completed an introduction of how light is absorbed, emitted and/or re-emitted, and the role that tissue plays in these instances is hopefully a bit clearer. What is left now is to consider what happens to that part of the incident radiation that after interacting with matter has not been absorbed and transformed into heat, light production, or any of the other mechanisms of energy conversion mentioned in the previous sections. Let us go back to the simplistic picture of Fig. 1.2, the classical description of electromagnetic radiation interacting with a dipole. When attempting to explain the basis of absorption, we basically covered how the dipole will oscillate and how these oscillations would be damped on certain directions depending on the environment of the dipole and therefore on the accessible energy levels. In the case where the secondary radiation is of the same frequency as the incident radiation or *incident field*, what we have is elastic *scattering*; this secondary radiation will be what is termed the *scattered field*. For the purpose of this book we will not consider any type of *inelastic scattering* as in the case of Raman scattering (which is due to the absorption and re-emission of light through an intermediate excited state with a virtual energy level) or Brillouin scattering (which takes place when the medium's optical density changes very fast with time resulting in

---

[23]This will be dealt with in Chap. 9, when studying the ill-posedness of fluorescence and bioluminescence tomography.

a Doppler shift of the incident radiation). In the common case of elastic scattering, the scattered radiation is of the same frequency and the total field measured will be the sum of the incident and scattered fields, always conserving both energy and momentum as long as the absorbed energy is taken into account. Don't forget that this more general definition of scattering is quite different to that used in many textbooks and specifically by Nassau, but as long as we bear this in mind we will hopefully avoid confusion.

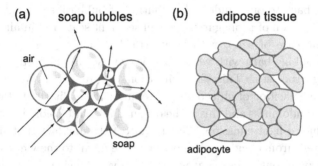

Fig. 1.13   Depiction of how light interacts with soap bubbles (a). The distribution of adipocytes (fat cells) in adipose tissue is shown for comparison in (b).

As mentioned at the beginning of this chapter, it will be the collective response of the atoms to the incident radiation that will determine the way light will propagate through matter. It is clear that if the medium is an 'ordered' one, i.e. one with a periodic or layered structure, for example, coherent effects such as interference will be clearly visible. Such would be the case of iridescence which is caused by interference from light reflected between different layers, as in abalone or a soap bubble: due to the irregular width of the soap layer, when illuminated with white light we see different colors on the surface of the bubble, similar to what we see in an oil slick. However, due to its transparency, most of the light will basically go through the bubble and barely change direction. If, on the other hand, what we have is a 'disordered' or random medium, depending on the amount of organization we will start to lose most of these interference effects to a more prominent contribution of an average intensity. Considering the soap bubble mentioned before, even though each individual bubble does not significantly change the propagation of the incident radiation, when we have many of them each of different size and at arbitrary positions, the overall radiation reflected or transmitted will contain the addition of many slight (and random, don't forget that it is a disordered medium) deviations from the incident direction of propagation (see Figure 1.13). It is in these cases where a *statistical description* of the medium is better suited, for example, by describing their size distribution and average distance between them. As an interesting fact, note that if these same bubbles were all of approximately the same size and ordered (like a stack of cannonballs for example), we would have an *opal*, which is a structure with very strong coherent effects that acts as a photonic band-gap material.

Going back to our random distribution of bubbles, since light loses its original direction of propagation we will see light scattered in all directions and in the case of soap, with very little lost due to absorption. If we view this bubble bath with white light we will see it white due to multiple scattering processes. This same situation will appear in any case where you have a collection of objects or scatterers that are distributed with some degree of randomness and this is the origin of all white colors in nature that are not due to light emission. This is quite a striking fact that should not be forgotten: from clouds, milk, cotton, white linen, to the white emulsions we use to paint our walls, what we have is a very efficient highly scattering medium (and don't forget hair; where else would gray hair get its color from?). To emphasize even further, there is no combination of pigments that will give white color as a result, since pigments are based on absorption of light. The only way of obtaining white is through multiple scattering[24] or light emission. Note that some emulsions and in particular some detergents use fluorophores in order to enhance the 'whiteness' we perceive, in which case your white clothes not only scatter light very efficiently but also emit a small amount of fluorescence.

Before moving on to how tissue scatters light, let me introduce a very visual example of how much the degree of order affects light propagation: the fried egg. Before heating the egg, the egg-white is transparent. This is because the strands of proteins in egg albumin (approximately 15% are proteins, the rest is water) are all aligned and in order, coexisting with water in a stable conformation which hides their hydrophobic regions and exposes their hydrophilic ones (don't forget proteins are amino-acid polymers). Once we start heating the egg, we denaturate these proteins altering their structure irreversibly. This breaks the order imposed by their stable

---

[24]Whenever using the term multiple scattering it will always refer to scattering in the presence of disorder, since rigorously speaking multiple scattering is present in any interaction of radiation with a macroscopic object.

interaction with their surrounding water, distributing them randomly and with different structures. In this case the egg white becomes more highly scattering the more the proteins are denatured, forming an interconnected solid mass.

1.5.1 *Tissue Scattering*

Table 1.7 Characteristic Index of Refraction of Cellular Components (from [Drezek *et al.* (1999)])

Structure	Index of Refraction
Extracellular Fluid	1.35-1.36
Cytoplasm	1.36-1.375
Nucleus	1.38-1.41
Mitochondria and Organelles	1.38-1.41
Melanin	1.6-1.7

As mentioned in the previous subsection, when dealing with a distribution of objects with a certain degree of randomness a statistical description is more suitable. By describing the average shape, size and spatial distribution, together with the composition of these objects we can describe how the medium will affect light propagation *on average*. The actual statistical coefficients that take all these parameters into account are:

(1) the *scattering coefficient*, expressed as μ_s, which describes how efficiently these objects scatter light;
(2) the *absorption coefficient*, expressed as μ_a, which defines how much light these objects absorb;
(3) and the *scattering anisotropy*, expressed as g, which is an indication of the transparency of the object.

These parameters will be derived from first principles in the next chapter and for the time being do not need to be explained further. The important thing to remember is that they are a statistical description of our complex original problem with objects of several sizes arranged arbitrarily.

Note: The Statistical Description of Light Propagation

Whenever dealing with statistical descriptions it is important to remember that we must always have a *statistically significant* number of objects — in our case, scatterers — within the volume of interest. With this in mind, it makes no sense describing the scattering coefficient of a sample with very few scatterers, or modeling the spatially-varying value of the scattering coefficient at distances which are much smaller than the average distance between particles. Note, however, that the true distance between particles is not always directly reflected in the overall scattering properties measured, since anisotropy plays a very important role.

Following the same example introduced previously, consider a single cell as a scatterer. Cells are quite transparent, as is understood from their optical properties shown in Table 1.7. Their average refractive index is in the order of 1.38, which is very close to that of water, 1.333, and really close to the index of refraction of the extracellular fluid, 1.35. For this reason, cells can barely be seen directly under the microscope and in order to see them clearly we need to fiddle with the direction of incident light (as in darkfield microscopy or in differential interference contrast microscopy), stain them with chromophores (the name chromosome comes from the fact that they are easily stained — apart from very specific cells such epidermal onion cells, the nucleus is pretty much invisible unless labeled with dyes), or label them with fluorophores. However, as in the case of the soap bubble, they do spread the incident electromagnetic field, even though slightly. Of course, different types of cells will do so differently, some of them also exhibiting strong absorption properties as in the case of red blood cells or in melanocytes as presented earlier. Following the same reasoning as before we may describe light propagation in different tissue types statistically, using the quantities defined previously: the scattering and absorption coefficients and the anisotropy factor. What should be remembered is that cells, in general terms, are more than an order of magnitude larger than the incident radiation wavelength (see Table 1.2 for some comparative values). This, together with the fact that they are quite transparent, is reflected in the angular distribution of scattered light which is by no means isotropic but rather forward scattered, enabling the visualization of thin tissue slices under a microscope.

So as to have an example of what the angular distribution of a fat cell would approximately look like, we may consider the fat cell as a perfect homogeneous sphere in which case we may use Mie Theory (see, for ex-

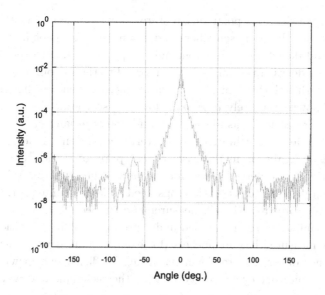

Fig. 1.14   Approximation of what the angular intensity distribution would look like for a fat cell, assuming it is a sphere of radius 50 times the wavelength of the incident light (i.e. 80μm diameter for a 800nm incident wavelength) with index of refraction 1.4 and surrounded by water.

ample [van de Hulst (1981)]) to solve exactly for the scattered intensity. Mie Theory is used to describe light scattering from spheres and cylinders, where closed form solutions to Maxwell's equations have been found. Figure 1.14 presents data generated using Mie Theory showing the angular distribution of scattered intensity for a homogeneous sphere with an index of refraction of 1.4 surrounded by water, with a diameter 100 times the incident wavelength. As can be clearly seen in this figure, most of the scattered radiation is forward scattered, as expected, with a difference of six orders of magnitude when compared to the back-scattered contribution. It is the distribution of fat cells in a disordered manner (see Figure 1.13) that yields the highly scattering properties. Note that, of course, the true angular distribution of scattered intensity of a fat cell will be slightly different due to the presence of the nucleus and other components of the cell's anatomy. Even so, in most cells (see [Watson *et al.* (2004)] for example) the difference between forward and backscattered light is in the vicinity of 4 orders of magnitude.

Similarly to what was said in the case of tissue absorption, each tissue type will have its own scattering properties: it will have a specific structure that will give rise to a polarization state, as in muscle due to the highly structured bundles of muscle fibers; a different scattering coefficient due to cell size and density; or a different anisotropy factor due to the variation in relative indexes of the cells when compared to the surrounding medium. This means that the 'real' or 'actual' scattering coefficient of different tissues is never known accurately *a priori*. However, scattering does not change *in-vivo* as dramatically as absorption, and one is usually on the safe-side assuming an average value for the scattering coefficient, at least for each tissue type. Again, as in the case of absorption of tissues, one can find in literature very different values for the same type of tissue. When doing such a literature search several things need to be accounted for: how the samples were prepared (i.e. if measurements are *in-vivo* or *ex-vivo*, and if so, if the samples were maintained under proper humidity conditions) or if the tissue measured has air pockets (for example lung is very tricky, since the presence of air may render a scattering coefficient which will change with the respiratory cycle), together with the technique and wavelengths used.

Some common approaches are to either assume an average scattering coefficient in the range of $\mu_s = 80 - 110cm^{-1}$ with $g = 0.8 - 0.95$, or use the following formula to fit for its value:

$$\mu_s = a \times \lambda^{-b},$$
$$g = c \times \lambda^{-d},$$

being a, b, c, and d constants one would need to fit for, being λ the wavelength. Typical values, for example for Intralipid-10% are [see Flock *et al.* (1992) for details][25]:

$$\mu_s \simeq 1.17 \times 10^9 \lambda^{-2.33}, \qquad \left(cm^{-1}\right)$$
$$g \simeq 2.25\lambda^{-0.155},$$

where λ is given in nanometers. This approximation is based on the fact that scattering changes smoothly for random distributions of particles, at least far from resonances. Note that in the case of very small scattering particles we have that μ_s behaves as λ^{-4}. This is termed Rayleigh scattering, which predicts scattering from particles much smaller than the incident

---

[25] Intralipid is a fat emulsion composed of lipid droplets and water which is used as a source of calories and essential fatty acids for intravenous administration to patients. It is a very common source of scattering particles that mimic tissue scattering, and as such is extensively used to prepare tissue phantoms. Intralipid-10% refers to 10g of lipid per 100ml of suspension.

wavelength and explains the blue color of our atmosphere. Rayleigh scattering will be considered briefly while describing the scattering phase function or scattering diagram in the next section.

1.6   Optical Molecular Imaging

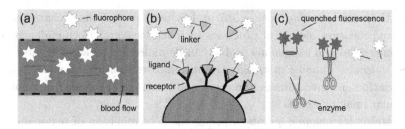

Fig. 1.15   Depiction of different types of fluorescent probes: (a) non-specific probes which go with the bloodstream, (b) targeted probes which target a specific receptor, and (c) activatable probes, which only fluoresce in the presence of a specific enzymatic activity. Note that combinations of targeted activatable probes are also possible.

Once we have established how light interacts with tissue, I believe an introduction to molecular imaging is called for, specifically since most of the applications that this book targets are framed within the novel developments in optical 'molecular imaging'. It is not clear exactly when the field of molecular imaging developed, since for several decades radiologists have been making use of specific markers *in-vivo*. It is clear, however, that this field emerged within the broader field of radiology and those working to improve imaging approaches with human health in mind. As such, it must not be confused with what would be understood in a more physical sense as 'imaging molecules'. The main goal of Molecular imaging *is not* to image molecules, but rather to image *function* at a molecular level. Following the definition approved by the Society of Nuclear Medicine (SNM) in 2005,

> *molecular imaging is the visualization, characterization and measurement of biological processes at the molecular and cellular levels in humans and other living systems.*

The definition further elaborates:

> *molecular imaging typically includes two- or three-dimensional imaging as well as quantification over time. The techniques used include ra-*

diotracer imaging/nuclear medicine, magnetic resonance imaging (MRI), magnetic resonance spectroscopy (MRS), optical imaging, ultrasound and others.

The different components involved in molecular imaging are mainly imaging agents and imaging instrumentation, always with quantification in mind. With regards to these, the SNM defines:

- *Molecular imaging agents* are probes used to visualize, characterize and measure biological processes in living systems. Both endogenous molecules and exogenous probes can be molecular imaging agents.
- *Molecular imaging instrumentation* comprises tools that enable visualization and quantification in space and over time of signals from molecular imaging agents.
- *Molecular imaging quantification* is the determination of regional concentrations of molecular imaging agents and biological parameters, and it provides measurements of processes at molecular and cellular levels. This quantification is a key element of molecular imaging data and image analysis, especially for inter- and intra-subject comparisons.

This definition of molecular imaging is general and includes optical imaging as long as it makes use of molecular imaging agents. In fact, it is so broad that it includes the current microscopy imaging approaches used in biology which make use of fluorescent proteins or probes as long as the experiments take place *in-vivo*. With respect to optical molecular imaging agents or probes, there are basically three categories:

- *Non-specific probes* which are delivered through the bloodstream and basically tend to accumulate at regions with slower clearance rate, such as tumors, for example. These probes fluoresce all the time and are thus not optimal due to a high background signal always present and their inherent non-specificity. An equivalent of these probes in X-ray imaging would be iodine, for example, which is commonly used for angiography. An optical imaging non-specific fluorescent probe currently used is indocyanine green (ICG).
- *Targeted probes* which are typically also delivered through the bloodstream, but in opposition to the non-specific probes these are attached to a ligand and therefore bind preferentially to the targeted receptor. As in the case of non-specific probes, these probes also fluoresce all the time and therefore present a high background signal. However,

since they target specific receptors, most of the background signal will
be cleared by the system if one waits 'enough' — how much time is
needed to obtain the maximum clearance while maintaining the probe
bound to the receptors must be tested for every specific application and
probe — leaving only those fluorophores attached to the target receptor (and inevitably some non-targeted receptors). Most of the optical
imaging probes used for immunofluorescence are of this sort.

- *Activatable probes* which are a new type of probes which do not fluoresce
  in their normal state, but emit light due to a change in their conformation that is caused by the specific function targeted. In particular, some
  probes use the fact that two fluorochromes (or a fluorochrome and a
  'quencher') when placed in very close proximity quench their emission
  (i.e. non-radiative processes are favored in this case): by engineering
  the probe in such a way that the link between the fluorochromes can be
  cleaved in the presence of a particular enzymatic activity, it is possible
  to have a specific probe with very high signal to background ratios.
  Of course, combination of activatable probes with targeted probes is
  also possible yielding high specificity and contrast (see [Blum (2008)]
  for a review). Similar probes exist in microscopy as would be Fluorescence (or Förster) Resonance Energy Transfer (FRET) [26] biosensors.
  Note that engineering this type of activatable probes for other imaging
  modalities such as nuclear imaging is a very difficult matter — you
  might understand that for this we would need a substance that can
  switch its positron emission properties on and off, for example.

With regards to this book, the type of optical probe used is not important as long as we understand its possible contribution to the background
(non-specific) signal. Some tips on how to improve the signal to noise
ratio in the presence of a background signal will be discussed in later chapters, however it is important to understand the importance of the optical
probes: good probes with high specificity, high brightness, and good ab-
sorption/emission characteristics will yield good results on practically any
optical imaging setup. Probes with low specificity and less than optimal
spectra will be very difficult to image, whatever the theoretical/engineering
approach used.

There is a comparison I like to present whenever introducing molecular

---

[26] Even though the terminology is similar, it must not be confused with bioluminescence
resonant energy transfer (BRET)

imaging: if one imagines molecular imaging as a stringed instrument, for example, the imaging setup would be the body, the theory used the tuning pegs, and the probes would be represented by the strings. Without strings vibrating, the instrument is useless. However, with the appropriate strings, the body (i.e. the imaging approach) can amplify the string's vibration, while the tuning pegs (i.e. the theory) ensure that the vibrations are in tune. With this in mind, given an optical imaging approach such as those used for small animal optical imaging, this book will focus on how to tune the strings to obtain the highest amount of information, always understanding what are the governing principles, the limitations, and possible improvements.

Key Points

In this introductory chapter many basic concepts were revised such as light emission, absorption and scattering. In particular, emphasis on how these phenomena are affected by tissue was presented, introducing a statistical description as a means to predict how light will propagate within randomly distributed media. From all the basic phenomena that we went through, the following key points will be very valuable in the following chapters and should not be forgotten:

- **Absorption**: absorption is the main limiting factor with regards to how deep we can image optically, much more than scattering. Multiple scattering does not convert radiative energy into non-radiative, it just spreads it in all directions. With this in mind, selecting the sources (be it the excitation sources or the fluorescent probes or proteins used) so that they fall within the near infra-red region will improve penetration depth, sensitivity and signal to noise ratio.
- **Statistical description**: the optical properties of tissue are described statistically by the scattering coefficient μ_s, the absorption coefficient μ_a, and the anisotropy factor g. As such, care must be taken to ensure that the volumes and/or distances considered are statistically relevant.
- **Average optical properties**: since we describe both the scattering and the absorption tissues statistically, we must not forget that an accurate description of the absorption properties of tissue *in-vivo* is extremely difficult to obtain *a priori*. In particular, blood volume and oxygen saturation change significantly the absorption spectra of tissue. Since in most cases it is not practical to fit for these values, we usually assume an average value. As will be shown in this book, assuming

the wrong value does not have such dramatic consequences as long as relative measurements are used. However, the fact that whenever dealing with *a priori* values we might be very far from the true ones should not be forgotten, specially for the absorption coefficient. With relation to the scattering coefficient, the variation - at least within tissue types - is not so dramatic, but it must not be forgotten either. What is expected, however, is that the average scattering coefficient will decrease with wavelength.

- **Background signal**: once the upper limit of sensitivity is dictated by the amount of absorption present in tissue (where, once again, the wavelengths used play a major role), the next limiting factor is the amount of tissue autofluorescence and non-specific fluorescence: choosing the appropriate probes with optimal excitation/emission spectra in the near infra-red, and more importantly with high specificity, is the key to high sensitivity.

- **Quantification**: As we have seen in this chapter, we have approximated — without much loss of generality — that the fluorescence emission depends linearly on the excitation radiation. Even though this is not a strong assumption, we must not forget that the quantum yield of our fluorophore/fluorescent protein does depend on its environment. Additionally, the precise amount of energy that reaches the fluorophore is not known accurately due to the issues mentioned above regarding the statistical values used for absorption and scattering. As will hopefully be clear throughout this book, it is still possible to obtain quite accurate quantitative results: in any case, we must not forget that we are working and building on these basic assumptions. With regards to bioluminescence, quantification is a much more difficult task as will be shown in the following chapters.

Further Reading

Since this chapter covered several very different areas and disciplines, I recommend the following books that will expand in greater detail the concepts that were presented here.

First of all, since this book is centered in biological and medical applications, for those that wish to get more acquainted with biology I strongly recommend the general book on biology by Alberts *et al.* (2009). With respect to issues more related to the basis of molecular biology a very good source is the book by Watson and Berry (2003), which is very well written for a general audience. Also, within the area of biology the book by Harvey (1957), even though more than half a century old, provides very good insight on the history of bioluminescence and its biology. From this author, [Harvey (1920)] is also a very interesting book which helps forming an idea of just how much we have advanced in understanding and controlling the emission of 'living light'.

A great source for understanding all the different origins of color in nature is the book by Nassau (1983). Additionally, for those that wish to know the scientific answer to why veins appear blue, I suggest reading the paper by Kienle *et al.* (1996). It is a simple question with quite a complicated answer that is directly related to what has been presented in this chapter. Similar to the reason of our perception of the blue color of bruises, the answer commonly given to this question is in most cases incorrect: it is not related in a direct and simple manner solely to the oxygen content of blood.

With regards to the subject of basic electromagnetic interaction with matter and the dipole approximation, I would recommend the book by Lakowicz (1999) which also covers the basics of fluorescence. A more rigorous description on the basis of scattering can be found in the book by Bohren and Huffman (1998) and, for those interested in scattering from single cells, I recommend the PhD dissertation of Dunn (1997). For those interested in generating their own scattering diagrams for homogeneous or coated spheres, a great source for Mie scattering codes is the internet; for the theoretical part I would recommend the famous book by van de Hulst (1981). Apart from these, a great description on the basis of absorption and fluorescence can be found in the book by Valeur (2002), which also explains very well what modifies the fluorescence emission.

As was introduced in this chapter, the amount of radiative and non-radiative de-excitation depends greatly on the environment of the atom. To gain further insight on this issue I recommend the works by R. Carminati and co-workers [Froufe-Pérez *et al.* (2007); Froufe-Pérez and Carminati (2008); Pierrat and Carminati (2010)] who study the changes in radiative emission depending on the amount of highly scattering particles present.

With respect to tissue optics, a great source is the book by Tuchin (2007), while the webpages of Scott Prahl and Steve Jacques provide very useful information on tissue spectra (see http://olmc.ogi.edu/spectra and references therein). If you are interested in the effect of polarization in skin, I recommend the chapter by Jacques and Ramella-Roman (2004).

Finally, for all those not familiar to molecular imaging I strongly recommend the books by Rudin (2005) and Ntziachristos *et al.* (2006). These books present in detail all imaging modalities (nuclear, magnetic resonance and optical), together with reporter probes and applications. A valuable source of information are the reviews on molecular imaging by Weissleder and Mahmood (2001) and Cherry (2004). More specific on the optical approach to molecular imaging, I suggest the reviews by Weissleder and Ntziachristos [Weissleder and Ntziachristos (2003); Ntziachristos *et al.* (2005); Ntziachristos (2006)].

Chapter 2

Scattering and Absorption

Summary. Once the basis of absorption and scattering has been understood from the previous chapter, we will more rigorously consider here the basic interactions of light with particles and define scattering and absorption using classical electromagnetic theory. We will first tackle the case of single particles and then move on to a collection of particles, introducing the definition of energy flow through the Poynting vector.

2.1 Definition of Scattering

Even though we have generally defined scattering as will be used in this book in the previous chapter, it is useful to understand the different definitions one might encounter elsewhere. If one looks up the word 'scatter' in the Oxford English dictionary the following definitions will appear: *1) throw in various random directions; 2) separate and move off in different directions; 3)(be scattered) occur or be found at various places rather than all together. 4) Physics: deflect or diffuse (electromagnetic radiation or particles).* In everyday life, 'scatter' is a word that is commonly used. You could use it, for example, to describe the amount of chaos in your room with clothes 'scattered' all over, or to describe how you have found pieces of glass 'scattered' around the kitchen floor. However, despite being a common word, when used in the context of physics it may still have slightly different connotations in different fields.

Scattering is indeed a difficult concept to pinpoint since it covers a wide range of interactions. If I had to choose one single definition for scattering from those found in the dictionary, it would be deflection. However, and here is the catch, this would hold only for particles: scattering by marbles, for example, can be explained as 'a change in direction due to interaction with a particle', but this would not apply to the more general case of scattering by electromagnetic radiation. As seen in Chapter 1, one reason for this is that there are special cases of scattering which occur without a change in the path of propagation but with a change in the polarization or phase of the wave, for example.

Another example of how deep the implications of scattering are consists on considering the effect of refraction of light as it enters a homogeneous medium with a different index of refraction. Macroscopically one can use Snell's law and predict how the original direction of light will change after entering the medium. However, if one wishes to fully understand this effect we need to reach down to the microscopic point of view. If you recall, in the previous chapter we discussed how it was the combined interaction of all oscillating dipoles with the incident radiation that determined the macro-scopical behavior of the homogeneous medium. As a specific example, in order to understand the implications of the General Extinction Theorem (see Born and Wolf (1999) and Nieto-Vesperinas (2006), for example), one needs to consider each molecule as a scattering center. That is, in order to explain the interaction of light with what seems to be a simple homogeneous macroscopic medium such as water, one has to reach down to the micro-scopic level and explain everything in terms of scattering from a collection of individual molecules.

Therefore, in order to avoid confusion I believe a more general definition is necessary; my personal favorite is the definition by Bohren and Huffman (1998) which I here rewrite for convenience: "*Matter is composed of discrete electric charges which are electrons and protons. If an obstacle, which could be a single electron, an atom or a molecule, a solid or liquid particle, is illuminated by an electromagnetic wave, electric charges in the obstacle are set into oscillatory motion by the electric field of the incident wave. Accelerated electric charges radiate electromagnetic energy in all directions; it is this secondary radiation that is called the radiation scattered by the obstacle*". Note, however, that absorption does not enter into this definition of scattering since, as we saw in Chap. 1, absorption is the process where the excited elementary charges transform part of the incident electromagnetic energy into other forms (mainly thermal energy in the case of tissue), i.e. it does not comprise re-radiation. Following this definition of scattering, it is clear that it refers uniquely to electromagnetic radiation and is not applicable to particles, be it marbles or any other object, and serves as a good reminder that light should not be seen as a collection of particles and as such one should be careful when using the term 'photons'. If the solution to a specific problem does require quantization of the electromagnetic field then we should refer to quantum electrodynamics. Since all applications we are interested in this book can be described by classical electromagnetism, I will avoid when possible the use of 'photon' to describe localized packets of radiation, and simply refer to it as light.

The goal of this book is to describe light propagation through media which contain a very large number of particles which are randomly arranged in space, such as cells in tissue. As has been mentioned previously, all cases considered in this book are linear interactions of light with matter, i.e. scattered and incident light have the same wavelength (unless we consider fluorescence, of course), in opposition to what happens in other non-linear interactions such as Raman scattering, frequency doubling, or higher order harmonic generation where there is a change of wavelength implied. In order to understand light propagation in tissue we need to understand the basic mechanisms of multiple scattering that lead to diffusion of light. And for that, we first need to define the interaction of light with a single one of those particles.

2.2   Poynting's Theorem and Energy Conservation

A fundamental equation that arises from applying Maxwell's equations directly is the equation of energy conservation. Assuming we have a locally isotropic medium (and by this we mean that its optical properties are the same locally, independent of the direction we are looking from) of properties given by its dielectric constant ϵ and magnetic permeability μ, given an electric field vector $\mathbf{E}$, a magnetic field vector $\mathbf{H}$ and a density current $\mathbf{j}$, Maxwell's equations can be expressed as:

$$\nabla \times \mathbf{H} - \frac{\epsilon}{c}\frac{\partial \mathbf{E}}{\partial t} = \frac{4\pi}{c}\mathbf{j}, \tag{2.1}$$

$$\nabla \times \mathbf{E} + \frac{\mu}{c}\frac{\partial \mathbf{H}}{\partial t} = 0, \tag{2.2}$$

where c is the speed of light in vacuum. Multiplying Eq. (2.1) by $\mathbf{E}$ and Eq. (2.2) by $\mathbf{H}$, subtracting and rearranging we obtain:

$$\mathbf{E} \cdot (\nabla \times \mathbf{H}) - \mathbf{H} \cdot (\nabla \times \mathbf{E}) = \frac{4\pi}{c}\mathbf{j} \cdot \mathbf{E} + \frac{1}{c}\left(\epsilon \mathbf{E} \cdot \frac{\partial \mathbf{E}}{\partial t} + \mu \mathbf{H} \cdot \frac{\partial \mathbf{H}}{\partial t}\right),$$

which using the vector identity $\mathbf{E} \cdot (\nabla \times \mathbf{H}) - \mathbf{H} \cdot (\nabla \times \mathbf{E}) = -\nabla \cdot (\mathbf{E} \times \mathbf{H})$ can be rewritten to the following expression which holds at any point in space $\mathbf{r}$ [Born and Wolf (1999)]:

$$\frac{1}{4\pi} \left(\epsilon \mathbf{E} \cdot \frac{\partial \mathbf{E}}{\partial t} + \mu \mathbf{H} \cdot \frac{\partial \mathbf{H}}{\partial t} \right) + \mathbf{j} \cdot \mathbf{E} + \frac{c}{4\pi} \nabla \cdot (\mathbf{E} \times \mathbf{H}) = 0. \qquad (2.3)$$

Let us now analyze what each of the terms of this equation represents. To begin with, the first term can be rewritten as:

$$\frac{1}{4\pi} \left(\epsilon \mathbf{E} \cdot \frac{\partial \mathbf{E}}{\partial t} + \mu \mathbf{H} \cdot \frac{\partial \mathbf{H}}{\partial t} \right) = \frac{1}{8\pi} \frac{\partial}{\partial t} \left(\epsilon \mathbf{E}^2 + \mu \mathbf{H}^2 \right).$$

What this term represents is the change in *energy density* with time, being energy density defined as:

$$W = \frac{1}{8\pi} (\epsilon \mathbf{E}^2 + \mu \mathbf{H}^2), \qquad \left(Joules/\text{cm}^3 \right) \qquad (2.4)$$

where we can clearly distinguish the electric and magnetic energy densities of the field.

The second term $\mathbf{j} \cdot \mathbf{E}$ represents Joule's heat, i.e. the resistive dissipation of energy. In all cases related to this book all loss of energy will be due to absorption. In our isotropic medium this loss of electromagnetic energy for a harmonic time dependence is defined by Ohm's law as $\mathbf{j} = \omega \epsilon^{(i)} \mathbf{E}$, where ω is the frequency of the electromagnetic wave and $\epsilon^{(i)}$ is the imaginary part of the permittivity:

$$\epsilon(\mathbf{r}) = \epsilon^{(r)}(\mathbf{r}) + i \epsilon^{(i)}(\mathbf{r}). \qquad (2.5)$$

As defined above, $\epsilon^{(i)}$ can have two contributions, one associated to the 'bound' charge current density that arises from a complex susceptibility and one associated to the 'free' charge current density. Absorption is the sum of both quantities and it is not possible to determine their relative contributions. Note that following this notation the conductivity would be defined as $\sigma = \omega \epsilon^{(i)}$ [Bohren and Huffman (1998)] in order for Ohm's law to take the more commonly used form $\mathbf{j} = \sigma \mathbf{E}$, but only if the 'free' charge current density is included. From now on we will avoid using σ to prevent confusion with the cross-sections that will appear in the following sections. This second term in Eq. (2.3) therefore represents the absorbed energy per unit volume:

$$\frac{dP_{abs}}{dV} = \mathbf{j} \cdot \mathbf{E}, \qquad \left(Watts/cm^3 \right) \qquad (2.6)$$

which for a harmonic time-dependence becomes:

$$\frac{dP_{abs}}{dV} = \omega \epsilon^{(i)} \mathbf{E}^2.$$

Finally, we arrive to the third and final term, $\nabla(\mathbf{E} \times \mathbf{H})$. This term represents the local net flow of energy, and it is usually written as:

$$\frac{c}{4\pi}\nabla(\mathbf{E} \times \mathbf{H}) = \nabla \cdot \mathbf{S},$$

where

$$\mathbf{S} = \frac{c}{4\pi}\mathbf{E} \times \mathbf{H}, \qquad \left(Watts/cm^2\right) \tag{2.7}$$

is the *Poynting vector*. This quantity is directly related to what is commonly understood in the lab as intensity: what one measures physically with a detector is not the electric or magnetic fields, but the flow of energy density through a detector area S for a given time interval. This quantity is characterized in electromagnetism by the integral over the detector area of the Poynting vector $\mathbf{S}$, $\int_S \mathbf{S} \cdot \mathbf{n}dS$, where $\mathbf{n}$ is the surface normal. It should be noted that the Poynting vector is not uniquely defined by Eq. (2.7) since any combination of a generic vector function $\mathbf{f}$ in the form $\nabla \times \mathbf{f}$ would still yield the same surface integral[1]. As long as we have this in mind, for all the derivations that follow we may safely assume that the Poynting vector represents the amount of energy which crosses per second a unit area normal to the directions of the electric vector field $\mathbf{E}$ and the magnetic vector field $\mathbf{H}$, or equivalently the magnitude and direction of the density of the flow of energy.

In terms of the above definitions, we may finally write the equation for conservation of energy as:

$$\frac{\partial W}{\partial t} + \frac{dP_{abs}}{dV} + \nabla \cdot \mathbf{S} = 0, \tag{2.8}$$

which is commonly termed *Poynting's Theorem*, and where it should be emphasized that it is valid only in those cases where $\mathbf{E}$ and $\mathbf{H}$ are mutually orthogonal, which is mainly in the far field.

---

[1]Remember that $\nabla \cdot (\nabla \times \mathbf{f}) = 0$. Therefore, the surface integral of $\nabla \times \mathbf{f}$ over a surface with normal $\mathbf{n}$ would be $\int_S (\nabla \times \mathbf{f}) \cdot \mathbf{n}dS = \int_V \nabla \cdot (\nabla \times \mathbf{f})dV = 0$ through the application of Gauss' theorem. Note that it is still not clear how much actual freedom exists in the addition of a rotational term: some books give freedom in its definition since it is of no consequence to the measurement, some consider it should be zero in order for momentum to be conserved, while on the other hand recent studies on forces on particles show that a term in the form of $\nabla \times \mathbf{L}$ is needed in order to account for spin angular momentum of light [Albaladejo *et al.* (2009)]. This might have direct implications in our definition of the Poynting vector.

Building Block: **Poynting's Theorem**

This expression for the conservation of energy through Poynting's Theorem will be later expanded and considered for rotational invariance in its time-averaged expression. This latter expression, a direct consequence of Poynting's Theorem, will be used as a starting point to derive the diffusion equation.

2.2.1 *The Time-Averaged Expressions*

We have now defined the equation for conservation of energy at any point in space by Eq. (2.8). However, all these expressions account for the fact that we are dealing with a time-harmonic electromagnetic field. Since the period of an electromagnetic wave ($T = 2\pi/\omega$) in the optical range is many orders of magnitude larger than any measurement time used in an optical imaging application (typically $T \sim 10^{-13}s$ versus $T' \sim 10^{-9}s$), the quantity of interest is their *time averaged* quantities. Writing the time-harmonic dependence of the electric and magnetic fields as $\mathbf{E}_t(\mathbf{r}, t) = \mathbf{E}(\mathbf{r}) \exp(-i\omega t)$ and $\mathbf{H}_t(\mathbf{r}, t) = \mathbf{H} \exp(-i\omega t)$, the time-averaged expression for the Poynting vector $\langle \mathbf{S} \rangle$ is found as:

$$\langle \mathbf{S} \rangle = \frac{1}{2T'} \int_{-T'}^{T'} \frac{c}{4\pi} \left[\mathbf{E}_t(\mathbf{r}, t) \times \mathbf{H}_t(\mathbf{r}, t) \right] dt \simeq \frac{c}{8\pi} \Re \left\{ \mathbf{E} \times \mathbf{H}^* \right\}, \quad (2.9)$$

where T' is the measurement time and now $\mathbf{E}$ and $\mathbf{H}$ are the complex vector functions of position and do not have any time-harmonic dependence ([Born and Wolf (1999)]). Defining the vector magnetic and electric fields in terms of their unitary vectors as $\mathbf{H} = H\mathbf{h}$ and $\mathbf{E} = E\mathbf{e}$, we will now make the following assumption: we will consider all measurements to be taken in the *far-field* where we will assume the fields have the structure of a *plane* electromagnetic wave which propagates in the direction of $\hat{\mathbf{s}}$. In this case we may use the orthogonal triad of vectors, $\hat{\mathbf{s}} = \hat{\mathbf{e}} \times \hat{\mathbf{h}}$ where $\hat{\mathbf{s}}$ represents the flow of energy, which assumes the overall behavior of the far field is that of an *outgoing spherical wave*. In this case we may use the solution of Maxwell Equations for mutually orthogonal $\mathbf{E}$ and $\mathbf{H}$ which relates their modulus as $\sqrt{\mu}H = \sqrt{\epsilon}E$, which yields the following expression for the time-averaged Poynting vector *valid only in the far-field* or for the special case of a plane wave:

$$\langle \mathbf{S} \rangle = \frac{\epsilon c_0}{8\pi} |\mathbf{E}|^2 \hat{\mathbf{s}}. \quad \left(Watts/cm^2 \right) \quad (2.10)$$

Building Block: **Time-Averaged Poynting Vector**

This expression, valid only where we may consider the fields to be mutually orthogonal, will be used to arrive to the scattering and absorption cross-sections and therefore to the final expressions for the absorption and scattering coefficients. Additionally, the Poynting vector will be the basis for the derivation of the diffusion equation as derived in this book.

In Eq. (2.10) $c_0 = c/n_0$ is the speed of light of the medium and we have made use of Maxwell's formula $n_0 = \sqrt{\epsilon\mu}$ where ϵ and μ are the dielectric permittivity and magnetic permeability. Finally, we have introduced $\hat{\mathbf{s}}$ which is a unitary vector which points in the direction of propagation $\hat{\mathbf{s}} = \mathbf{e} \times \mathbf{h}$ (see Fig. 2.1), being $\mathbf{e}$ and $\mathbf{h}$ the unit vectors for the electric and magnetic fields, respectively. Eq. (2.10) is a very important expression that will be used in several instances in this book, specially due to its direct relation with the specific intensity that will be defined in Chap. 3.

Proceeding in exactly the same manner as for the time-averaged Poynting vector, the *time averaged* expression for the density of electromagnetic energy in the *far-field* is, using $\sqrt{\mu}H = \sqrt{\epsilon}E$:

$$\langle W \rangle = \frac{1}{8\pi}\epsilon|\mathbf{E}|^2. \quad \left(Joules/cm^3 \right)$$

Comparing this expression to Eq. (2.10) it is clear that:

$$\langle W \rangle = \frac{|\langle \mathbf{S} \rangle|}{c_0} = \frac{\langle \mathbf{S} \rangle \cdot \hat{\mathbf{s}}}{c_0}. \tag{2.11}$$

Finally, the *time-averaged* expression for the absorbed power per unit volume is [Tsang *et al.* (2000)]:

$$\left\langle \frac{\mathrm{d}P_{abs}}{\mathrm{d}V} \right\rangle = \frac{1}{2}\omega\epsilon^{(i)}|\mathbf{E}|^2, \quad \left(Watts/cm^3 \right) \tag{2.12}$$

which rewritten with the aid of Eq. (2.10) becomes:

$$\left\langle \frac{\mathrm{d}P_{abs}}{\mathrm{d}V} \right\rangle = \frac{4\pi}{\epsilon c_0}\omega\epsilon^{(i)}\langle \mathbf{S} \rangle \cdot \hat{\mathbf{s}}. \tag{2.13}$$

Making use of the time-averaged expressions derived in this subsection, we can now express the time-averaged conservation of energy as:

$$\frac{1}{c_0}\frac{\partial \langle \mathbf{S}(\mathbf{r}) \rangle \cdot \hat{\mathbf{s}}}{\partial t} + \left\langle \frac{\mathrm{d}P_{abs}}{\mathrm{d}V}(\mathbf{r}) \right\rangle + \nabla \cdot \langle \mathbf{S}(\mathbf{r}) \rangle = 0, \tag{2.14}$$

which is the time-averaged expression of Poynting's Theorem and represents the conservation of energy flux along the direction of the Poynting vector $\hat{\mathbf{s}}$.

However, energy conservation should hold for any projection of this vector in any arbitrary direction. If we can write the projection of $\langle \mathbf{S} \rangle$ onto an arbitrary direction $\hat{\mathbf{s}}_J$ as:

$$\langle \mathbf{S}_J(\mathbf{r}) \rangle = S_J(\mathbf{r})\hat{\mathbf{s}}_J = S(\mathbf{r})(\hat{\mathbf{s}} \cdot \hat{\mathbf{s}}_J)\hat{\mathbf{s}}_J,$$

it is then clear that Poynting's theorem for $\langle \mathbf{S} \rangle_J$ can be expressed as:

$$\frac{1}{c_0}\frac{\partial \langle \mathbf{S}(\mathbf{r}) \rangle \cdot \hat{\mathbf{s}}_J}{\partial t} + \left\langle \frac{dP_{abs}}{dV}(\mathbf{r}) \right\rangle (\hat{\mathbf{s}} \cdot \hat{\mathbf{s}}_J) + \hat{\mathbf{s}}_J \cdot \nabla(\langle \mathbf{S}(\mathbf{r}) \rangle \cdot \hat{\mathbf{s}}_J) = 0, \quad (2.15)$$

where we have made use of the fact that:

$$\nabla \cdot \langle \mathbf{S}_J(\mathbf{r}) \rangle = \hat{\mathbf{s}}_J \cdot \nabla \left[S(\mathbf{r})(\hat{\mathbf{s}} \cdot \hat{\mathbf{s}}_J) \right] = \hat{\mathbf{s}}_J \cdot \nabla \left(\langle \mathbf{S}(\mathbf{r}) \rangle \cdot \hat{\mathbf{s}}_J \right).$$

Eq. (2.15) ensures rotational invariance of the flow of energy (see Fig. 2.1) which in terms of something we can relate to in the laboratory, represents that the power should be conserved no matter the angle our detector holds with respect to the direction of power flow. Note that Eq. (2.15) may be reached by multiplying Eq. (2.14) consecutively by $\hat{\mathbf{s}}$ and $\hat{\mathbf{s}}_J$, which proves that no *a priori* assumption needs to be used in its derivation.

 Building Block: **Conservation of Energy Flow under Rotation**

This expression, valid only in the far-field, will be the starting point to derive the diffusion approximation in this book since the radiative transfer equation will be obtained directly from this expression, as is shown in Appendix C.

Fig. 2.1 Rotation of the main axis conformed by $\mathbf{e}$, $\mathbf{h}$ and $\hat{\mathbf{s}}$ which hold the relationship, $\hat{\mathbf{s}} = \mathbf{e} \times \mathbf{h}$.

Given the expression for the time-averaged flow of energy density by $\langle \mathbf{S} \rangle$, we can finally find the total power measured by a detector at position $\mathbf{r}$ of area A with surface normal $\hat{\mathbf{n}}$ as:

$$P(\mathbf{r}) = \int_A \langle \mathbf{S}(\mathbf{r}') \rangle \cdot \hat{\mathbf{n}} \mathrm{d}S'. \qquad \left(Watts \right) \qquad (2.16)$$

Eq. (2.16) is the physically measurable quantity and will thus reappear constantly throughout the book.

Finally, as a reminder it is important to consider that in the case of a *non-absorbing medium* (that is, $\mathrm{d}P_{abs}/\mathrm{d}V = 0$) which contains no sources or scatterers the time-averaged expression for the conservation of energy states that (note that for continuous illumination[2] we have that $\langle \partial W/\partial t \rangle = 0$):

$$\nabla \cdot \langle \mathbf{S} \rangle = 0. \qquad (2.17)$$

If we now apply Gauss' theorem,

$$\int_V \nabla \cdot \langle \mathbf{S} \rangle \mathrm{d}V = \int_S \langle \mathbf{S} \rangle \cdot \hat{\mathbf{n}} \mathrm{d}S, \qquad (2.18)$$

we obtain that in a medium containing no sources or absorbers the averaged total flux of energy through any *closed surface*[3] is zero:

$$\int_S \langle \mathbf{S} \rangle \cdot \hat{\mathbf{n}} \mathrm{d}S = 0, \qquad (2.19)$$

being $\hat{\mathbf{n}}$ the outward normal to the surface S which encloses volume V.

2.3 Single Scattering

Now that we have defined the main quantities involved in energy conservation, we will now study what happens when in this previously homogeneous medium we introduce an object. Let us first consider a single particle of arbitrary size and shape, which we illuminate by a distant[4] light source. For simplicity, let us consider that this isolated particle is surrounded by a homogeneous medium, which we illuminate with a laser of a certain wavelength λ.

---

[2]I include here this explicitly, since we will see the case in future chapters where the intensity (instead of the field) is modulated.

[3]A closed surface is a surface that is compact and without boundary, examples of which are the sphere and the torus. Note that a closed surface does not need to be smooth.

[4]Whenever speaking about distances, length is usually compared to the wavelength of the incident light. In this particular case, a distant light source refers to a source which is to all practical effects placed at infinity when compared to the wavelength of light or to the size of the particle.

2.3.1   *The Scalar Theory of Scattering*

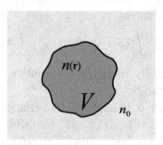

Fig. 2.2   An arbitrary particle of spatially varying index of refraction $n(\mathbf{r})$ occupying a volume V in an otherwise homogeneous medium of refractive index n_0.

Let us consider now the case where we have no charges or currents (i.e. $\mathbf{j} = 0$). If we take Maxwell's equations and apply a time-derivative onto Eq. (2.1) we can substitute the expression for $\partial \mathbf{H}/\partial t$ obtained from Eq. (2.2) giving:

$$-\frac{c}{\mu}\nabla \times \left(\nabla \times \mathbf{E}\right) - \frac{\epsilon}{c}\frac{\partial^2 \mathbf{E}}{\partial t^2} = 0,$$

where we have assumed that μ has no spatial dependence so that we do not need to apply the rotational to it. If we now re-arrange terms and include the fact that $\nabla \times (\nabla \times \mathbf{E}) = \nabla \cdot \nabla \mathbf{E} - \nabla^2 \mathbf{E}$ we obtain:

$$\nabla^2 \mathbf{E} - \frac{\mu\epsilon}{c^2}\frac{\partial^2 \mathbf{E}}{\partial t^2} - \nabla \cdot \nabla \mathbf{E} = 0.$$

Introducing again Maxwell's formula, $n = \sqrt{\epsilon\mu}$, we obtain the wave equation for the electric field[5]:

$$\nabla^2 \mathbf{E} - \frac{n^2}{c^2}\frac{\partial^2 \mathbf{E}}{\partial t^2} - \nabla \cdot \nabla \mathbf{E} = 0.$$

Assuming a time-harmonic dependence of the field:

$$\mathbf{E}(\mathbf{r}, t) = \mathbf{E}(\mathbf{r}) \exp(-i\omega t),$$

we obtain the equation:

$$\nabla^2 \mathbf{E}(\mathbf{r}) + n^2\frac{\omega^2}{c^2}\mathbf{E}(\mathbf{r}) - \nabla \cdot \nabla \mathbf{E}(\mathbf{r}) = 0.$$

---

[5]Detailed derivation of this equation and the similar one of the magnetic field, $\mathbf{H}$, can be found in most optics textbooks, see for example Nieto-Vesperinas (2006), Sec. 3.10 or Born and Wolf (1999) Sec 13.1.1. It is important to remember that we reached this equation assuming μ constant and no charges or currents present.

Considering now a particle of spatially varying index of refraction given by $n(\mathbf{r})$ that occupies a volume V (see Fig. 2.2) in an otherwise infinite and homogeneous medium of refractive index n_0, if the scatterer is nonmagnetic (i.e. $\mu(\mathbf{r}) = \mu_0 = 1$, as will always be the case from now on) the resulting differential equation for the electric vector $\mathbf{E}$ is given by :

$$\nabla^2 \mathbf{E}(\mathbf{r}) + k^2 \mathbf{E}(\mathbf{r}) = -k^2 \left(\frac{n^2(\mathbf{r})}{n_0^2} - 1 \right) \mathbf{E}(\mathbf{r}) + \nabla[\nabla \cdot \mathbf{E}(\mathbf{r})], \qquad (2.20)$$

where $k = n_0 \omega / c = 2\pi/\lambda$ is the wavenumber, being ω the frequency of the electromagnetic wave of wavelength λ. The term $F(\mathbf{r}) = k^2[n^2(\mathbf{r})/n_0^2 - 1]$ is usually termed the *scattering potential*. Note that Eq. (2.20) will be zero for points in space outside of the particle where $n(\mathbf{r}) = n_0$:

$$F(\mathbf{r}) = \begin{cases} k^2 \left[n^2(\mathbf{r})/n_0^2 - 1 \right], & \text{if } \mathbf{r} \text{ is inside } V \\ 0, & \text{if } \mathbf{r} \text{ is outside } V \end{cases} \qquad (2.21)$$

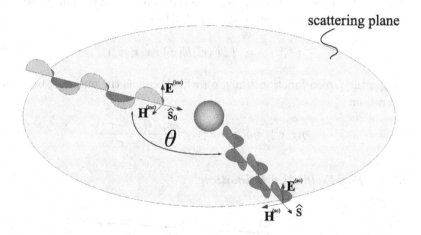

Fig. 2.3 Representation of the incident electromagnetic field scattered by an object. The scattering plane is defined by the incident and outgoing directions $\hat{\mathbf{s}}_0$ and $\hat{\mathbf{s}}$.

Additionally, Eq. (2.20) represents the complete problem to the electric field vector, and shows that the change in polarization of $\mathbf{E}$ as a result of scattering with the particle (or its interaction with the scattering potential) is given by the source term $\nabla[\nabla \cdot \mathbf{E}(\mathbf{r})]$. This term couples the Cartesian components of $\mathbf{E}$, and for this reason rigorous treatment of scattering based on this equation is rather complicated. However, if we assume the refractive index n varies so slowly with position that it is effectively constant over distances of the order of the wavelength, this term can be neglected yielding the uncoupled *scalar* differential equations:

$$\nabla^2 E(\mathbf{r}) + k^2 E(\mathbf{r}) = -F(\mathbf{r})E(\mathbf{r}), \qquad (2.22)$$

where we would have to solve Eq. (2.22) *for each Cartesian component* [6] of **E**. Note that this does not imply that the electric and magnetic fields are scalar, obviously: the electric and magnetic fields will always remain vectors, with the added simplicity that we can solve the wave equation for each cartesian component individually and do not have to deal with coupled equations. For now on we will solve for any one of the Cartesian components of **E** and **H** and use the notation E and H respectively.

The solution to Eq. (2.22) for any point outside the particle can be written as (see Fig. 2.3):

$$E(\mathbf{r}) = E^{(inc)}(\mathbf{r}) + E^{(sc)}(\mathbf{r}), \qquad (2.23)$$

where $E^{(inc)}(\mathbf{r})$ is the incident field (i.e the value of the field in the absence of the particle) and where the scattered field is:

$$E^{(sc)}(\mathbf{r}) = \int_V F(\mathbf{r}')E(\mathbf{r}')g(\mathbf{r}, \mathbf{r}')d^3\mathbf{r}', \qquad (2.24)$$

being g the Green function which we will choose as the outgoing free-space expression:

$$g(\mathbf{r}, \mathbf{r}') = g(|\mathbf{r} - \mathbf{r}'|) = \frac{\exp{(ik|\mathbf{r} - \mathbf{r}'|)}}{4\pi|\mathbf{r} - \mathbf{r}'|}. \qquad (2.25)$$

2.3.2 Far-Field Approximation

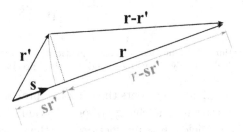

Fig. 2.4 Far field approximation. $\mathbf{r}'$ is projected onto $\mathbf{r}$ through $\hat{\mathbf{s}} \cdot \mathbf{r}'$, in which case if r is much larger than r' we can approximate $r - \hat{\mathbf{s}} \cdot \mathbf{r}' \sim |\mathbf{r} - \mathbf{r}'|$ shown by the gray curve.

---

[6]Note that this can be done only in Cartesian coordinates

So far we have not made any approximations yet on the scattering from the particle, apart from the validity of the scalar approximation to the electric and magnetic field vectors, and Eq. (2.24) together with Eq. (2.25) represent the complete solution. Let us now consider that we are measuring the scattered intensity at a position very far away from our particle (as will always be the case in all instances in this book). Under such conditions we may say we are in the *far field* regime where we may assume that far away from the scatterer the scattered field behaves as an outgoing spherical wave. By expressing vector $\mathbf{r}$ as $\mathbf{r} = r\hat{\mathbf{s}}$ being $\hat{\mathbf{s}}$ a unit vector pointing in the direction of $\mathbf{r}$, when $r \gg r'$ we may write (see Fig. 2.4):

$$|\mathbf{r} - \mathbf{r}'| \sim r - \hat{\mathbf{s}} \cdot \mathbf{r}', \qquad (2.26)$$

in which case the far-field expression of the Green function is:

$$g(|\mathbf{r} - \mathbf{r}'|) \sim \frac{\exp(ikr)}{4\pi r} \exp(-ik\hat{\mathbf{s}} \cdot \mathbf{r}'). \qquad (2.27)$$

In order to arrive to the far-field expression of the scattered electric field, let us now consider that we are illuminating our particle with a monochromatic plane wave of amplitude $|E_0|$ defined by its wavevector $\mathbf{k_i}$ which is pointing in the direction of propagation given by the unit vector $\hat{\mathbf{s}}_0$, $\mathbf{k_i} = k\hat{\mathbf{s}}_0$:

$$E^{(inc)}(\mathbf{r}) = |E_0| \exp(ik\hat{\mathbf{s}}_0 \cdot \mathbf{r}). \qquad (2.28)$$

Taking into account Eq. (2.28) and substituting the far-field expression for the Green function Eq. (2.27) into Eq. (2.24) yields the scattered electric field in the far-field as:

$$E^{(sc)}(\mathbf{r}) = |E_0| f(\hat{\mathbf{s}}, \hat{\mathbf{s}}_0) \frac{\exp(ikr)}{4\pi r}, \qquad (2.29)$$

where we have introduced the *scattering amplitude* f as:

$$f(\hat{\mathbf{s}}, \hat{\mathbf{s}}_0) = \frac{1}{4\pi} \int_V F(\mathbf{r}') \frac{E(\mathbf{r}')}{|E_0|} \exp(-ik\hat{\mathbf{s}} \cdot \mathbf{r}') d^3 r'. \qquad (2.30)$$

 Building Block: **Scattering Amplitude**

This expression contains all information relevant to the particle: its size, shape, and through the inhomogeneous index of refraction, its composition. It will be used to derive all cross-sections in order to finally arrive to the expressions for the absorption and scattering coefficients that will be used within the diffusion approximation.

In Eq. (2.30), f is defined with respect to the incident direction $\hat{s}_0$ and we have introduced the factor $|E_0|$ in order to ensure $f(\hat{s}, \hat{s}_0)$ is independent of the amplitude of the incident field.

Going back to our main quantity of interest, i.e. the time-averaged Poynting vector $\langle \mathbf{S} \rangle$, the scattered time-averaged Poynting vector $\langle \mathbf{S}^{(sc)} \rangle$ once introduced the far-field expression for $E^{(sc)}$ (see Eq. (2.29)) is:

$$\langle \mathbf{S}^{(sc)} \rangle = \frac{\epsilon c_0}{8\pi} |E_0|^2 \frac{|f(\hat{s}, \hat{s}_0)|^2}{r^2} \hat{s}. \tag{2.31}$$

Similarly, the incident time-averaged Poynting vector is expressed as:

$$\langle \mathbf{S}^{(inc)} \rangle = \frac{\epsilon c_0}{8\pi} |E_0|^2 \hat{s}_0, \tag{2.32}$$

in which case we may rewrite the scattered flow of energy $\langle \mathbf{S}^{(sc)} \rangle$ as:

$$\langle \mathbf{S}^{(sc)} \rangle = |\langle \mathbf{S}^{(inc)} \rangle| \frac{|f(\hat{s}, \hat{s}_0)|^2}{r^2} \hat{s}. \tag{2.33}$$

Note that the total time-averaged Poynting vector $\langle \mathbf{S} \rangle$ (i.e the Poynting vector due to the sum of $E^{(inc)}$ and $E^{(sc)}$) will include an additional term due to interference between scattered and incident waves: this is the basis of the *optical theorem*.

2.4 Main Optical Parameters of a Particle

So far, we have found how to describe the field that has been scattered by a particle when we measure it in the far-field, i.e. very far away from it when compared to its size and the wavelength of the incident light, and we have seen that the quantity measured will be directly related to the time-averaged Poynting vector $\langle \mathbf{S} \rangle$. We have also found an expression for the electromagnetic field which depends exclusively on the optical properties of the particle (the refractive index), and the incident and outgoing wave-vectors. By using Eq. (2.29) we are able also to define several very important quantities that refer directly to the properties of the particle and will be the basis for describing optical properties of more complex media such as tissues: the total (or extinction) cross-section, the scattering cross-section, the absorption cross-section, the phase function and the anisotropy factor.

2.4.1 *The Absorption Cross-Section*

The total amount of energy lost by the interaction of the incident light with the particle we defined in the previous section will be only due to absorption, in which case the total power lost is expressed as:

$$\bar{P}_{abs} = \int_V \left\langle \frac{\mathrm{d}P_{abs}}{\mathrm{d}V} \right\rangle \mathrm{d}V = \frac{1}{2}\omega \int_V \epsilon^{(i)}(\mathbf{r})|E(\mathbf{r})|^2 \mathrm{d}V. \qquad \left(Watts\right) \qquad (2.34)$$

This expression represents the amount of energy lost per second due to absorption, and as such depends on the incident flux of energy. The relative amount of energy lost due to the interaction with the particle will be then the ratio of the absorbed power $\bar{P}_{abs}$ to the total power incident on the particle P_{inc}, which is commonly termed *absorption efficiency*:

$$Q_a^{eff} = \frac{\bar{P}_{abs}}{P_{inc}}.$$

Since our illumination source was a plane wave (see Eq. (2.28)), assuming the particle has a geometrical cross-section A the total incident power on the particle per unit area will be $\langle \mathbf{S}^{(inc)} \rangle = P_{inc}/A$. Substituting in our expression for the absorption efficiency we obtain:

$$Q_a^{eff} = \frac{\bar{P}_{abs}}{\langle \mathbf{S}^{(inc)} \rangle} \frac{1}{A}. \qquad (2.35)$$

From Eq. (2.35) we see a direct relation between the relative amount of energy lost due to absorption and the geometrical cross-section of the particle A. What relates these two quantities has units of area and thus is termed the *absorption cross-section* σ_a, so that $Q_a^{eff} = \sigma_a/A$, or equivalently $\sigma_a = Q_a^{eff}A$. In terms of $\bar{P}_{abs}$ it can be quantified as:

$$\sigma_a = \frac{\bar{P}_{abs}}{|\langle \mathbf{S}^{(inc)} \rangle|} = 4\pi k \int_V \frac{\epsilon^{(i)}(\mathbf{r})|E(\mathbf{r})|^2}{\epsilon|E_0|^2} \mathrm{d}V, \qquad \left(cm^2\right) \qquad (2.36)$$

where we have made use of $k = \omega/c_0$.

 Building Block: **Absorption Cross-section**

This relationship will be used to arrive to the final expression for the absorption coefficient, which is one of the fundamental parameters that define the optical properties of diffuse media.

As mentioned before, the absorption cross-section σ_a has units of area, and typically in the bio-optics community it is expressed in cm^2. What it represents is the effective cross-section that the particle has when compared to its geometrical cross-section with respect to the amount of power absorbed. A highly transparent particle, for example, will have an absorption cross-section much smaller than its geometrical cross-section. In any case, the important fact is that the absorption cross-section is independent of the incident fields and is defined uniquely by its material properties and geometry.

Even though Eq. (2.36) will give us the absorption cross-section for a particle of known shape and refractive index distribution it is seldom used and will not be used in this book. What is traditionally used is a value for σ_a which is obtained through direct measurement, for example by the use of tables which have been obtained by measuring a suspension of such particles with a spectrophotometer. In any case, this formula has been included here in order to clearly define the absorption cross-section in terms of the complex index of refraction (or permittivity as shown in Eq. (2.36)) of the particle.

2.4.2   The Scattering Cross-Section

Following the same approach used to derive the absorption cross-section, we can arrive to an expression for scattering efficiency Q_{sc}^{eff} as:

$$Q_{sc}^{eff} = \frac{\bar{P}_{sc}}{\langle \mathbf{S}^{(inc)} \rangle} \frac{1}{A}, \qquad (2.37)$$

where we recall that A is the geometrical cross-section of the particle. The scattering cross-section, σ_{sc}, is expressed as:

$$\sigma_{sc} = \frac{\bar{P}_{sc}}{|\langle \mathbf{S}^{(inc)} \rangle|} = \int_V \frac{\nabla \cdot \langle \mathbf{S}^{(sc)} \rangle}{|\langle \mathbf{S}^{(inc)} \rangle|} dV = \int_S \frac{\langle \mathbf{S}^{(sc)} \rangle \cdot \mathbf{n}}{|\langle \mathbf{S}^{(inc)} \rangle|} dS, \qquad (2.38)$$

where we have made use of Gauss' Theorem Eq. (2.18), in which case it is important to note that *no sources or scatterers* can be outside V. Introducing the far-field expressions for $\langle \mathbf{S}^{(inc)} \rangle$ and $\langle \mathbf{S}^{(sc)} \rangle$ we obtain (note that here we are selecting the center of our surface of integration at the center of the particle) :

$$\sigma_{sc} = \int_S |f(\hat{\mathbf{s}}, \hat{\mathbf{s}}_0)|^2 \frac{dS}{r^2}. \qquad (2.39)$$

Considering that dS/r^2 is in this case the solid angle $d\Omega$, integrating over all angles, (4π), we obtain:

$$\sigma_{sc} = \int_{(4\pi)} |f(\hat{\mathbf{s}}, \hat{\mathbf{s}}_0)|^2 d\Omega. \qquad \left(cm^2 \right) \qquad (2.40)$$

Building Block: **Scattering Cross-section**

This relationship will be used to arrive to the final expression of the scattering coefficient, one of the key parameters that define the optical properties of diffuse media.

It is important to note that even though one might expect the scattering efficiency to be smaller than one, this is by no means the case: due to constructive interference, and more so close to resonances of the particle, we may have scattering efficiencies larger than one.

2.4.3 The Total or Extinction Cross-Section and the Optical Theorem

Once we have defined the absorption and scattering cross-sections, the total or extinction cross-section is defined as:

$$\sigma_{tot} = \sigma_a + \sigma_{sc}. \quad \left(cm^2\right) \tag{2.41}$$

and equivalently, the extinction efficiency:

$$Q_{ext}^{eff} = Q_a^{eff} + Q_{sc}^{eff}. \tag{2.42}$$

Additionally, σ_{tot} can be found using the *optical theorem* which in the case of unpolarized light is expressed as:

$$\sigma_{tot} = \frac{4\pi}{k}\Im\{f(\hat{s}_0, \hat{s}_0)\}. \quad \left(cm^2\right) \tag{2.43}$$

A formal derivation of Eq. (2.43) is out of the scope of this book, but a few words about it are called for. The optical theorem is an important theorem in optics since it relates the properties of the particle (its shape and refractive index dependence) through the scattering amplitude f (see Eq. (2.30)) to a quantity that can be measured in the far-field: the total or extinction cross-section. Even more importantly, this quantity, even though derived from integration over a closed surface, depends *only* on the incident direction of propagation. A traditional way of interpreting the physical meaning of the optical theorem is that the extinction of the incident wave when measured at the direction of incidence is caused by interference between incident and scattered waves [see Bohren and Huffman (1998) for example]. In any case, Eq. (2.43) implies that the amount of light that is removed from the incident wave measured at the same direction of incidence is due to both absorption and scattering and depends only on the imaginary part of the scattering amplitude f as shown in Eq. (2.43).

With respect to the extinction efficiency, as was mentioned previously we may have scattering efficiencies larger than one. This means that we may have extinction efficiencies quite larger than one and, even though it might seem unphysical at first sight (and as such it has been termed the *extinction paradox*), it is directly related to interference effects. Further insight on this effect and on cross-sections in general can be found in [Bohren and Huffman (1998)].

Going back to Eq. (2.43), it is important to remember that f is not a solution on its own: it depends *non-linearly* on the incident wave and therefore each particle will have its own function f depending on its shape and optical properties (see Eq. (2.30)). The good news is that for all purposes of this book and, more generally, in the field of biomedical optics a generic value for f (or more accurately, for $|f|^2$) is used. We will now introduce this quantity as the *phase function*.

2.4.4 The Phase Function

So far, based on the scattering amplitude f we have defined the main parameters that will give us information about the optical properties of a particle: the scattering and absorption cross-sections. Since the scattering cross-section depends only on the modulus-squared of the scattering amplitude, at this point it is convenient to introduce a new function $p(\hat{s}, \hat{s}_0)$ as:

$$p(\hat{s}, \hat{s}_0) = \frac{1}{\sigma_{tot}} |f(\hat{s}, \hat{s}_0)|^2. \qquad \left(no\ units\right) \qquad (2.44)$$

 Building Block: **Phase Function**

The phase function is the key parameter that will be used to describe statistically light interaction with particles.

Eq. (2.44) represents the *scattering diagram* or *phase function*. In biomedical optics the term most commonly employed is *phase function* even though this nomenclature has its origins in astronomy and bears no relationship whatsoever with the actual phase of the electromagnetic wave (I personally prefer 'scattering diagram', but in order to avoid confusion I will stick to the most commonly used expression 'phase function'). Eq. (2.44) can also be seen statistically as the probability of light incident on the particle from direction $\hat{s}_0$ to be scattered in direction $\hat{s}$, thus the normalization factor. Introducing Eq. (2.44) into the expression for the scattering cross-section σ_{sc}, Eq. (2.40), we obtain the following important relationship for the phase function:

$$\int_{(4\pi)} p(\hat{s},\hat{s}_0)d\Omega = \frac{\sigma_{sc}}{\sigma_{tot}} = W_0, \tag{2.45}$$

where W_0 is the *albedo*[7] or *whiteness* of the particle which in the presence of absorption gives $W_0 < 1$. Note that when no absorption is present $\int_{(4\pi)} p(\hat{s},\hat{s}_0)d\Omega = 1$. Before continuing, it is convenient to express the phase function explicitly, in order to better understand the approximations that will follow. This expression can be readily found from the squared modulus of Eq. (2.30), in which case it takes the following form:

$$p(\hat{s},\hat{s}_0) = \frac{1}{\sigma_{tot}}|f(\hat{s},\hat{s}_0)|^2 = \left(\frac{1}{4\pi\sigma_{tot}|E_0|}\right)^2 \times$$

$$\int_V \int_V F(\mathbf{r}')F^*(\mathbf{r}'')E(\mathbf{r}')E^*(\mathbf{r}'')\exp\left(-ik\hat{s}\cdot(\mathbf{r}'-\mathbf{r}'')\right)d^3r'd^3r''. \tag{2.46}$$

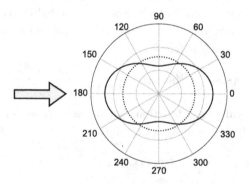

Fig. 2.5 Polar plot of the scattering diagram or phase function for isotropic (dotted line) and Rayleigh (solid line) scattering. In this figure light would be incident from the left.

Eq. (2.46) includes all the interference effects that might occur *inside* the particle. Clearly, solving this integral for each particle is not practical. Therefore, when dealing with large collections of particles which are statistically equivalent it is more convenient to use an approximation to this expression. The following are examples of phase functions commonly used:

[7]From the Latin *albus* which means white.

- **Isotropic phase function:** In the isotropic case (see Fig. 2.5) we have that $p(\hat{\mathbf{s}}, \hat{\mathbf{s}}_0) = p_0$ and therefore:

$$p(\hat{\mathbf{s}}, \hat{\mathbf{s}}_0) = \frac{W_0}{4\pi}. \qquad (2.47)$$

  This formula is reached by considering that in the scattering amplitude f in Eq. (2.30) that the total electric field inside the particle is the incident field $E \simeq E^{(inc)}(\mathbf{r})$ *and* that the particle is smaller than the wavelength so that we do not need to integrate. Since this approximation is not too accurate for describing light interaction with tissue components it is not commonly used.

- **Rayleigh's phase function:** Rayleigh's phase function is defined as:

$$p(\hat{\mathbf{s}}, \hat{\mathbf{s}}_0) = p(\hat{\mathbf{s}} \cdot \hat{\mathbf{s}}_0) = \frac{3}{16\pi}(1 + (\hat{\mathbf{s}} \cdot \hat{\mathbf{s}}_0)^2)W_0, \qquad (2.48)$$

  where we must remember that $\hat{\mathbf{s}} \cdot \hat{\mathbf{s}}_0 = \cos\theta$, being θ the angle subtended by the directions $\hat{\mathbf{s}}$ and $\hat{\mathbf{s}}_0$. Its polar plot is represented in Fig. 2.5.

- **Henyey-Greenstein's phase function:** Originally derived for interstellar scattering [Henyey and Greenstein (1941)] it is the most common phase function used in biomedical optics since it accurately describes, at least in a *statistical* manner the average interaction of light with tissue[8]. It depends on both the angle subtended between the incident and measurement vectors $\hat{\mathbf{s}} \cdot \hat{\mathbf{s}}_0 = \cos\theta$ *and* the anisotropy of the phase function $g = \langle \hat{\mathbf{s}} \cdot \hat{\mathbf{s}}_0 \rangle = \langle \cos\theta \rangle$:

$$p(\hat{\mathbf{s}}, \hat{\mathbf{s}}_0) = p(\hat{\mathbf{s}} \cdot \hat{\mathbf{s}}_0) = \frac{1}{4\pi} \frac{W_0(1 - g^2)}{(1 + g^2 - 2g\hat{\mathbf{s}} \cdot \hat{\mathbf{s}}_0)^{3/2}}. \qquad (2.49)$$

A more general phase function can be defined as series of Legendre polynomials with constants a_m of the form:

$$p(\hat{\mathbf{s}} \cdot \hat{\mathbf{s}}_0) = \sum_{m=0}^{n} a_m P_m(\hat{\mathbf{s}} \cdot \hat{\mathbf{s}}_0) =$$

$$a_0 + a_1(\hat{\mathbf{s}} \cdot \hat{\mathbf{s}}_0) + a_2\frac{1}{2}(3(\hat{\mathbf{s}} \cdot \hat{\mathbf{s}}_0)^2 - 1)\ldots \qquad (2.50)$$

Note that this more general expression for the phase function includes the assumption $p(\hat{\mathbf{s}}, \hat{\mathbf{s}}_0) = p(\hat{\mathbf{s}} \cdot \hat{\mathbf{s}}_0)$.

---

[8]I would like to emphasize here the fact that it is overall on *average* a good approximation in biomedical optics. There are cases where it does not work too well and additional factors need to be included. A good example of where its accuracy is questionable is in scattering by individual cells. I recommend reading the PhD thesis dissertation of A. Dunn where the accuracy of the Henyey-Greenstein phase function is studied in detail [Dunn (1997)].

Finally, due to *reciprocity relation* of the scattering amplitude $f(\hat{s}, \hat{s}_0) = f(-\hat{s}_0, -\hat{s})$ we obtain the following important reciprocity property of the phase function:

$$p(\hat{s}, \hat{s}_0) = p(\hat{s}_0, \hat{s}) = p(-\hat{s}, -\hat{s}_0) = p(-\hat{s}_0, -\hat{s}).$$

In terms of the phase function, the expression for the *scattered flow of energy in the far field* as defined in Eq. (2.31) becomes:

$$\langle \mathbf{S}^{(sc)} \rangle = \frac{\epsilon c_0}{8\pi} |E_0|^2 \sigma_{tot} \frac{p(\hat{s}, \hat{s}_0)}{r^2} \hat{s},$$

where we recall that $|E_0|$ is the modulus of the electric field *incident on the particle*. As can be seen from this expression, if we integrate over a closed surface we obtain once again the expression for the scattering cross-section, $\int_S \langle \mathbf{S}^{(sc)} \rangle \hat{s} dS = \sigma_{sc} |\langle \mathbf{S}^{(inc)} \rangle|$.

Before moving on it is important to note that there are several ways of defining the phase function. In particular, you will also find quite regularly in the literature an alternative definition as $p(\hat{s}, \hat{s}_0) = |f(\hat{s}, \hat{s}_0)|^2$, in which case it is always normalized to unity, instead of the albedo as in Eq. (2.45). Another common expression is to define the phase function multiplied by 4π, i.e. $p(\hat{s}, \hat{s}_0) = 4\pi |f(\hat{s}, \hat{s}_0)|^2 / \sigma_{tot}$. Which definition you use is not important as long as it is used consistently, since it will not affect the derivation of the radiative transfer equation or the diffusion equation.

2.4.5 *The Anisotropy Factor*

Up to this point we have been able to define how much a particle scatters and absorbs light, and we can now define precisely how it does this through the recently introduced phase function which relates the direction of incidence with the direction of measurement. In terms of these definitions we can also now quantify how efficiently a particle scatters light in terms of its albedo $W_0 = \sigma_{sc}/\sigma_{tot}$. However, the albedo informs us of the *overall* properties of the particle. A very useful parameter that was briefly introduced in the previous section in the context of the Henyey-Greenstein phase function is the *anisotropy factor* which will also tell us how much scattering occurs in the forward direction compared to the backward direction. In more practical terms this can be understood as how 'transparent' a particle is, at least when dealing with tissues[9]: the more transparent the

---

[9]In some special cases you might have enhancement of the forward direction due to interference effects. This, however, is generally not the case for the type of scatterers we might find in tissue.

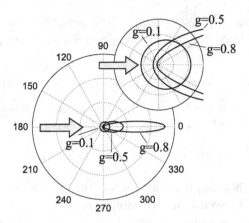

Fig. 2.6 Polar plot of the scattering diagram or phase function for the Henyey-Greenstein expression for anisotropy values of $g = 0.1$, $g = 0.5$ and $g = 0.8$. The inset shows in greater detail the difference for backscattering angles (note that light is incident from the left).

particle the more light is distributed in the forward direction (note that within this relaxed definition transparency is only affected by scattering, not absorption). If for example we considered a fat particle, we could easily predict that its scattering will be highly anisotropic in the forward direction, which is actually very close to reality. This can be seen in Fig. 2.6, where different scattering diagrams are shown for this function for several values of the anisotropy. A way to express this in terms of the phase function is by defining the anisotropy factor as the average of the cosine of the angle subtended between the incident and scattering directions as:

$$g = \langle \cos\theta \rangle = \langle \hat{\mathbf{s}} \cdot \hat{\mathbf{s}}_0 \rangle = \frac{\int_{(4\pi)} p(\hat{\mathbf{s}}, \hat{\mathbf{s}}_0)\hat{\mathbf{s}} \cdot \hat{\mathbf{s}}_0 d\Omega}{\int_{(4\pi)} p(\hat{\mathbf{s}}, \hat{\mathbf{s}}_0)d\Omega}, \qquad (2.51)$$

where it is clear that g can take values from $g = 1$ (all light is in the forward direction) to $g = -1$ (all light is back-scattered). $g = 0$ of course represents the isotropic scattering case. Introducing the expression from Eq. (2.45) we obtain:

$$g = \frac{1}{W_0} \int_{(4\pi)} p(\hat{\mathbf{s}}, \hat{\mathbf{s}}_0)\hat{\mathbf{s}} \cdot \hat{\mathbf{s}}_0 d\Omega. \qquad \left(no\ units\right) \qquad (2.52)$$

Building Block: **Anisotropy Factor**

The anisotropy factor represents the average of the cosine between the incident and scattered fields and will appear whenever defining the average optical properties of diffuse media.

Typical values of the anisotropy factor g in tissue range from $g \simeq 0.8$ to $g \simeq 0.9$ (see [Cheong *et al.* (1990)] for example). For all examples shown in this book, $g = 0.8$ has been chosen.

2.5 Multiple Scattering

In the previous section we introduced all the parameters that characterize light interaction with a particle of arbitrary shape, and following the derivations presented we should be able to predict how much light (i.e. what power) would be collected at a detector of area A located at a certain point in space $\mathbf{r}$ (remember, always far away from the particle since all derivations were done in the far-field region) when we shine light from a specific direction $\hat{\mathbf{s}}_0$ onto our particle. It is time now then to move on to a more practical scenario with relevance in biomedical optics: what happens if we have a collection of these particles, distributed randomly?

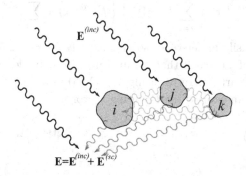

Fig. 2.7 Representation of multiple scattering from three generic particles, where only up to second order of scattering is shown. The total field will be the sum of *all* multiple scattering contributions. Note that this figure represents our approach to solving the multiple scattering problem (as a sum of a homogeneous and particular solutions), and *does not* represent what physically takes place in a complex scattering problem.

To begin with, we will assume that all our particles have the same optical properties. For the time being we will also assume that each particle has a different size and shape. For a collection of N such particles at positions $\mathbf{r}_i$, and ignoring de-polarization effects, we can write the scattered field in the far-field at $\mathbf{r}$ as (see Fig. 2.7):

$$E^{(sc)}(\mathbf{r}) = \sum_{i=1}^{N} E_i^{(sc)}(\mathbf{r}).$$

being (see Eq. (2.29) for comparison):

$$E_i^{(sc)}(\mathbf{r}) = |E_0| f_i(\hat{\mathbf{s}}_i, \hat{\mathbf{s}}_0) \frac{\exp(ik|\mathbf{r} - \mathbf{r}_i|)}{|\mathbf{r} - \mathbf{r}_i|}, \tag{2.53}$$

assuming an incident field $E^{(inc)}(\mathbf{r}) = E_0 \exp(ik\hat{\mathbf{s}}_0 \cdot \mathbf{r})$, and where we have defined f_i as:

$$f_i(\hat{\mathbf{s}}_i, \hat{\mathbf{s}}_0) = \frac{1}{4\pi} \int_{V_i} F_i(\mathbf{r}'_i) \frac{E(\mathbf{r}'_i)}{|E_0|} \exp(-ik\hat{\mathbf{s}}_i \cdot \mathbf{r}'_i) \mathrm{d}^3 r'_i, \tag{2.54}$$

where $\hat{\mathbf{s}}_i$ is now the unit vector pointing from particle i in the direction of $\mathbf{r}$, i.e. $\hat{\mathbf{s}}_i = (\mathbf{r} - \mathbf{r}_i)/|\mathbf{r} - \mathbf{r}_i|$, and $E(\mathbf{r}_i) = E^{(inc)}(\mathbf{r}_i) + E^{(sc)}(\mathbf{r}_i)$ is the *total* field at $\mathbf{r}_i$. The following step is to retrieve the expression for the flow of energy density at point $\mathbf{r}$ expressed by the Poynting vector $\langle \mathbf{S} \rangle$. For this, we first need to find the intensity $|E|^2$ which expressed in terms of the contribution of each particle is:

$$|E(\mathbf{r})|^2 = \left(E^{(inc)}(\mathbf{r}) + \sum_{i=1}^{N} E_i^{(sc)}(\mathbf{r}) \right) \left(E^{(inc)}(\mathbf{r}) + \sum_{i=1}^{N} E_i^{(sc)}(\mathbf{r}) \right)^*. \tag{2.55}$$

As can be easily understood, solving exactly for Eq. (2.55) is a colossal task, specially if we recall that we also need to solve for the field inside *each* individual particle. Introducing Eq. (2.55) into the expression for $\langle \mathbf{S} \rangle$ Eq. (2.10), we can express the Poynting vector as:

$$\langle \mathbf{S} \rangle = \langle \mathbf{S}^{(inc)} \rangle + \sum_{i=1}^{N} \langle \mathbf{S}^{(sc)} \rangle_i + \sum_{\substack{i,j=1 \\ i \neq j}}^{N} \langle \mathbf{S}^{(sc)} \rangle_{ij} + \dots, \tag{2.56}$$

where $\langle \mathbf{S}^{(sc)} \rangle_i$ is the contribution to the flow of energy at $\mathbf{r}$ due to particle i, $\langle \mathbf{S}^{(sc)} \rangle_{ij}$ is the contribution of the interference between particle i and particle j; all other orders of scattering (i.e. the interference term between particles i,j and k for example) have not been explicitly written. This situation is schematically shown in Fig. 2.7. In any case, it is clear this series carries on to as many combinations as we can obtain between N

particles[10]. In order to simplify this extremely complex problem and arrive to the diffusion equation, at this point we will neglect *all* interference effects that might take place due to scattering between particles (see Fig. 2.8):

$$\langle \mathbf{S} \rangle \simeq \langle \mathbf{S}^{(inc)} \rangle + \sum_{i=1}^{N} \langle \mathbf{S}^{(sc)} \rangle_i. \tag{2.57}$$

 Note: **About the Interference Contribution**

From this point on we will not consider any more interference phenomena. For all practical purposes the wavelength of light will be considered much smaller than the distances and sizes we are interested in. By means of this approximation we will not consider any effects that arise from the addition of fields such as interference but will rather consider only the addition of intensities. Note that polarization of light can still be included.

Following Eq. (2.53), $\langle \mathbf{S}^{(sc)} \rangle_i$ are represented as:

$$\langle \mathbf{S}^{(sc)}(\mathbf{r}) \rangle_i = \frac{\epsilon c_0}{8\pi} |E^{(inc)}(\mathbf{r}_i)|^2 |f_i(\hat{\mathbf{s}}_i, \hat{\mathbf{s}})|^2 \frac{1}{|\mathbf{r} - \mathbf{r}_i|^2} \hat{\mathbf{s}}_i,$$

which can be rewritten in terms of $\langle \mathbf{S}^{(inc)} \rangle$ and the phase function p as:

$$\langle \mathbf{S}^{(sc)}(\mathbf{r}) \rangle_i = \sigma_{tot} \langle \mathbf{S}^{(inc)}(\mathbf{r}_i) \rangle \cdot \hat{\mathbf{s}} \frac{p(\hat{\mathbf{s}}_i, \hat{\mathbf{s}})}{|\mathbf{r} - \mathbf{r}_i|^2} \hat{\mathbf{s}}_i. \tag{2.58}$$

In Eq. (2.57), depending on how we approximate the field inside each particle (i.e. we can approximate it to the incident field[11], or write it explicitly as the sum of the field contribution from all particles), we will be accounting for higher orders of scattering. However, as was discussed earlier, typically the scattering diagram is approximated by a phase function p defined *a-priori* by fitting different expressions on experimental measurements.

Let us now study in detail the implications that the approximation present in Eq. (2.57) has. To begin with, using exactly the same approach used in the previous section to derive the scattering cross-section we find that the *total* scattering cross-section is defined as:

$$\bar{\sigma}_{sc} = \sum_{i=1}^{N} \sigma_{sc}^{(i)}, \qquad \left(cm^2 \right)$$

---

[10]And bear in mind that even if we wrote this series down explicitly we wouldn't even be close to solving the complete problem: remember that each of these $\langle \mathbf{S}^{(sc)} \rangle_i$ terms needs the *total* electric field in order for us to reach an exact solution.

[11]This is the Born approximation which will be dealt with in detail in the context of light diffusion.

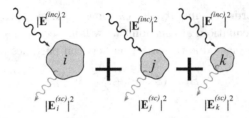

Fig. 2.8 If we neglect all interference effects between particles, we can approximate the total intensity as the sum of intensities scattered from each particle.

being $\sigma_{sc}^{(i)}$ the scattering cross-section for each individual particle, and the *total* absorption cross-section takes the expression:

$$\bar{\sigma}_a = \sum_{i=1}^{N} \sigma_a^{(i)}. \qquad \left(cm^2 \right)$$

That is, if we neglect interference effects between particles, their cross-sections are *additive*.

2.5.1 *The Scattering and Absorption Coefficients*

At this point we have found an approximation that gives us the total scattering and absorption cross-sections of an ensemble of N particles. However, in all practical situations we do not know how many particles are present, let alone what are their individual sizes or shapes. Therefore, it is convenient to introduce an approximation with statistical significance. This is usually done assuming that we can describe the total distribution of particles as a collection of particles of *average* radius R_a. Being $\xi(R_a)dR_a$ the number of particles per cm^3 with radii in the interval dR_a, the average scattering cross-section per unit volume of the medium will be given by:

$$\mu_s = \int_0^\infty \xi(R_a)\sigma_{sc}(R_a)dR_a, \qquad \left(cm^{-1} \right) \qquad (2.59)$$

being

$$\int_0^\infty \xi(R_a)dR_a = \frac{N}{V},$$

where N is the total number of particles and V the volume they occupy. The quantity defined in Eq. (2.59) is the *scattering coefficient* and will be the quantity used throughout this book. If all particles have the same average radius R_o, $\xi = \rho\delta(R_a - R_o)$ and we retrieve the well-known relationship:

$$\mu_s = \rho\sigma_{sc}, \qquad \left(cm^{-1} \right) \qquad (2.60)$$

where $\rho = N/V$ is the density in particles per cm^3 and σ_{sc} is the scattering cross-section of a single particle (see Fig. 2.9). Proceeding the same way for the total absorption cross-section, the average absorption cross-section per unit volume is defined as the *absorption coefficient* and is expressed as:

$$\mu_a = \int_0^\infty \xi(R)\sigma_a(R)dR, \qquad \left(cm^{-1}\right)$$

or, in the case of identical particles, $\xi = \rho\delta(R_a - R_o)$, we obtain:

$$\mu_a = \rho\sigma_a. \qquad \left(cm^{-1}\right) \qquad (2.61)$$

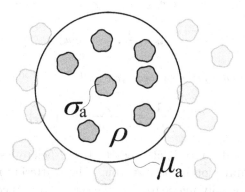

Fig. 2.9 Representation of the relationship between the absorption cross-section of a single particle σ_a and the absorption coefficient μ_a, for a specific density of particles ρ. The same relationship holds for the scattering coefficient.

With respect to the other important quantities of interest, namely the phase function and the anisotropy factor, the following relations hold: since we have assumed spherical particles *on average*, there will be no preferential scattering with respect to the direction of incidence:

$$p(\hat{\mathbf{s}}, \hat{\mathbf{s}}_0) = p(\hat{\mathbf{s}} \cdot \hat{\mathbf{s}}_0).$$

From now on we will assume that the Henyey-Greenstein phase function Eq. (2.49) yields a scattering diagram similar to that of the constituents of tissue, as has been shown experimentally to yield accurate results (see [Cheong *et al.* (1990)] and references therein). With regards to the anisotropy factor g, once the above mentioned approximations are taken, the anisotropy of our medium must either be measured experimentally or must be known *a priori*. However, in order to account for the average anisotropy of scattering, another quantity used which relates the scattering coefficient of an anisotropic particle to that of an isotropic particle is the *reduced scattering coefficient*:

$$\mu_s' = \mu_s(1 - g). \quad \left(cm^{-1} \right) \tag{2.62}$$

With respect to the above mentioned coefficients it is also common to define the following distances:

- The *scattering mean free path*:

$$l_{sc} = \frac{1}{\mu_s}, \quad \left(cm \right) \tag{2.63}$$

- and the *reduced scattering mean free path*:

$$l_{sc}' = \frac{1}{\mu_s'} = \frac{l_{sc}}{(1 - g)}, \quad \left(cm \right) \tag{2.64}$$

which is the distance light needs to travel before losing memory of its original direction of propagation. The scattering mean free path is also traditionally understood as the *average* distance light needs to travel until it suffers the next scattering event. Great care should be taken when using this definition. Consider the following case: we have a collection of particles with density ρ, with scattering cross-section σ_{sc}. Their scattering coefficient and, in turn, l_{sc} will be *identical* to that from a collection of particles with σ_{sc}' and ρ' as long as $\rho' = \rho\sigma_{sc}/\sigma_{sc}'$. Clearly, if $\rho \neq \rho'$, the particles are located at different average distances from one another. Therefore, speaking about 'distances' to the next scattering event does not hold a true physical meaning; the only relevant physical parameter is the statistical distance needed to travel before the direction of the scattered light is independent of the direction of incidence. This issue will be further studied in Chap. 4.

Additionally, other distances that are traditionally defined are the transport mean free path, $l_{tr} = 1/(\mu_a + \mu_s)$, the reduced transport free path $l_{tr}' = 1/(\mu_a + \mu_s')$ and the absorption length $l_{abs} = 1/\mu_a$. The transport mean free path can be seen as the distance light has to travel to have its intensity reduced due to extinction by a factor of e.

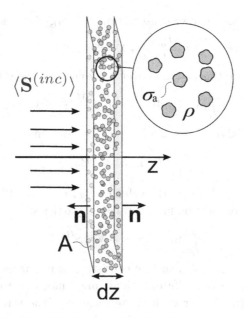

Fig. 2.10   Light incident on a thin slab of absorbing particles.

2.6   Extinction by a Slab of Absorbing Particles

Before moving on, in order to get a clearer picture of the effect that the recently introduced scattering and absorption coefficients have on the incident radiation we will now attempt to solve for the situation depicted in Fig. 2.10. This represents the case of a thin slab which contains a certain number of *non-scattering* particles with density ρ whose optical properties are represented by their absorption coefficients. In order to avoid contributions from specular reflection we also assume that the index of refraction is constant throughout space. Let us now consider a plane wave with normal incidence on the slab, whose dimensions are given by a surface of area A and normals **n** pointing in the z direction (see Fig. 2.10). To simplify matters, we will assume that the width of the slab dz is much smaller than the area of each of the faces of the slab, i.e. $dz \ll A$, in which case we do not need to account for the sides of the slab. In order to solve for this, we will resort to the equation of energy conservation, Eq. (2.14). The approach that follows to solve for the slab of particles is not the common one found in text books but, since it makes use of most of the concepts recently introduced, I believe it is quite helpful. Assuming we have a constant illumination, the time-averaged expression for conservation of energy becomes:

$$\left\langle \frac{\mathrm{d}P_{abs}}{\mathrm{d}V} \right\rangle + \nabla \cdot \langle \mathbf{S} \rangle = 0.$$

If we now integrate this equation over the volume of the slab V we obtain:

$$\int_V \nabla \cdot \langle \mathbf{S} \rangle \mathrm{d}V = -\int_V \left\langle \frac{\mathrm{d}P_{abs}}{\mathrm{d}V} \right\rangle \mathrm{d}V.$$

Applying Gauss' theorem on the left term, we obtain:

$$\int_S \langle \mathbf{S} \rangle \cdot \mathbf{n}\mathrm{d}S = -\bar{P}_{abs},$$

where we can now substitute the value of $\bar{P}_{abs}$ from Eq. (2.36) in terms of the absorption cross-section for a total of N particles:

$$\int_S \langle \mathbf{S} \rangle \cdot \mathbf{n}\mathrm{d}S = -N\sigma_a S^{(inc)},$$

Since what we have is a surface consisting of two faces of area A with normals pointing in opposite directions, and since due to the symmetry of the problem (remember we have an incident plane wave) the intensity should depend only on z (i.e. $S(\mathbf{r}) = S(z)$), we may write the surface integral as:

$$[S(z = 0) - S(z = dz)]\,A = -N\sigma_a \|\mathbf{S}^{(inc)}\|.$$

which, considering dz infinitesimally small, and accounting for the fact that the number of particles in the slab is $N = \rho V = \rho A dz$, we may write as:

$$\frac{dS}{dz} = -\rho\sigma_a S, \tag{2.65}$$

where we have also assumed that the flow of energy on the slab of particles is $S^{(inc)} \simeq S$ since dz is infinitesimally small. Eq. (2.65) enables us to calculate the absorption due to a thicker slab of width L as a combination of these slices, in which case we must integrate Eq. (2.65) yielding:

$$S(z = L) = S^{(inc)} \exp\left(-\mu_a L\right), \tag{2.66}$$

where we have used the boundary condition $S(z = 0) = S^{(inc)}$. Eq. (2.66) is commonly referred to as *Beer-Lambert's law*. In the case where *each slice* has a different absorption coefficient, Eq. (2.66) results in:

$$S(z = L) = S^{(inc)} \exp\left(-\int_0^L \mu_a(z)dz\right), \tag{2.67}$$

which is commonly used in X-ray computed tomography and in the recently developed optical projection tomography, its optical equivalent.

Similar to the derivation we just saw for an absorbing slab, it is possible to introduce scattering by considering the scattering contribution through the surface integral, $\int_S S^{(sc)} dS = N\sigma_{sc} S^{(inc)}$. Alternatively, the optical theorem accounts for extinction of the incident wave on the direction of propagation and would take into account scattering. However, a great deal of approximations need to be introduced since we cannot account properly for scattering using this simple approach: the intensity of the beam will decay exponentially proportional to the total or extinction coefficient, but will not take into account the *scattering anisotropy*. For scattering volumes there will be a significant positive contribution from scattered intensity to the total flow of energy. This is usually accounted for by using the reduced transport coefficient, $\mu'_{tr} = \mu'_s + \mu_a$. We will reach this expression from the RTE in Chap. 3.

2.7 Polarization Effects

In all derivations presented up to now, we have not discussed the effect of polarization of the incident wave $\mathbf{E}^{(inc)}$. In particular, we have presented the optical theorem in the context of *unpolarized waves*. Even though exploiting the use of polarization to probe biological media is indeed very promising, up to date there are very few instances where polarization is taken into account when dealing with diffuse light. I have therefore opted for not including a thorough derivation of the effects of polarization, and directing the reader to Bohren and Huffman (1998), Chap. 3.3, for example. However, I do believe a few basic concepts are necessary. First, we need to define the changes in polarization upon scattering with an object. For that, what is commonly used is the *scattering matrix* , which relates an incident polarization given by $\mathbf{E}^{(inc)} = E_\parallel^{(inc)} \mathbf{e}_\parallel^{(i)} + E_\perp^{(inc)} \mathbf{e}_\perp^{(i)}$ to a scattered polarization given by $\mathbf{E}^{(sc)} = E_\parallel^{(sc)} \mathbf{e}_\parallel^{(sc)} + E_\perp^{(sc)} \mathbf{e}_\perp^{(sc)}$ as:

$$\begin{pmatrix} E_\parallel^{(sc)} \\ E_\perp^{(sc)} \end{pmatrix} = \frac{\exp(ikr)}{r} \begin{pmatrix} S_2 & S_3 \\ S_4 & S_1 \end{pmatrix} \begin{pmatrix} E_\parallel^{(inc)} \\ E_\perp^{(inc)} \end{pmatrix}, \qquad (2.68)$$

where all $S_{1,2,3,4}$ depend on the outgoing, $\hat{s}$, and incident, $\hat{s}_0$, directions of propagation[12]. It is clear from Eq. (2.68) that this definition is in the *far-field* and should be compared with Eq. (2.29). Since the quantity of interest is the resulting scattered flow of energy or scattered Poynting vector, $\langle \mathbf{S}^{(sc)} \rangle$,

---

[12]In most text books the notation for the scattering matrix is as shown in Eq. (2.68), i.e. the S_1 term is the one that relates the perpendicular polarization contributions.

we must characterize the interaction between the incident flow of energy and the particle while maintaining information on the polarization state. Note that if de-polarization needs to be properly accounted for, the scalar approximation to the wave equation cannot be used (see Eq. (2.22)). The polarization state of the intensity can be described in terms of the Stokes parameters which, using the above notation, are expressed for the scattered component as:

$$I_{sc} = \langle E_{\parallel}{}^{(sc)}(E_{\parallel}{}^{(sc)})^* + E_{\perp}{}^{(sc)}(E_{\perp}{}^{(sc)})^* \rangle,$$
$$Q_{sc} = \langle E_{\parallel}{}^{(sc)}(E_{\parallel}{}^{(sc)})^* - E_{\perp}{}^{(sc)}(E_{\perp}{}^{(sc)})^* \rangle,$$
$$U_{sc} = \langle E_{\perp}{}^{(sc)}(E_{\parallel}{}^{(sc)})^* + E_{\parallel}{}^{(sc)}(E_{\perp}{}^{(sc)})^* \rangle,$$
$$V_{sc} = i\langle E_{\perp}{}^{(sc)}(E_{\parallel}{}^{(sc)})^* - E_{\parallel}{}^{(sc)}(E_{\perp}{}^{(sc)})^* \rangle,$$

where $\langle \cdot \rangle$ denotes time-averaging as in Eq. (2.9). The Stokes parameters are real numbers which satisfy the relation:

$$I_{sc}^2 \geq Q_{sc}^2 + U_{sc}^2 + V_{sc}^2,$$

and have been derived in order to define the most general state of polarization, which is elliptic polarization (by changing the ratio of the axis of the ellipse we can define all states of polarization: linear, circular and elliptical). In the above expression we would have $I_{sc}^2 = Q_{sc}^2 + U_{sc}^2 + V_{sc}^2$ only in the case of completely polarized light. In terms of the Stokes parameters, the description of the scattered flow of energy which includes the state of polarization is given by:

$$\begin{pmatrix} I_{sc} \\ Q_{sc} \\ U_{sc} \\ V_{sc} \end{pmatrix} = \frac{1}{r^2} \begin{pmatrix} S_{11} & S_{12} & S_{13} & S_{14} \\ S_{21} & S_{22} & S_{23} & S_{24} \\ S_{31} & S_{32} & S_{33} & S_{34} \\ S_{41} & S_{42} & S_{43} & S_{44} \end{pmatrix} \begin{pmatrix} I_{inc} \\ Q_{inc} \\ U_{inc} \\ V_{inc} \end{pmatrix}.$$

This 4×4 matrix is the *Mueller matrix* for scattering by a single object, which more conveniently we can write in matrix form as:

$$\left[\mathbb{I}^{(sc)} \right]_{4 \times 1} = \frac{1}{r^2} \left[\mathbb{M} \right]_{4 \times 4} \left[\mathbb{I}^{(inc)} \right]_{4 \times 1}$$

From the 16 matrix elements of $[\mathbb{M}]$ only 7 can be independent, corresponding to the diagonal components which represent the moduli of the components of the scattering matrix Eq. (2.68), $S_{ii=1..4} = |S_i|^2$ and the three differences in phase between the $S_{i=1..4}$.

In the case of a distribution of particles, as long as we may use the approximation Eq. (2.57) the Stokes parameters of the light scattered can be considered the sum of the Stokes parameters of the light scattered by each individual particle:

$$\left[\mathbb{I}^{(sc)} \right]_{4 \times 1} = \frac{1}{r^2} \sum_{i=1}^{N} \left[\mathsf{M}^{(i)} \right]_{4 \times 4} \left[\mathbb{I}^{(inc)} \right]_{4 \times 1},$$

where $\left[\mathsf{M}^{(i)} \right]$ is the Muller matrix for particle i and we have N particles.

We will not dwell anymore on describing polarized light but, before moving on, there are a few things to keep in mind:

- **Unpolarized Light**: In the case in which we have completely unpolarized light, the incident Stokes parameters Q, U, and V will be zero: $Q_{inc} = U_{inc} = V_{inc} = 0$. Assuming we have an incident intensity I_{inc}, the Stokes parameters for the scattered light are:

$$I_{sc} = \frac{S_{11}}{r^2} I_{inc}, \qquad Q_{sc} = \frac{S_{21}}{r^2} I_{inc},$$

$$U_{sc} = \frac{S_{31}}{r^2} I_{inc}, \qquad V_{sc} = \frac{S_{41}}{r^2} I_{inc},$$

which yields a degree of polarization:

$$\delta = \sqrt{\frac{S_{21}^2 + S_{31}^2 + S_{41}^2}{S_{11}^2}}.$$

That is, scattering polarizes light. As an example, single Rayleigh scattering is known to polarize the scattered light in a cloudless blue sky. Bear in mind that since the S_{ij} factors depend on the direction we measure the scattered light $\hat{s}$ so will the degree of polarization.

- **Scattering by a Collection of particles**: In the case where we have a collection of identical particles all receiving light from the same direction, the scattered light will be completely polarized. Note that this does not mean that it will keep the same polarization necessarily; generally, linearly polarized light will be transformed into elliptically polarized light. Possibly due to this reason circular polarization is maintained for larger distances into highly scattering media such as tissue[MacKintosh *et al.* (1989)]. However, in the general case of a collection of particles with different orientations and sizes, scattering will depolarize light.

Even though, in my personal opinion, polarization has not been yet fully exploited in the field of diffuse optical imaging in biology and medicine it is however quite extensively used in Optical Coherence Tomography (OCT), and detailed derivation and the use of the Mueller matrix can be found in Wang and Wu (2007). With respect to highly scattering media with applications in biology and medicine, there was great interest in the 90's, since it is clear that using polarized light we can discriminate between different scattering contributions (see Schmitt *et al.* (1992) for example). After that, some work can be found in diffuse-reflectance imaging (see for example Li and Yao (2009) for a visual representation of the Mueller matrix in tissue). Additionally, several works can be found based on Monte Carlo simulations, such as in [Yao and Wang (2000)] and Guo *et al.* (2010) where the authors predict the propagation of polarized light in turbid media by means of Monte Carlo simulations.

2.8   Self-Averaging

Before closing this chapter, there are a few things that need to be understood. First of all, since we are dealing with biological tissue, there will never be the case where all particles are identical. Additionally, all distances to our detectors can be considered in the far-field, since simply by the use of a lens we are effectively in that range. Also, considering the size of a single cell, which is between 10 and 100 μm, we can be almost certain to be studying media where the scatterers are much larger than the wavelength, even though sub-cellular components are in the wavelength range or smaller (it seems that organelles have an important contribution at large scattering angles, since being smaller that the wavelength they act as isotropic scatterers [Dunn and Richards-Kortum (1996)]). And, finally and most importantly, we will always be probing or studying a *dynamic* medium. Unless we study frozen sections of tissue, there will always be movement present, causing what is called *self-averaging*. Let me elaborate on this issue a bit further. Consider the case where we have a collection of scatterers embedded in a solid. If we shine coherent light (a laser, for example) on this sample we will have multiple scattering effects as discussed in this chapter, and will measure a great deal of speckle, which arises from coherent interference between the scattered waves (i.e. the terms we neglected in Eq. (2.56)). This will happen whatever the sizes of our scatterers, and is due to the simple fact that differences in phase are maintained. However, if these scatterers were moving, *on average* most of these phase differences would cancel out and we would measure practically no speckle[13]. This is exactly what happens in tissue, and for that I recommend you try the following simple experiment: shine a laser pointer at a white surface which is static, a white-painted wall for instance. You will see speckle in this case. However, if you shine this laser pointer onto your hand you will not see any speckle patterns, they have all been effectively canceled out due to self-averaging caused by the movement of your cells, motility and blood flow[14]. This very basic difference is what relaxes most of the assumptions we make in order to use the diffusion approximation and it is what renders it so useful for *in-vivo* imaging applications.

Key Points

From this chapter the following key points need to be remembered:

- **Far-field approximation**: In all our derivation, we have assumed that all distances (including distances between particles) are much larger than the dimensions of the scattering particles, and much larger than the incident wavelength of light.

- **Self-Averaging**: As mentioned in the previous section, one of the most important facts is that the media we are studying have self-averaging properties. This means that the important quantities are the statistical properties of the ensemble of particles.

- **Loosely packed approximation**: Greatly alleviated by the fact that tissues are self-averaging, by considering that we can add the intensities scattered from each particle rather than the fields, we are not taking into account interference contributions due to scattering between particles. This approximation holds in a loosely packed medium.

- **Statistical description of scattering**: By using *a priori* defined phase functions and considering that the scattering diagram or phase function is independent on the orientation of the particle (we approximated $p(\hat{s}, \hat{s}_0) = p(\hat{s} \cdot \hat{s}_0)$), we are describing the scattering amplitude in a statistical manner independent on the actual size and properties of each individual particle. This allows us to study the average properties of a collection of particles, independent on their orientation with respect to the incident beam.

---

[13]Depending on the measurement times and how sensitive our measurements are we may still measure coherent effects, which are the basis of correlation studies, but these will not be considered in this book.

[14]If you look carefully enough you will see the actual speckle pattern moving.

Further Reading

For those of you interested in understanding single particle scattering I strongly recommend the book by Bohren and Huffman (1998), and the classic book by van de Hulst (1981). In the specific case of light scattering by cells, I suggest the thesis dissertation of Dunn (1997). For general electromagnetic theory I would recommend Born and Wolf (1999), Tsang *et al.* (2000) and the book by Nieto-Vesperinas (2006). Finally, in order to gain insight into how to solve exactly multiple scattering from a collection of scatterers I would recommend reading the work by Mishchenko, in particular Mishchenko *et al.* (2007) and references therein.

On the subject of polarization, as it was already mentioned, even though it is an extremely interesting subject, it will not be covered in this book. For applications of polarization in tissues and the use of the Mueller matrix in general I recommend the book by Wang and Wu (2007).

Finally, regarding the use the term 'photon' to describe light propagation I would suggest reading the special issue in the October 2003 supplement to Optics and Photonics News titled 'The Nature of Light: What is a Photon?', edited by Roychoudhuri and Roy (2003), and the article 'Anti-photon' by Lamb (1995).

Chapter 3

The Radiative Transfer Equation (RTE)

Summary. In this chapter the radiative transfer equation (RTE) and basic radiosity concepts such as the *specific intensity* are presented. The RTE will be used as starting point to derive the expression for the diffusion equation and gain a better insight of the approximations that it implicitly involves. Important quantities such as the average intensity and the energy flux will be introduced.

3.1 Radiative Transfer

The Radiative Transfer Equation (from now on the RTE) is an energy balance equation that has been used to model transport of energy for over 100 years. It represents the energy conservation law for a quantity coined by Max Planck as *specific intensity* in his *The Theory of Heat Radiation* published originally in 1906, [Planck (1906, 1914)]. In his book Planck introduced the revolutionary concept of energy 'quanta' while establishing the theory of heat radiation. What is usually not attributed to him is the fact that, in describing the propagation of heat radiation, he introduced the basic concept behind the equation of radiative transfer and for radiometry in general through the definition of the specific intensity. In the very first page of the English version of his classic book [Planck (1914)], he introduces the concept of *specific intensity*:

> *Generally speaking, radiation is a far more complicated phenomenon that conduction of heat. The reason for this is that the state of radiation at a given instant and at a given point of the medium cannot be represented, as can the flow of heat by conduction[1], by a single vector (that*

---

[1] The flow of heat by conduction or the flow of diffuse light in our case can be described,

is, a single directed quantity). All heat rays which at a given instant pass through the same point of the medium are perfectly independent of one another, and in order to specify completely the state of the radiation the intensity of radiation must be known in all the directions, infinite in number, which pass through the point in question; for this purpose two opposite directions must be considered as distinct, because the radiation in one of them is quite independent of the radiation in the other.

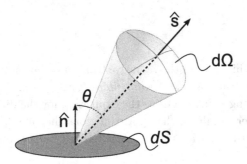

Fig. 3.1 Representation of the specific intensity $I(\mathbf{r}, \hat{\mathbf{s}})$ at the point $\mathbf{r}$ flowing in the direction of unit vector $\hat{\mathbf{s}}$ within the solid angle $d\Omega$.

Shortly after this brief introduction, Planck introduces the amount of *energy radiated* in time dt into a solid angle $d\Omega$ from a small differential area dS as:

$$dtdS \cos\theta d\Omega I(\mathbf{r}, \hat{\mathbf{s}}),$$

where $\cos\theta$ is the angle subtended by the normal $\hat{\mathbf{n}}$ of the area dS and the direction of the cone $d\Omega$ given by $\hat{\mathbf{s}}$ (see Fig. 3.1). The quantity $I(\mathbf{r}, \hat{\mathbf{s}})$ he termed the *specific intensity* or the *brightness*, with units of energy per area per unit solid angle[2]. This definition, or more importantly, this concept for describing energy radiation set the foundations for the classical works of Milne[Milne (1930)], Chandrasekhar[Chandrasekhar (1960)], and finally Ishimaru[Ishimaru (1978)] whose book has become one of the main references for the RTE in biomedical optics.

Even though the RTE has since then been well established in many areas of research, its original derivation is purely phenomenological: it was derived by establishing what would be expected to be the energy balance

as we shall see, by a single vector represented by the flux, $\mathbf{J}(\mathbf{r})$.

[2]Note that in his original derivation he used K to define the specific intensity.

when accounting for absorption and scattering. Its rigorous derivation directly from Maxwell's equations (see Chap. 2) had not been established until Mishchenko recently did so through the microphysical derivation of the *vector* RTE (VRTE) [Mishchenko (2002)] and the Poynting-Stokes tensor [Mishchenko (2010)]. As such, the traditionally used RTE (i.e the scalar RTE, which is the one used in this book) can be seen as a specific case of the microphysical VRTE where we neglect de-polarization[3] and coherence effects, and place all sources and detectors in the far-field of the particles. However, rigorous derivation of the RTE (and more relevant to us, of the diffusion equation) from the VRTE and the Poynting-Stokes tensor requires a great deal of non-trivial steps and a solid background in tensor algebra which deviates from our purpose. Since this book by no means intends to be a treatise in radiative transfer I refer the reader to these original references for the detailed and rigorous derivation of the RTE.

One of the main problems faced when relating the RTE with the solution to Maxwell's equation has been the relationship between the specific intensity (a heuristically defined quantity, with no physical measurement) to the Poynting vector or the density flow of energy (which strictly speaking is not a measurable quantity either[4], but it is a direct consequence of Maxwell's equations). This main discrepancy lays in the fact that for each position in space there is a *single value* of the Poynting vector and thus energy flow is represented by a *unique direction*: the concept introduced by Planck for an angular-dependent flow of energy for a specific point in space r has no direct representation in electromagnetic theory. To overcome this, Ishimaru (1978) introduced in his book the relationship between the Poynting vector and the specific intensity by assigning a probability of the Poynting vector to point in specific direction when considering a small volume [see Ishimaru (1978), Sec. 7-8]. It is this picture, which unfortunately has received very little attention in the RTE community, which we will slightly modify and expand.

We shall therefore attempt to overcome the seemingly unconnected relationship between specific intensity and Poynting vector by defining the energy distribution for each angle in terms of the average of the energy flux in a differential volume δV. In defining a volume averaged Poynting vector we shall see that a quantity more directly related to the specific intensity is

---

[3]Note that we can still solve for different polarization states through the Stokes vector presented in Chap. 2.

[4]Don't forget that it is the integral of the Poynting vector over a finite region what is an observable.

recovered. The justification for this approach, however, requires two mayor approximations to be introduced: the addition of intensities rather than fields will hold within this differential volume, and the electric and magnetic field vectors are mutually orthogonal, which requires all distances to be considered in the far-field.

3.1.1 *Volume Averaged Flow of Energy*

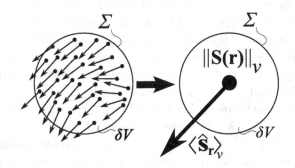

Fig. 3.2 Representation of the Poynting vector $\langle \mathbf{S}(\mathbf{r}') \rangle$ averaged over the differential volume δV. This will yield a volume averaged flow of energy $\|\mathbf{S}(\mathbf{r})\|_v$ pointing in an average direction of energy flow $\hat{\mathbf{s}}_v$. The flow of energy at any point in space $\mathbf{r}'$ has been represented in such a way that it is understood that the Poynting vector is continuous.

In order to have a better insight on the specific intensity, I will now introduce a picture which might have more physical significance due to its direct connection to the Poynting vector, following an approach similar to that of Ishimaru ([Ishimaru (1978)], Sec. 7-8), where we shall consider the volume averaged Poynting vector.

Let us consider the Poynting vector $\langle \mathbf{S} \rangle = S\hat{\mathbf{s}}$ averaged over a small (how small is 'small' will be dealt with later) volume δV. This volume will have *in each point in its volume* a Poynting vector defined by $\langle \mathbf{S}(\mathbf{r}') \rangle$ (see Fig. 3.2), which will be defined by a magnitude S and unit vector $\hat{\mathbf{s}}$. Since the Poynting vector at each point in space is defined by a single direction $\hat{\mathbf{s}}$, the resulting integral could have contributions over all directions in space $\hat{\mathbf{s}}_J$. In this case, we can define the normalized contribution of the flow of energy pointing in a specific direction $\hat{\mathbf{s}}_J$ within δV as (remember that both $\hat{\mathbf{s}}$ and $\hat{\mathbf{s}}_J$ are unitary):

$$w_{\mathbf{r}}(\hat{\mathbf{s}}_J) = \frac{1}{\delta V \|\mathbf{S}(\mathbf{r})\|_v} \int_{\delta V} S(\mathbf{r} - \mathbf{r}')(\hat{\mathbf{s}}' \cdot \hat{\mathbf{s}}_J) \mathrm{d}^3 r', \qquad \left(sr^{-1} \right) \qquad (3.1)$$

where now $\|\mathbf{S}(\mathbf{r})\|_v$ represents the total *volume averaged* flow of energy (see Fig. 3.2):

$$\|\mathbf{S}(\mathbf{r})\|_v = \frac{1}{\delta V} \int_{\delta V} S(\mathbf{r} - \mathbf{r}') \mathrm{d}^3 r'. \qquad (3.2)$$

As defined above, $w_\mathbf{r}(\hat{\mathbf{s}}_J)$ can also be seen as the probability at point $\mathbf{r}$ of having the flow of energy in a specific direction $\hat{\mathbf{s}}_J$ and clearly:

$$\frac{1}{4\pi} \int_{(4\pi)} w_\mathbf{r}(\hat{\mathbf{s}}_J) \mathrm{d}\Omega = 1. \qquad (3.3)$$

Note the similarity between this expression and the definition of the phase function in Chap. 2: in the limit of a single particle $w_\mathbf{r}(\hat{\mathbf{s}}_J)$ and $p(\hat{\mathbf{s}}, \hat{\mathbf{s}}')$ should be equivalent. By using $w_\mathbf{r}(\hat{\mathbf{s}}_J)$ we can now define an *average direction of energy flow* $\hat{\mathbf{s}}_v$ as:

$$\langle \hat{\mathbf{s}}_\mathbf{r} \rangle_v = \frac{1}{4\pi} \int_{(4\pi)} w_\mathbf{r}(\hat{\mathbf{s}}_J) \hat{\mathbf{s}}_J \mathrm{d}\Omega, \qquad (3.4)$$

with which we could now express a *volume averaged* flow of energy as:

$$\langle \mathbf{S}(\mathbf{r}) \rangle_v = \|\mathbf{S}(\mathbf{r})\|_v \langle \hat{\mathbf{s}}_\mathbf{r} \rangle_v. \qquad (3.5)$$

Note that in all the previous equations I have explicitly included the $\mathbf{r}$ dependence: the average flow of energy will depend on the location in space, both in its magnitude and in its direction. This expression in Eq. (3.5) is very useful, since we may associate it directly with the specific intensity. It conveys what one would expect in a more general case: if one considers the average flow of energy within a small volume it will depend on the average magnitude of flow and point in a spectrum of directions $\hat{\mathbf{s}}_J$, each one with a different weight (see Fig. 3.3).

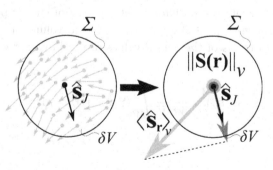

Fig. 3.3 Representation of the volume averaged flow of energy $\|\mathbf{S}(\mathbf{r})\|_v$ projected onto a generic direction of energy flow $\hat{\mathbf{s}}_J$ (compare with Fig. 3.2).

3.2 Specific Intensity, Average Intensity and Flux

In the previous section we introduced a way of relating the quantity first introduced by Planck as the specific intensity with the Poynting vector or the density of flux of energy *within a certain volume* δV. Before introducing the RTE which will account for scattering and absorption within the medium, we can introduce now several quantities which can be deduced directly from our *volume averaged* expressions for the Poynting vector: the specific intensity, average intensity, energy density and total flux density. These will be the quantities with which we will model light diffusion. It should not be forgotten, however, that this approach for accounting for diffuse light propagation is not unique and that the definition of these main quantities involves some important approximations: as will be discussed in the next chapter, the diffusion approximation itself can also be derived directly from vector electromagnetic waves.

3.2.1 *The Specific Intensity*

As mentioned before, the specific intensity represents the amount of power per area that flows in a certain direction defined by a unit solid angle[5]. In terms of the average energy flow represented by the volume averaged Poynting vector introduced the previous section, we can define a specific intensity $I(\mathbf{r}, \hat{\mathbf{s}})$ at point $\mathbf{r}$ flowing in direction $\hat{\mathbf{s}}$ as (see Fig. 3.1):

$$I(\mathbf{r}, \hat{\mathbf{s}}) = \frac{1}{4\pi} \|\mathbf{S}(\mathbf{r})\|_v w_{\mathbf{r}}(\hat{\mathbf{s}}), \qquad \left(Watts/cm^2 sr \right) \qquad (3.6)$$

where $w_{\mathbf{r}}$ represents the probability of the flow of energy pointing in direction $\hat{\mathbf{s}}$ as shown Eq. (3.1) and $\|\mathbf{S}(\mathbf{r})\|_v$ is the volume averaged flow of energy at $\mathbf{r}$. In terms of the actual values of the Poynting vector $\langle \mathbf{S} \rangle$ it is defined as:

$$I(\mathbf{r}, \hat{\mathbf{s}}_J) = \frac{1}{4\pi\delta V} \int_{\delta V} \langle \mathbf{S}(\mathbf{r} - \mathbf{r}') \rangle \cdot \hat{\mathbf{s}}_J \mathrm{d}^3 r'. \qquad (3.7)$$

---

[5]Usually $I(\mathbf{r}, \hat{\mathbf{s}})$ and all the related quantities are defined per unit frequency interval ω and represented as I_ω. Throughout this book we will assume we are dealing with monochromatic waves and drop this notation.

Building Block: **The Specific Intensity**

The specific intensity will be the main function to model light propagation in tissues, and the RTE will be derived based on its definition. It represents the average flow of energy in a certain direction $\hat{\mathbf{s}}$ at a position $\mathbf{r}$.

Note that we have not yet established what volume δV represents. This will be discussed towards the end of this chapter; for the time being we shall consider δV as a small volume where the number of particles (scatters and absorbers) that it contains are enough so that their average optical properties are equivalent to that of the surrounding medium.

At this point it is important to remember one key aspect of this definition of the specific intensity, namely that in this manner it is *not uniquely defined*. Exactly as in the case of the Poynting vector, as mentioned in Sec. 2.2 of Chap. 2, the measurable quantity is the integral of the Poynting vector over a finite region. This means that any expression in the form of $I(\mathbf{r}, \hat{\mathbf{s}}) = I(\mathbf{r}, \hat{\mathbf{s}}) + \nabla \times \mathbf{f}(\mathbf{r}, \hat{\mathbf{s}})$, with $\mathbf{f}$ being a generic *vector* function, would yield the same measured power or flux density. In any case, what should always be remembered is that *neither the Poynting vector nor the specific intensity can be measured*. The measurable quantity is the total power that traverses a surface (i.e. the flux), in which case the power and average intensities obtained through the specific intensity and the time and volume-averaged Poynting vector should be identical.

3.2.2   The Average Intensity

The average intensity at a certain point in space $\mathbf{r}$ can be easily understood as the total flow of energy when considering all directions, or equivalently the volume-averaged magnitude of the Poynting vector. In terms of the specific intensity this can be expressed as[6]:

$$U(\mathbf{r}) = \int_{(4\pi)} I(\mathbf{r}, \hat{\mathbf{s}}) d\Omega. \qquad \left(Watts/cm^2 \right) \qquad (3.8)$$

---

[6]When defining the average intensity, it is also common to find it normalized to the unit solid angle (4π) (see [Ishimaru (1978)], [Chandrasekhar (1960)]). However, note that this normalization has been included in the definition of $I(\mathbf{r}, \hat{\mathbf{s}})$.

Building Block: **The Average Intensity**

The average intensity will be used throughout this book both in the context of the RTE and in the context of the diffusion approximation. It represents the volume-averaged magnitude of the flow of energy.

In terms of the volume-averaged Poynting vector $\|\mathbf{S}(\mathbf{r})\|_v$, Eq. (3.8) can be expressed as:

$$U(\mathbf{r}) = \frac{1}{4\pi}\|\mathbf{S}(\mathbf{r})\|_v \int_{(4\pi)} w_{\mathbf{r}}(\hat{\mathbf{s}})\mathrm{d}\Omega,$$

which yields:

$$U(\mathbf{r}) = \|\mathbf{S}(\mathbf{r})\|_v, \tag{3.9}$$

where we have made use of Eq. (3.3). As expected, the average intensity represents the magnitude of the total volume-averaged flow of energy.

3.2.3   *The Energy Density*

Another quantity commonly used in radiative transfer formalism is the energy density. The way to arrive to the definition of energy density in radiative transfer books is by assuming the energy occupies a volume which can be given in terms of the speed of light, c_0. However, a more convenient way is to go directly to the definition of energy density we obtained from Maxwell's equations, Eq. (2.11), and relate it to our expression for the volume averaged Poynting vector. Since we are dealing with volume averaged quantities, the energy density as used in radiative transfer formalism is defined as:

$$u(\mathbf{r}) = \frac{1}{c_0 \delta V} \int_{\delta V} |\langle \mathbf{S}(\mathbf{r} - \mathbf{r}')\rangle|\mathrm{d}^3 r',$$

or equivalently,

$$u(\mathbf{r}) = \frac{1}{c_0}\|\mathbf{S}(\mathbf{r})\|_v, \tag{3.10}$$

which defined in terms of the specific intensity becomes:

$$u(\mathbf{r}) = \frac{1}{c_0} \int_{(4\pi)} I(\mathbf{r}, \hat{\mathbf{s}})\mathrm{d}\Omega = \frac{U(\mathbf{r})}{c_0}. \qquad \left(Joules/cm^3\right) \tag{3.11}$$

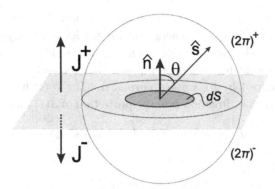

Fig. 3.4 Schematic view of the geometry with the upward, $J^+(\mathbf{r})$ and downward, $J^-(\mathbf{r})$, density flux at an interface.

3.2.4 *The Total Flux Density*

So far, we have defined the average intensity, U, and the energy density, u, quantities which both relate to the average flow of energy at a certain point when considering all angles, with no information on the average direction of propagation. In order to specify in which average direction does the overall flow of energy propagate we will refer to the total flux density. In terms of the specific intensity, the *total flux density* or simply *flux* $\mathbf{J}(\mathbf{r})$ is defined as:

$$\mathbf{J}(\mathbf{r}) = \int_{(4\pi)} I(\mathbf{r}, \hat{s})\hat{s}d\Omega. \quad \left(Watts/cm^2\right) \tag{3.12}$$

 Building Block: **The Total Flux Density**

The total flux density will be used throughout this book to account for light interaction at an interface, either between two diffusive media, or between a diffusive and a non-diffusive medium. It represents the power that a detector measures per unit area.

In terms of the volume-averaged Poynting vector, it can be clearly seen that this expression represents the total flux of energy averaged over all angles:

$$\mathbf{J}(\mathbf{r}) = \frac{1}{4\pi}\|\mathbf{S}(\mathbf{r})\|_v \int_{(4\pi)} w_{\mathbf{r}}(\hat{s})\hat{s}d\Omega,$$

yielding,

$$\mathbf{J}(\mathbf{r}) = \|\mathbf{S}(\mathbf{r})\|_v\langle\hat{s}_{\mathbf{r}}\rangle_v, \tag{3.13}$$

which is a vector representing the magnitude and direction of the average flow of electromagnetic energy, which we have already encountered in Eq. (3.5). Following this definition, let us consider a more practical case where we want to measure the power that reaches a detector of area $A = \int_A dS$ defined by its surface normal $\hat{n}$. The total flux density that flows in a particular direction $\hat{n}$ is the projection of $\mathbf{J}(\mathbf{r})$ onto $\hat{n}$:

$$J_n(\mathbf{r}) = \mathbf{J}(\mathbf{r}) \cdot \hat{n} = \int_{(4\pi)} I(\mathbf{r}, \hat{s}) \hat{s} \cdot \hat{n} d\Omega, \qquad (3.14)$$

or equivalently,

$$J_n(\mathbf{r}) = \|\mathbf{S}(\mathbf{r})\|_v \langle \hat{s}_\mathbf{r} \rangle_v \cdot \hat{n}. \qquad (3.15)$$

Additionally, the total flux passing through dS can also be defined as a sum of the forward flux, $J^+(\mathbf{r})$, and the backward flux, $J^-(\mathbf{r})$ (see Fig. 3.4), these quantities being (note the minus sign to account for the backward flux):

$$J^+(\mathbf{r}) = \int_{(2\pi)^+} I(\mathbf{r}, \hat{s}) \hat{s} \cdot \hat{n} d\Omega, \qquad (3.16)$$

$$J^-(\mathbf{r}) = \int_{(2\pi)^-} I(\mathbf{r}, \hat{s}) \hat{s} \cdot (-\hat{n}) d\Omega, \qquad (3.17)$$

where, as shown in Fig. 3.4, $(2\pi)^+$ stands for integration in $0 \leq \theta \leq \frac{\pi}{2}$ and $(2\pi)^-$ corresponds to integration in $\frac{\pi}{2} \leq \theta \leq \frac{3\pi}{2}$. We can represent this using our volume-averaged expressions for the Poynting vector as:

$$J^+(\mathbf{r}) = \frac{1}{4\pi} \|\mathbf{S}(\mathbf{r})\|_v \int_{(2\pi)^+} w_\mathbf{r}(\hat{s}) \hat{s} \cdot \hat{n} d\Omega, \qquad (3.18)$$

$$J^-(\mathbf{r}) = \frac{1}{4\pi} \|\mathbf{S}(\mathbf{r})\|_v \int_{(2\pi)^-} w_\mathbf{r}(\hat{s}) \hat{s} \cdot (-\hat{n}) d\Omega, \qquad (3.19)$$

In terms of the upward and downward flux, the total flux density in the direction of $\hat{n}$ is expressed as:

$$J_n(\mathbf{r}) = J^+(\mathbf{r}) - J^-(\mathbf{r}). \qquad \left(Watts/cm^2 \right) \qquad (3.20)$$

3.3   The Detected Power

We have now defined the main quantities that represent the amount of energy that traverses an interface. We are now in the position to predict what a measurement device (detector, from now on) will obtain as a readout, i.e. what the detected power should be. Let us consider the situation shown in Fig. 3.5 where we have a detector at a point $\mathbf{r}$, defined by a certain area A,

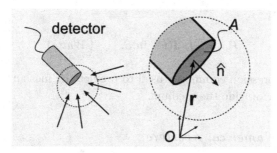

Fig. 3.5 An arbitrary medium (could be scattering and absorbing) where we have placed a detector at a certain point **r**. The inset shows the area, A, and normal, $\hat{n}$, of the surface of the detector.

and its surface normal $\hat{n}$. The medium can be homogeneous or inhomogeneous, for the time being. Since we have previously defined in Eq. (3.14) the total flux at a certain point **r** flowing in a specific direction $\hat{n}$, the power traversing a differential area dS of our detector defined by its unit normal $\hat{n}$ would be:

$$dP(\mathbf{r}) = J_n(\mathbf{r})dS. \qquad \Big(Watts\Big) \qquad (3.21)$$

However, this accounts for flux traversing the area in both directions. In order to account for the flux traversing *inward* into our detector, we must consider only the $(2\pi)^-$ contribution, $J^-(\mathbf{r})$. In this case, the *total power* measured at our detector of area A would be:

$$P_{det} = \int_A J^-(\mathbf{r})dS = \int_A dS \int_{(2\pi)^-} I(\mathbf{r},\hat{s})\hat{s}\cdot\hat{n}d\Omega. \qquad (3.22)$$

In terms of the Poynting vector, by substituting Eq. (3.6) we obtain that the total power at the detector is defined as:

$$P_{det} = \frac{1}{4\pi} \int_A \|\mathbf{S}(\mathbf{r})\|_v dS \int_{(2\pi)^-} w_{\mathbf{r}}(\hat{s})\hat{s}\cdot\hat{n}d\Omega. \qquad (3.23)$$

Note that in practice it is not physically possible to place a detector and not alter the distribution of energy. This means that by placing the detector we are effectively removing the $J^+(\mathbf{r})$ contribution to the flux, and therefore we may say that the boundary conditions *at the detector* are well approximated by $J_n(\mathbf{r}) = J^-(\mathbf{r})$. In this case, Eq. (3.23) with the aid of Eq. (3.13) becomes:

$$P_{det} = \int_A \|\mathbf{S}(\mathbf{r})\|_v \langle\hat{s}_{\mathbf{r}}\rangle_v \cdot \hat{n}dS, \qquad \Big(Watts\Big) \qquad (3.24)$$

or equivalently,

$$P_{det} = \int_A \mathbf{J}(\mathbf{r}) \cdot \hat{\mathbf{n}} dS. \quad \left(Watts \right) \qquad (3.25)$$

This is the expression commonly used to predict the measured power anywhere inside or outside the medium.

3.3.1   *The Numerical Aperture*

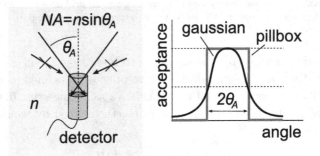

Fig. 3.6   Definition of numerical aperture for a detector depicting an aperture and the photo-sensible component, which could represent an optical fiber, for example. n represents the average index of refraction of the medium. To the left of the figure two types of angular acceptance are shown: a pillbox function (i.e. a step function), and a Gaussian function.

Before moving on, there is one issue regarding the formulas we just derived which should be carefully considered, namely the receiving *angular acceptance* or *numerical aperture (NA)* of our detector. If we go back to Eq. (3.23) it is easy to see that we are assuming that *all angles* contribute to the measured signal (each weighted by its cosine through the $\hat{\mathbf{s}} \cdot \hat{\mathbf{n}}$ factor). In reality no detector has a perfect angular acceptance and only a limited number of angles in fact account for the measured power. The actual form of the angular acceptance curve depends of course on the design of the detector. It is convenient, however, to define the acceptance of a detector through its numerical aperture, *NA*, as (see Fig. 3.6):

$$NA = n \sin \theta_A, \qquad (3.26)$$

where n is the average index of refraction outside the detector. Low numbers of the *NA* will therefore indicate very small acceptance angles, while $NA \sim 1$ indicate almost all light incident on the detector is measured. It is also convenient for our purposes to describe the cosine of the acceptance angle in terms of the NA as:

$$\cos\theta_A = \sqrt{1 - (NA/n)^2}, \tag{3.27}$$

where we have simply made use of the trigonometric formula $\sin^2\theta + \cos^2\theta = 1$. Once defined the acceptance angle, for $-\pi/2 \le \theta \le \pi/2$ we may now model different angular acceptance curves, as shown Fig. 3.6, the simplest being:

- *Pillbox or Step function:*

$$f(\theta) = H(|\theta| - |\theta_A|), \tag{3.28}$$

  where H is the Heavy-side step function. This function can also be written more conveniently as $f(\cos\theta) = H(|\cos\theta| - |\cos\theta_A|)$ since if $|\cos\theta| = |\cos\theta_A|$ then $|\theta| = |\theta_A|$, as long as $-\pi/2 \le \theta \le \pi/2$.
- *Gaussian function:*

$$f(\theta) = \frac{1}{\theta_A\sqrt{2\pi}} \exp\left(-\frac{\theta^2}{2\theta_A^2}\right), \tag{3.29}$$

  which is a simple normalized Gaussian function centered at zero.

As can be seen from Eq. (3.29), one problem arises with this expression, namely that it does not go to zero at $\theta = \pm\pi/2$ as it should in a realistic situation. In order to properly account for this behavior, I have found that a convenient expression is given by a *Hanning function*, which is commonly used for apodizing signals and thus reduce aliasing effects in Fourier transforms [Blackman and Tukey (1959)]. We can thus model the acceptance angle distribution through the following Hanning function:

$$f(\theta) = \frac{1}{2}\left[1 + \cos\left[\pi\left(\frac{\theta}{2\theta_A}\right)^k\right]\right], \tag{3.30}$$

where I have included the factor k which can account for the 'flatness' of the response (for $k = 1$ we have the original Hanning function, as k increases we slowly move towards a step function with smooth sides).

Since these functions only depend on the angle of incidence θ, they can be written in terms of the detector normal $\hat{n}$ and the unit vector that defines the direction of incidence, $\hat{s}$, through the relation $\hat{s} \cdot \hat{n} = \cos\theta$. In this way we can account for the detector's acceptance angle when measuring the power as in Eq. (3.22) as:

$$P_{det} = \int_A dS \int_{(2\pi)^-} I(\mathbf{r}, \hat{s})\hat{s} \cdot \hat{n} f(\hat{s} \cdot \hat{n}) d\Omega. \tag{3.31}$$

Accounting for the acceptance angle of the detector will be important when describing the measurements from diffusive media when taken from a non-scattering medium (free space), as we shall see in Sec. 8.6 of Chap. 8.

3.4   Isotropic Emission and its Detection

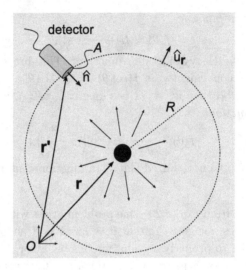

Fig. 3.7   Example of detection of radiation from an isotropic source, located at a distance $R = |\mathbf{r} - \mathbf{r}'|$ from the detector.

At this point it is interesting to include a specific case for the angular dependence of the emission, namely isotropic emission. Let us consider the situation depicted in Fig. 3.7, where we have our detector of area A defined by its surface normal $\hat{\mathbf{n}}$, located at a distance R from an isotropic source. This isotropic source, of an arbitrary shape, is emitting a total power of P_0 *Watts*. Since our light source radiates isotropically, there will be no angular dependence on the emitted power and we may assume that the Poynting vector depends only on the distance from the source, $\langle \mathbf{S}(\mathbf{r}) \rangle = \langle \mathbf{S}(r) \rangle$[7], where we may also assume radiation in the radial direction given by unit vector $\hat{\mathbf{u}}_R$. Assuming there is no absorption, the total power that will

---

[7]Note that we have gone back to the original definition of the Poynting vector, since what follows does not need the volume-averaged expressions.

traverse a sphere of radius R centered in the source would be:

$$P_{total} = \int_{4\pi R^2} \langle \mathbf{S}(R) \rangle \cdot \hat{\mathbf{u}}_R \mathrm{d}S = \int_{4\pi R^2} \|\mathbf{S}(R)\| \mathrm{d}S = \|\mathbf{S}(R)\| 4\pi R^2,$$

which accounting for the fact that in the absence of absorption $P_{total} = P_0$, we obtain:

$$\|\mathbf{S}(R)\| = \frac{P_0}{4\pi R^2}. \quad \left(Watts/cm^2 \right)$$

Proceeding in a similar way, the total power received at the detector would be:

$$P_{det} = \int_A \|\mathbf{S}(R)\| \hat{\mathbf{u}}_R \cdot \hat{\mathbf{n}} \mathrm{d}S = \int_A \|\mathbf{S}(R)\| \cos\theta \mathrm{d}S = \|\mathbf{S}(R)\| \cos\theta A,$$

which on substitution for $\|\mathbf{S}(R)\|$ results in:

$$P_{det}(\mathbf{r}) = P_0 \cos\theta \frac{A}{4\pi R^2}. \quad \left(Watts \right) \quad (3.32)$$

Eq. (3.32) can be seen as the total power emitted by the source weighted by the contribution of the detector area to the total area. The same approach can be used to arrive to this expression using the volume averaged Poynting vector, $\|\mathbf{S}(\mathbf{r})\|_v$, or the specific intensity. Note also that for very small detector areas, $A \simeq \mathrm{d}A$, we may use the expression for the solid angle $\mathrm{d}\Omega = \mathrm{d}A/R^2$.

We can follow a similar argument for an emitting bounded surface (note that in the previous case no need for defining the surface of the emitter was necessary, only the condition that its emission had no angular dependence). At this point, however, it is very important to understand that the concept of an isotropic emitting surface is not an object that appears in electromagnetic theory. This is something we already encountered in the definition of the specific intensity at the beginning of this chapter: if we consider an arbitrarily small surface $\mathrm{d}A'$ at an arbitrary point in space $\mathbf{r}'$, the flow of energy at that specific point in space can only flow in one single direction which is given by the Poynting vector $\mathbf{E} \times \mathbf{H}$. Therefore, to account for an isotropic emitting surface we must make use once again of the concept of a volume averaged Poynting vector, or equivalently, to the specific intensity. Isotropic radiation in terms of these quantities can be defined as:

$$I(\mathbf{r}, \hat{\mathbf{s}}) = I_0(\mathbf{r}) = \frac{1}{4\pi} \|\mathbf{S}(\mathbf{r})\|_v, \quad \left(Watts/cm^2 sr \right) \quad (3.33)$$

where now of course the angular contribution of the average flow of energy is unity, $w_{\mathbf{r}}(\hat{\mathbf{s}}_J) = 1$. Assuming we have a small emitting surface of area dA' at $\mathbf{r}'$, defined by its normal $\hat{\mathbf{n}}'$, the total power *emitted* by the surface would be (see Eq. (3.22) for comparison):

$$P_{em}(\mathbf{r}') = \int_{A'} J^+(\mathbf{r})\mathrm{d}A' = \int_{A'} \mathrm{d}A' \int_{(2\pi)^+} I(\mathbf{r}, \hat{\mathbf{s}})\hat{\mathbf{s}} \cdot \hat{\mathbf{n}}\mathrm{d}\Omega,$$

where substituting for $I(\mathbf{r}, \hat{\mathbf{s}})$ we obtain:

$$P_{em}(\mathbf{r}') = \int_{A'} \mathrm{d}A' \int_{(2\pi)^+} I_0(\mathbf{r}')\hat{\mathbf{s}} \cdot \hat{\mathbf{n}}\mathrm{d}\Omega.$$

This equation can be rewritten in terms of the average power emitted per unit solid angle, $dp_{em}(\mathbf{r}, \hat{\mathbf{s}})$ as:

$$dp_{em}(\mathbf{r}, \hat{\mathbf{s}}) = I_0(\mathbf{r})\cos\theta \mathrm{d}A', \qquad \left(Watts/sr\right) \qquad (3.34)$$

which is *Lambert's cosine law* [Ishimaru (1978)], frequently used to describe light radiating from a strong scattering medium into a non-scattering medium. Of course, on integrating Eq. (3.34) over the surface A' and over $(2\pi)^+$ we should obtain the total emitted power, P_0:

$$P_{em}(\mathbf{r}') = \int_{A'} \int_{(2\pi)^+} dp_{em}(\mathbf{r}, \hat{\mathbf{s}})\mathrm{d}\Omega,$$

which accounting for the fact that (see Appendix B),

$$\int_{(2\pi)^+} \cos\theta \mathrm{d}\Omega = \int_0^1 2\pi \cos\theta d\cos\theta = \pi,$$

results in:

$$P_{em}(\mathbf{r}') = \pi I_0 A' = P_0.$$

This expression will re-appear in Chap. 8 when modeling diffuse light propagation in free space, due to the fact that by means of Eq. (3.14) we have that if $I(\mathbf{r}, \hat{\mathbf{s}})$ represents an isotropic source then the total flux $J_n(\mathbf{r})$ emitted from the surface is $J_n(\mathbf{r}) = \pi I_0$ or, equivalently, $J_n(\mathbf{r}) = P_0/A'$.

In the above discussion we presented the total power traversing areas in both directions, inward and outward, where we specifically separated detection from emission. As mentioned before in Sec. 3.2.4, the total flux that traverses an interface, $J_n(\mathbf{r})$, given by Eq. (3.14), can be separated into two contributions, $J^+(\mathbf{r})$ and $J^-(\mathbf{r})$. In radiosity theory, it is common to name the flux of energy that would be received by a detector as *irradiance*, while the flux of energy emitted by a surface is usually termed *emittance*. Expressed in these terms we can say that the emitted power in direction $\hat{\mathbf{s}}$ from a surface dS defined by its surface normal $\hat{\mathbf{n}}$ is given by:

$$dp_{em}(\mathbf{r}, \hat{\mathbf{s}}) = I(\mathbf{r}, \hat{\mathbf{s}})\hat{\mathbf{s}} \cdot \hat{\mathbf{n}}\mathrm{d}S, \forall \hat{\mathbf{s}} \cdot \hat{\mathbf{n}} > 0, \qquad \left(Watts/sr\right) \qquad (3.35)$$

and, equivalently, the received power from direction $\hat{\mathbf{s}}$ is:

$$dp_{det}(\mathbf{r}, \hat{\mathbf{s}}) = I(\mathbf{r}, \hat{\mathbf{s}})\hat{\mathbf{s}} \cdot (-\hat{\mathbf{n}})dS. \forall \hat{\mathbf{s}} \cdot \hat{\mathbf{n}} < 0, \qquad \left(Watts/sr\right) \qquad (3.36)$$

In general, it is convenient to describe the power emitted in an arbitrary direction $\hat{\mathbf{s}}$, irrespective of whether it is received or emitted power, as:

$$dp(\mathbf{r}, \hat{\mathbf{s}}) = I(\mathbf{r}, \hat{\mathbf{s}})\hat{\mathbf{s}} \cdot \hat{\mathbf{n}}dS. \qquad \left(Watts/sr\right) \qquad (3.37)$$

Of course, if we integrate the above expressions over 4π we obtain:

$$dP_{em}(\mathbf{r}) = \int_{(2\pi)^+} dp_{em}(\mathbf{r}, \hat{\mathbf{s}})d\Omega = J^+(\mathbf{r})dS, \qquad \left(Watts\right)$$

$$dP_{det}(\mathbf{r}) = \int_{(2\pi)^-} dp_{det}(\mathbf{r}, \hat{\mathbf{s}})d\Omega = J^-(\mathbf{r})dS, \qquad \left(Watts\right) \qquad (3.38)$$

and the total power at $\mathbf{r}$ will thus be $dP = dP_{det} - dP_{em}$. From now on, when speaking about surfaces I will only refer to emitted and received or detected power, and will drop the use of emittance and irradiance.

3.5 Reflectivity and Transmissivity

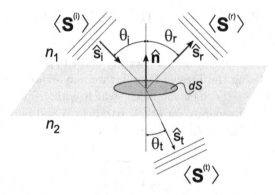

Fig. 3.8 Reflection and transmission of the flow of energy at an interface between homogeneous media of different index of refraction n_1, n_2.

In the previous section we considered how we can measure the flow of energy with a detector, and in general how we can predict the flow of energy through a surface defined by its surface normal $\hat{\mathbf{n}}$. In all the derivations presented we always assumed we were working in a homogeneous medium with constant index of refraction, n. Once understood how the flow of energy through a surface may be accounted for, the next step is to consider how the flow of energy will interact with a plane surface which separates two otherwise homogeneous media with different index of refraction n_1 and n_2 (see Fig. 3.8). In order to solve for this, we may use directly Maxwell's equations and consider a plane wave propagating in direction $\hat{\mathbf{s}}_i$ defined by its electric and magnetic fields, $\mathbf{E}^{(i)}$ and $\mathbf{H}^{(i)}$ respectively. In this case, the density flow of energy will be defined by its time-averaged Poynting vector as (see Eq. (2.10) from Chap. 2):

$$\langle \mathbf{S}^{(i)} \rangle = \frac{c}{8\pi} \Re\{\mathbf{E}^{(i)} \times \mathbf{H}^{(i)*}\} = \frac{c}{8\pi} \sqrt{\frac{\epsilon_1}{\mu_1}} |\mathbf{E}^{(i)}|^2 \hat{\mathbf{s}}_i,$$

where c is the speed of light in vacuum, ϵ_1 and μ_1 represent the dielectric constant and magnetic permeability of the medium, and where we have made use of the fact that for plane waves $\sqrt{\epsilon}E = \sqrt{\mu}H$. Equivalently, we may represent the Poynting vector related to the reflected and transmission plane waves as:

$$\langle \mathbf{S}^{(r)} \rangle = \frac{c}{8\pi} \Re\{\mathbf{E}^{(r)} \times \mathbf{H}^{(r)*}\} = \frac{c}{8\pi} \sqrt{\frac{\epsilon_1}{\mu_1}} |\mathbf{E}^{(r)}|^2 \hat{\mathbf{s}}_r,$$

$$\langle \mathbf{S}^{(t)} \rangle = \frac{c}{8\pi} \Re\{\mathbf{E}^{(t)} \times \mathbf{H}^{(t)*}\} = \frac{c}{8\pi} \sqrt{\frac{\epsilon_2}{\mu_2}} |\mathbf{E}^{(t)}|^2 \hat{\mathbf{s}}_t,$$

respectively, the subindex 1 and 2 referring to the upper and lower medium (see Fig. 3.8). Using the Fresnel formulas we may express these Poynting vectors in terms of the incident fields, which with respect to the plane of incidence we can separate into their parallel ($\parallel$) and perpendicular ($\perp$) polarization components, $\mathbf{E}^{(i)} = E_\parallel \mathbf{e}_\parallel + E_\perp \mathbf{e}_\perp$:

$$\langle \mathbf{S}^{(r)} \rangle_\parallel = \frac{c}{8\pi} \sqrt{\frac{\epsilon_1}{\mu_1}} |r_\parallel|^2 E_\parallel^2 \hat{\mathbf{s}}_r,$$

$$\langle \mathbf{S}^{(r)} \rangle_\perp = \frac{c}{8\pi} \sqrt{\frac{\epsilon_1}{\mu_1}} |r_\perp|^2 E_\perp^2 \hat{\mathbf{s}}_r,$$

$$\langle \mathbf{S}^{(t)} \rangle_\parallel = \frac{c}{8\pi} \sqrt{\frac{\epsilon_2}{\mu_2}} |t_\parallel|^2 E_\parallel^2 \hat{\mathbf{s}}_t,$$

$$\langle \mathbf{S}^{(t)} \rangle_\perp = \frac{c}{8\pi} \sqrt{\frac{\epsilon_2}{\mu_2}} |t_\perp|^2 E_\perp^2 \hat{\mathbf{s}}_t, \tag{3.39}$$

where $r_{\parallel,\perp}$ and $t_{\parallel,\perp}$ are the Fresnel formulas (see [Born and Wolf (1999)] for their derivation):

$$r_{\parallel} = \frac{\tan (\theta_i - \theta_t)}{\tan (\theta_i + \theta_t)},$$

$$r_{\perp} = -\frac{\sin (\theta_i - \theta_t)}{\sin (\theta_i + \theta_t)},$$

$$t_{\parallel} = \frac{2 \sin (\theta_t) \cos (\theta_i)}{\sin (\theta_i + \theta_t) \cos (\theta_i - \theta_t)},$$

$$t_{\perp} = \frac{2 \sin (\theta_t) \cos (\theta_i)}{\sin (\theta_i + \theta_t)}, \tag{3.40}$$

with the angles θ_i and θ_t given by $\hat{\mathbf{s}}_i \cdot \hat{\mathbf{n}} = \cos \theta_i$, $\hat{\mathbf{s}}_t \cdot (-\hat{\mathbf{n}}) = \cos \theta_t$.

It is convenient to express the reflected and transmitted Poynting vectors in terms of the incident density flow of energy. To that end, we may separate the incident Poynting vector into the contributions due to the parallel and perpendicular components of the electric field, $\langle \mathbf{S}^{(i)} \rangle = \left(S_{\parallel}^{(i)} + S_{\perp}^{(i)} \right) \hat{\mathbf{s}}_i$, in which case:

$$\langle \mathbf{S}^{(r)} \rangle_{\parallel} = |r_{\parallel}|^2 S_{\parallel}^{(i)} \hat{\mathbf{s}}_r,$$

$$\langle \mathbf{S}^{(r)} \rangle_{\perp} = |r_{\perp}|^2 S_{\perp}^{(i)} \hat{\mathbf{s}}_r,$$

$$\langle \mathbf{S}^{(t)} \rangle_{\parallel} = \frac{n_2}{n_1} |t_{\parallel}|^2 S_{\parallel}^{(i)} \hat{\mathbf{s}}_t,$$

$$\langle \mathbf{S}^{(t)} \rangle_{\perp} = \frac{n_2}{n_1} |t_{\perp}|^2 S_{\perp}^{(i)} \hat{\mathbf{s}}_t, \tag{3.41}$$

where we have assumed $\mu_1 = \mu_2$ in which case $n_2/n_1 = \sqrt{\epsilon_2/\epsilon_1}$ through Maxwell's formula, $n = \sqrt{\epsilon \mu}$.

Let us now consider the differential surface dS defined by its unit normal $\hat{\mathbf{n}}$ shown in Fig. 3.8. The incident power on dS is given by:

$$dP^{(i)} = \langle \mathbf{S}^{(i)} \rangle \cdot \hat{\mathbf{n}} dS = S^{(i)} \cos \theta_i dS. \quad \left(Watts \right) \tag{3.42}$$

Since the total power at dS must be conserved it is clear that:

$$dP_{\parallel}^{(i)} = dP_{\parallel}^{(r)} + dP_{\parallel}^{(t)},$$

$$dP_{\perp}^{(i)} = dP_{\perp}^{(r)} + dP_{\perp}^{(t)},$$

where:

$$dP_{\parallel}^{(r)} = \langle \mathbf{S}^{(r)} \rangle_{\parallel} \cdot \hat{\mathbf{n}} dS = S_{\parallel}^{(r)} \cos \theta_i dS,$$

$$dP_{\perp}^{(r)} = \langle \mathbf{S}^{(r)} \rangle_{\perp} \cdot \hat{\mathbf{n}} dS = S_{\perp}^{(r)} \cos \theta_i dS,$$

$$dP_{\parallel}^{(t)} = \langle \mathbf{S}^{(t)} \rangle_{\parallel} \cdot (-\hat{\mathbf{n}}) dS = S_{\parallel}^{(t)} \cos \theta_t dS,$$

$$dP_{\perp}^{(t)} = \langle \mathbf{S}^{(t)} \rangle_{\perp} \cdot (-\hat{\mathbf{n}}) dS = S_{\perp}^{(t)} \cos \theta_t dS. \tag{3.43}$$

Once understood the amount of power distributed from the incident field into the reflected and transmitted fields, the ratios

$$R_\parallel = \frac{dP_\parallel^{(r)}}{dP_\parallel^{(i)}} = |r_\parallel|^2; \quad T_\parallel = \frac{dP_\parallel^{(t)}}{dP_\parallel^{(i)}} = \frac{n_2 \cos\theta_t}{n_1 \cos\theta_i}|t_\parallel|^2; \qquad (3.44)$$

$$R_\perp = \frac{dP_\perp^{(r)}}{dP_\perp^{(i)}} = |r_\perp|^2; \quad T_\perp = \frac{dP_\perp^{(t)}}{dP_\perp^{(i)}} = \frac{n_2 \cos\theta_t}{n_1 \cos\theta_i}|t_\perp|^2, \qquad (3.45)$$

represent the *reflectivity* and *transmissivity* for each polarization respectively. In terms of the incident and transmitted angles, they can be written as:

$$R_\parallel = \frac{\tan^2(\theta_i - \theta_t)}{\tan^2(\theta_i - \theta_t)}; \quad R_\perp = \frac{\sin^2(\theta_i - \theta_t)}{\sin^2(\theta_i - \theta_t)},$$
$$T_\parallel = \frac{\sin 2\theta_i \sin 2\theta_t}{\sin^2(\theta_i + \theta_t)\cos^2(\theta_i - \theta_t)}; \quad T_\perp = \frac{\sin 2\theta_i \sin 2\theta_t}{\sin^2(\theta_i + \theta_t)}. \qquad (3.46)$$

Assuming α_i represents the angle which the electromagnetic field $\mathbf{E}$ makes with the plane of incidence, we may write $E_\parallel = E\cos\alpha_i$ and $E_\perp = E\sin\alpha_i$, and equivalently $S_\parallel^{(i)} = S^{(i)}\cos^2\alpha_i$, and $S_\perp^{(i)} = S^{(i)}\sin^2\alpha_i$, in which case:

$$R = \frac{dP^{(r)}}{dP^{(i)}} = \frac{dP_\parallel^{(r)} + dP_\perp^{(r)}}{dP^{(i)}} = R_\parallel \cos^2\alpha_i + R_\perp \sin^2\alpha_i, \qquad (3.47)$$

$$T = \frac{dP^{(t)}}{dP^{(i)}} = \frac{dP_\parallel^{(t)} + dP_\perp^{(t)}}{dP^{(i)}} = T_\parallel \cos^2\alpha_i + T_\perp \sin^2\alpha_i. \qquad (3.48)$$

R and T represent the total reflectivity and transmissivity of the interface, which hold the following relations:

$$R_\parallel + T_\parallel = 1; \quad R_\perp + T_\perp = 1; \quad R + T = 1;$$
$$R = R_\parallel \cos^2\alpha_i + R_\perp \sin^2\alpha_i;$$
$$T = T_\parallel \cos^2\alpha_i + T_\perp \sin^2\alpha_i.$$

Using this notation, the case of *unpolarized light* can be easily described through $\alpha_i = \pi/4$ in which case $\cos^2\alpha_i = \sin^2\alpha_i = 1/2$.

Assuming we may define a volume-averaged Poynting vector we could proceed in a similar manner for each of the angular components of $w_\mathbf{r}(\hat{\mathbf{s}})$ (see Sec. 3.1.1). In this case, the incident, reflected and transmitted powers in terms of the volume-averaged Poynting vector can be expressed as:

$$J^{(i)}(\mathbf{r}) = \|\mathbf{S}(\mathbf{r})\|_v \int_{(2\pi)^-} w_{\mathbf{r}}(\hat{\mathbf{s}})\hat{\mathbf{s}} \cdot (-\hat{\mathbf{n}})d\Omega, \tag{3.49}$$

$$J^{(r)}(\mathbf{r}) = \|\mathbf{S}(\mathbf{r})\|_v \int_{(2\pi)^+} R(\hat{\mathbf{s}} \cdot \hat{\mathbf{n}})w_{\mathbf{r}}(\hat{\mathbf{s}})\hat{\mathbf{s}} \cdot \hat{\mathbf{n}}d\Omega, \tag{3.50}$$

$$J^{(t)}(\mathbf{r}) = \|\mathbf{S}(\mathbf{r})\|_v \int_{(2\pi)^-} T(-\hat{\mathbf{s}} \cdot \hat{\mathbf{n}})w_{\mathbf{r}}(\hat{\mathbf{s}})\hat{\mathbf{s}} \cdot (-\hat{\mathbf{n}})d\Omega. \tag{3.51}$$

Note, however, that the incident and reflected powers do not represent a flow of energy *through* the surface. For the purposes of this book, specially in relation to finding appropriate boundary conditions, it is more convenient to describe the total flux through the interface in terms of the *transmitted* intensities considering light might be incident from both media. In terms of the specific intensity, through Eq. (3.46) we can now relate the transmissivity and reflectivity with our definitions for the power that traverses by surface dS, as we previously did in Eq. (3.16) and Eq. (3.17), assuming light might be incident from below and above surface dS and accounting for the reflected light:

$$J^+(\mathbf{r}) = \int_{(2\pi)^+} [1 - R_{2\to1}(\hat{\mathbf{s}} \cdot \hat{\mathbf{n}})]\, I(\mathbf{r},\hat{\mathbf{s}})\hat{\mathbf{s}} \cdot \hat{\mathbf{n}}d\Omega, \tag{3.52}$$

$$J^-(\mathbf{r}) = \int_{(2\pi)^-} [1 - R_{1\to2}(-\hat{\mathbf{s}} \cdot \hat{\mathbf{n}})]\, I(\mathbf{r},\hat{\mathbf{s}})\hat{\mathbf{s}} \cdot (-\hat{\mathbf{n}})d\Omega, \tag{3.53}$$

where $R_{i\to j}$ is the reflectivity on going from medium i to medium j through direction $\hat{\mathbf{s}}$, in which case the angle of incidence is defined through $\hat{\mathbf{n}} \cdot \hat{\mathbf{s}} = \cos\theta$ and $(-\hat{\mathbf{n}}) \cdot \hat{\mathbf{s}} = \cos\theta$, respectively. These equations account for the differences in index of refraction when determining the total upward and downward flux and will be the ones used when modeling diffuse light at interfaces.

 Building Block: **The Refractive Index Dependent Flux**

The expressions which account for the differences in index of refraction when describing flux traversing an interface will be basic for the derivation of the boundary conditions, both at diffusive/diffusive (D-D) and diffusive/non-diffusive (D-N) interfaces.

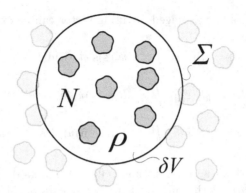

Fig. 3.9   Representation of the surface Σ which encloses the differential volume δV with a total of N particles distributed with density ρ.

3.6   Derivation of the Radiative Transfer Equation

At this point, all the necessary concepts and main formulas that enable us to predict the measurement at a detector or the emission from a surface have been briefly introduced. Hopefully a clearer picture of how a measurement takes place and what elements are involved in this measurement is available. We shall now attempt to model how the flux of energy, after it leaves our emitting surface, will behave in a homogeneous scattering and absorbing medium until it reaches our detector. As discussed at the beginning of this chapter, we shall model this propagation through the RTE. It was also mentioned then that a rigorous derivation of the RTE from the microphysical vector RTE (VRTE), is out of the scope of this book. On the other hand, having at least some level of understanding to how the RTE is derived is necessary if we need to somehow get a grasp of the level of approximation involved in moving away from the rigorous solution given by Maxwell's equations towards a simpler and statistical representation offered by the RTE. For those interested in the rigorous approach I recommend the works of Mischenko [Mishchenko (2002, 2010)]. Additionally, more related to the formulations presented in this book, I have included an alternative derivation of the scalar RTE directly from the equation of energy conservation (Poynting's Theorem) in Appendix C. This is the derivation that I will follow, which deviates from the classical phenomenological approach used to derive the RTE. For those interested in the classic derivation of the RTE I suggest [Ishimaru (1978); Chandrasekhar (1960)] and in the context of diffuse light in tissues [Martelli *et al.* (2010)].

In order to derive the Radiative Transfer Equation (RTE), lets first go back to the equation of energy conservation Eq. (2.8), and assume for generality that we have a time-varying dependence of the energy density (i.e. we modulate the intensity of our source, for example) which is much slower than the frequency of the electromagnetic oscillation ω. In this case, we may use the time-averaged expressions, which for a generic direction $\hat{s}_J$ were defined as (see Eq. (2.15)):

$$\frac{1}{c_0}\frac{\partial \langle \mathbf{S}(\mathbf{r}) \rangle \cdot \hat{s}_J}{\partial t} + \left\langle \frac{dP_{abs}}{dV}(\mathbf{r}) \right\rangle (\hat{s} \cdot \hat{s}_J) + \hat{s}_J \cdot \nabla(\langle \mathbf{S}(\mathbf{r}) \rangle \cdot \hat{s}_J) = 0, \quad (3.54)$$

which, as mentioned previously, basically ensures invariance under rotation, or equivalently, energy conservation in any particular direction $\hat{s}_J$.

In order to arrive to the RTE we first need to consider the equation for energy conservation shown above in a small differential volume δV, equivalent to that used for the derivation of the volume averaged Poynting vector. However, we shall now assume that this differential volume contains N particles such as those defined in Chapter 2 (see Fig. 3.9), with absorption cross-section σ_a and scattering cross-section σ_{sc}. If we thus integrate Eq. (3.54) over δV we obtain:

$$\frac{1}{c_0}\frac{\partial}{\partial t}\int_{\delta V}(\hat{s} \cdot \hat{s}_J)S(\mathbf{r} - \mathbf{r}')d^3r'+$$

$$\int_{\delta V}(\hat{s} \cdot \hat{s}_J)\left\langle \frac{dP_{abs}}{dV}(\mathbf{r} - \mathbf{r}') \right\rangle d^3r'+$$

$$\int_{\delta V}(\hat{s} \cdot \hat{s}_J)\hat{s}_J \cdot \nabla_{\mathbf{r}'}S(\mathbf{r} - \mathbf{r}')d^3r' = 0,$$

where we have written $\langle \mathbf{S}(\mathbf{r}) \rangle = S(\mathbf{r})\hat{s}$ and which, after several approximations, will result in the RTE. The detailed steps of this alternative derivation can be found in Appendix C, of which the final results derive directly from the three main terms of Eq. (3.54) and may be represented as:

- *Volume Averaged Change in Energy Density*

$$\frac{1}{c_0}\frac{\partial}{\partial t}\int_{\delta V}(\hat{s} \cdot \hat{s}_J)S(\mathbf{r} - \mathbf{r}')d^3r' \rightarrow \frac{1}{c_0}\frac{\partial}{\partial t}\|\mathbf{S}(\mathbf{r})\|_v w_{\mathbf{r}}(\hat{s}_J)\delta V, \quad (3.55)$$

- *Volume Averaged Absorbed Power*

$$\int_{\delta V}(\hat{s} \cdot \hat{s}_J)\left\langle \frac{dP_{abs}}{dV}(\mathbf{r} - \mathbf{r}') \right\rangle d^3r' \rightarrow N\sigma_a\|\mathbf{S}(\mathbf{r})\|_v w_{\mathbf{r}}(\hat{s}_J), \quad (3.56)$$

- *Volume Averaged Change in Energy Flow*

$$\int_{\delta V} (\hat{\mathbf{s}} \cdot \hat{\mathbf{s}}_J) \hat{\mathbf{s}}_J \cdot \nabla_{\mathbf{r}'} S(\mathbf{r} - \mathbf{r}') \mathrm{d}^3 r' \rightarrow$$

$$\hat{\mathbf{s}}_J \cdot \nabla \left[\|\mathbf{S}(\mathbf{r})\| w_{\mathbf{r}}(\hat{\mathbf{s}}_J) \right] \delta V + N\sigma_{tot} \|\mathbf{S}(\mathbf{r})\|_v w_{\mathbf{r}}(\hat{\mathbf{s}}_J) -$$

$$N\sigma_{tot} \int_{(4\pi)} \|\mathbf{S}(\mathbf{r})\|_v w_{\mathbf{r}}(\hat{\mathbf{s}}') p(\hat{\mathbf{s}}_J, \hat{\mathbf{s}}') \mathrm{d}\Omega'. \quad (3.57)$$

At this point, taking into account that the density of particles within δV will be given by $\rho = N/\delta V$, and making use of the definitions presented in Chap. 2 for the absorption, scattering and transport coefficients, namely $\mu_a = \rho\sigma_a$, $\mu_s = \rho\sigma_{sc}$ and $\mu_t = \rho\sigma_{tot}$, respectively, we may now identify $\|\mathbf{S}(\mathbf{r})\| w_{\mathbf{r}}(\hat{\mathbf{s}}_J)/4\pi$ as the specific intensity (see Eq. (3.6)). Introducing this change and adding all contributions we obtain the equation for radiative transfer with its well-known appearance (details in Appendix C):

$$\frac{1}{c_0} \frac{\partial}{\partial t} I(\mathbf{r}, \hat{\mathbf{s}}) + \hat{\mathbf{s}} \cdot \nabla I(\mathbf{r}, \hat{\mathbf{s}})$$

$$+ (\mu_a + \mu_s) I(\mathbf{r}, \hat{\mathbf{s}}) - \mu_t \int_{(4\pi)} I(\mathbf{r}, \hat{\mathbf{s}}') p(\hat{\mathbf{s}}, \hat{\mathbf{s}}') \mathrm{d}\Omega' = 0, \quad (3.58)$$

where $p(\hat{\mathbf{s}}, \hat{\mathbf{s}}')$ is the phase function as defined in Chap. 2.

It is very important to underline the fact that the μ_t factor in Eq. (3.58) appears *through our definition of the phase function*, Eq. (2.46), in Chap. 2. Since our choice of the phase function is quite arbitrary, we can also express it as $p(\hat{\mathbf{s}}, \hat{\mathbf{s}}') = |f(\hat{\mathbf{s}}, \hat{\mathbf{s}}')|^2$, in which case the μ_t factor will be absent. It is the $\mu_t p(\hat{\mathbf{s}}, \hat{\mathbf{s}}')$ product which yields the contribution which is due solely to scattering, i.e. the absorption term related to the phase function that appears through $(\mu_a + \mu_s) \int I(\mathbf{r}, \hat{\mathbf{s}}') p(\hat{\mathbf{s}}, \hat{\mathbf{s}}') \mathrm{d}\Omega'$ has no direct physical meaning.

Clearly, the RTE simply represents an energy balance for the flow of energy at $\mathbf{r}$ in a specific direction $\hat{\mathbf{s}}_J$, $I(\mathbf{r}, \hat{\mathbf{s}}_J)$ of the form:

[Change in $I(\mathbf{r}, \hat{\mathbf{s}}_J)$] + [Absorbed $I(\mathbf{r}, \hat{\mathbf{s}}_J)$] + [Scattered $I(\mathbf{r}, \hat{\mathbf{s}}_J)$] = 0,

where one of the main simplifications has been to write the scattered contribution as:

[Scattered $I(\mathbf{r}, \hat{\mathbf{s}}_J)$] = [Spatial Change in $I(\mathbf{r}, \hat{\mathbf{s}}_J)$] +

[Scattered $I(\mathbf{r}, \hat{\mathbf{s}}_J \rightarrow \hat{\mathbf{s}})$] − [Scattered $I(\mathbf{r}, \hat{\mathbf{s}} \rightarrow \hat{\mathbf{s}}_J)$],

where [Spatial Change in $I(\mathbf{r}, \hat{\mathbf{s}}_J)$] is clearly $\hat{\mathbf{s}} \cdot \nabla I(\mathbf{r}, \hat{\mathbf{s}})$, $I(\mathbf{r}, \hat{\mathbf{s}}_J \rightarrow \hat{\mathbf{s}})$ represents the flow of energy lost from direction $\hat{\mathbf{s}}_J$ into all other directions $\hat{\mathbf{s}}$ (i.e. the $\mu_s I(\mathbf{r}, \hat{\mathbf{s}})$ factor), and $I(\mathbf{r}, \hat{\mathbf{s}} \rightarrow \hat{\mathbf{s}}_J)$ the *gain* on energy flow on direction $\hat{\mathbf{s}}_J$ due to scattering from the flow of energy from all other directions $\hat{\mathbf{s}}$ into $\hat{\mathbf{s}}_J$ (i.e. the $\int I(\mathbf{r}, \hat{\mathbf{s}}') p(\hat{\mathbf{s}}, \hat{\mathbf{s}}') \mathrm{d}\Omega'$ factor).

3.6.1 The Source Term

You might have noticed that in all the previous derivation of the RTE I have not included explicitly the source term, assuming in all cases our incident source is placed at infinity. In order to account for the existence of sources, we shall consider two types: a) those due to an external illumination device (i.e. a fiber inside the scattering medium or in contact with the tissue, a laser source impinging on the surface, etc.) or that do not need radiation to produce light (such as bioluminescence) and b) those sources due to production of light within the scattering medium that need external radiation for this light production, i.e. fluorescent sources. Fluorescence sources are clearly distinguishable in the sense that light production is of a different wavelength (longer wavelength, as we saw in Chap. 1) compared to the excitation source. It is convenient to distinguish between them for future purposes, and will come in handy when modeling the fluorescence contribution.

When deriving the equation for energy conservation, Eq. (2.14), we included a term which accounted for the amount of energy lost due to absorption:

$$u_{abs} = \left\langle \frac{\mathrm{d}P_{abs}}{\mathrm{d}V} \right\rangle, \qquad \left(Watts/cm^3 \right) \tag{3.59}$$

which represents the amount of power absorbed per unit volume, that we also saw that when integrated over a small volume δV gave:

$$\|u_{abs}\|_v = \frac{1}{\delta V} \int_{\delta V} (\hat{\mathbf{s}}' \cdot \hat{\mathbf{s}}_J) \left\langle \frac{\mathrm{d}P_{abs}}{\mathrm{d}V}(\mathbf{r}' - \mathbf{r}) \right\rangle \mathrm{d}^3 r', \tag{3.60}$$

which in terms of the absorption coefficient and the specific intensity we expressed as:

$$\|u_{abs}\|_v = \mu_a I(\mathbf{r}, \hat{\mathbf{s}}). \qquad \left(Watts/cm^3 sr \right) \tag{3.61}$$

Clearly, we can proceed in a similar way and account for fluorescence using the formulas developed in Chap. 1, in particular Eq. (1.7) in which case:

$$\|u_{fl}(\mathbf{r}, \hat{\mathbf{s}}; \lambda_{em})\|_v = -\mu_a^{fl} \Phi(\lambda_{em}) I(\mathbf{r}, \hat{\mathbf{s}}), \qquad \left(Watts/cm^3 \right) \tag{3.62}$$

where μ_a^{fl} is the absorption coefficient of the fluorophore at the excitation wavelength, $\mu_a^{fl} = [C]\sigma_a$, being $[C]$ the concentration and σ_a the absorption cross-section, Φ is the quantum yield, and we have explicitly introduced the difference in wavelength through λ_{em}. Note the negative sign in Eq. (3.62), which reflects the fact that fluorescence can be seen as negative absorption *at the emission wavelength.*

In a more general case where we can consider the amount of energy introduced on the system to be delivered externally, we can introduce the source term as:

$$\|u_{src}\|_v = -\epsilon(\mathbf{r}, \hat{\mathbf{s}}), \qquad \left(Watts/cm^3 sr\right) \qquad (3.63)$$

where ϵ represents the amount of power per unit volume delivered to the system in direction $\hat{\mathbf{s}}$.

Accounting for the source term, we can therefore write the RTE as:

$$\frac{1}{c_0}\frac{\partial}{\partial t}I(\mathbf{r}, \hat{\mathbf{s}}) + \hat{\mathbf{s}} \cdot \nabla I(\mathbf{r}, \hat{\mathbf{s}})$$

$$+ \mu_t I(\mathbf{r}, \hat{\mathbf{s}}) - \mu_t \int_{(4\pi)} I(\mathbf{r}, \hat{\mathbf{s}}')p(\hat{\mathbf{s}}, \hat{\mathbf{s}}')d\Omega' = \epsilon(\mathbf{r}, \hat{\mathbf{s}}). \quad (3.64)$$

 Building Block: **The Radiative Transfer Equation (RTE)**

The radiative transfer equation will be the starting point to derive the diffusion equation in this book.

In the cases where we need to account for fluorescence, due to the fact that the emitted intensity depends on the excitation intensity, we must solve a set of coupled equations of the form:

$$\frac{1}{c_0}\frac{\partial}{\partial t}I(\mathbf{r}, \hat{\mathbf{s}}; \lambda_{ex}) + \hat{\mathbf{s}} \cdot \nabla I(\mathbf{r}, \hat{\mathbf{s}}; \lambda_{ex}) + \mu_t I(\mathbf{r}, \hat{\mathbf{s}}; \lambda_{ex})$$

$$- \mu_t \int_{(4\pi)} I(\mathbf{r}, \hat{\mathbf{s}}'; \lambda_{ex})p(\hat{\mathbf{s}}, \hat{\mathbf{s}}')d\Omega' = \epsilon(\mathbf{r}, \hat{\mathbf{s}}; \lambda_{ex}); \quad (3.65)$$

and

$$\frac{1}{c_0}\frac{\partial}{\partial t}I(\mathbf{r}, \hat{\mathbf{s}}; \lambda_{em}) + \hat{\mathbf{s}} \cdot \nabla I(\mathbf{r}, \hat{\mathbf{s}}; \lambda_{em}) + \mu_t I(\mathbf{r}, \hat{\mathbf{s}}; \lambda_{em})$$

$$- \mu_t \int_{(4\pi)} I(\mathbf{r}, \hat{\mathbf{s}}'; \lambda_{em})p(\hat{\mathbf{s}}, \hat{\mathbf{s}}')d\Omega' = \mu_a^{fl}(\mathbf{r})\Phi(\lambda_{em})I(\mathbf{r}, \hat{\mathbf{s}}; \lambda_{ex}), \quad (3.66)$$

where for the time being the wavelength dependence of the scattering and absorption coefficients have been omitted. We shall not deal with solving this set of coupled integro-differential equations for the RTE, but will come back to this set of equations when dealing with the diffusion approximation in the presence of fluorescence. For the time being, we shall only focus on approximating the RTE at a single wavelength, i.e. finding an approximate solution to Eq. (3.64).

3.6.2 The Equation of Energy Conservation

Once we have defined the RTE, a very useful relation results when integrating Eq. (3.64) over (4π):

$$\frac{1}{c_0}\frac{\partial}{\partial t}\int_{(4\pi)} I(\mathbf{r},\hat{\mathbf{s}})d\Omega + \nabla \cdot \int_{(4\pi)} \hat{\mathbf{s}}I(\mathbf{r},\hat{\mathbf{s}})d\Omega$$

$$+ \mu_t \int_{(4\pi)} I(\mathbf{r},\hat{\mathbf{s}})d\Omega - \mu_t \int_{(4\pi)}\int_{(4\pi)} I(\mathbf{r},\hat{\mathbf{s}}')p(\hat{\mathbf{s}},\hat{\mathbf{s}}')d\Omega'd\Omega =$$

$$\int_{(4\pi)} \epsilon(\mathbf{r},\hat{\mathbf{s}})d\Omega. \quad (3.67)$$

Expressing the source energy density as:

$$S_0(\mathbf{r}) = \int_{(4\pi)} \epsilon(\mathbf{r},\hat{\mathbf{s}})d\Omega, \quad \left(Watts/cm^3\right) \quad (3.68)$$

identifying terms and using the following relation:

$$\mu_t \int_{(4\pi)}\int_{(4\pi)} I(\mathbf{r},\hat{\mathbf{s}}')p(\hat{\mathbf{s}},\hat{\mathbf{s}}')d\Omega'd\Omega =$$

$$\mu_t U(\mathbf{r}) \int_{(4\pi)} p(\hat{\mathbf{s}},\hat{\mathbf{s}}')d\Omega' = \mu_s U(\mathbf{r}), \quad (3.69)$$

we obtain the continuity equation for the *energy conservation*:

$$\frac{1}{c_0}\frac{\partial}{\partial t}U(\mathbf{r}) + \nabla \cdot \mathbf{J}(\mathbf{r}) + \mu_t U(\mathbf{r}) - \mu_s U(\mathbf{r}) = S_0(\mathbf{r}),$$

where the contribution of scattering has been written explicitly. From the above equation we can now identify clearly each contribution related to energy conservation: the first term represents the overall change in average intensity in time, the second term represents the spatial change due to energy flux, the third the loss of average intensity due to local absorption and scattering, and finally the fourth term represents the *gain* in average intensity due to scattering contributions. As expected, the total energy is not lost due to scattering and the equation for flux conservation is thus independent of scattering:

$$\frac{1}{c_0}\frac{\partial}{\partial t}U(\mathbf{r}) + \mu_a U(\mathbf{r}) + \nabla \cdot \mathbf{J}(\mathbf{r}) = S_0(\mathbf{r}). \quad (3.70)$$

 Building Block: **Energy Conservation**

The equation for energy conservation in combination with Fick's law will produce the diffusion equation.

It is important to remember that this equation for energy conservation is exact as long as μ_a, $U(\mathbf{r})$ and $\mathbf{J}(\mathbf{r})$ can be defined, or equivalently, as long as $I(\mathbf{r}, \hat{\mathbf{s}})$ is an accurate representation of the density flow of energy and μ_a holds statistical significance. This equation for energy conservation is a consequence of the RTE, but it obviously holds a direct relationship with the equation for energy conservation derived from Poynting's theorem, introduced in Chapter 2, Eq. (2.8):

$$\frac{\partial W}{\partial t} + \frac{dP_{abs}}{dV} + \nabla \cdot \mathbf{S} = 0,$$

which, introducing the expressions in terms of $\langle \mathbf{S} \rangle$, becomes:

$$\frac{1}{c_0} \frac{\partial \|\mathbf{S}(\mathbf{r})\|}{\partial t} + \frac{4\pi}{\epsilon c_0} \omega \epsilon^{(i)} \|\mathbf{S}(\mathbf{r})\| + \nabla \cdot \langle \mathbf{S}(\mathbf{r}) \rangle = 0, \tag{3.71}$$

where now the absorbed energy cannot be described by the statistical average given by μ_a but rather by the local imaginary part of the dielectric constant of the medium, $\epsilon^{(i)}$. Additionally, note how U and $\mathbf{J}(\mathbf{r})$ reduce to $\|\mathbf{S}(\mathbf{r})\|$ and $\langle \mathbf{S}(\mathbf{r}) \rangle$ as $\delta V \to 0$.

3.6.3 *Summary of Approximations: How Small is 'Small Enough'?*

Before continuing it is convenient to discuss the main approximations that took place in order to reach the equation for the radiative transfer for $I(\mathbf{r}, \hat{\mathbf{s}}) = \|\mathbf{S}\|_v w_{\mathbf{r}}(\hat{\mathbf{s}}_J)/4\pi$, Eq. (3.64), and in particular, the meaning of the differential volume δV. As mentioned previously, detailed description of the derivation can be found in Appendix C, together with a thorough list of the approximations in Sec. C.1.6. I here include a list of the main approximations introduced during its derivation:

(1) **Far-field approximation:** The far-field approximation is the main and most relevant approximation introduced throughout the whole derivation of the RTE, implicit even in the definition of the specific intensity. This approximation has been used to account for the fact that the contributions from each of the scatterers can be seen as an outgoing spherical wave, with the magnetic and electric fields always mutually orthogonal (note that this should hold also *within* δV). The far-field approximation has also been taken in order to account for a slowly varying Poynting vector within δV, which enabled simplifications when dealing with volume integrals.

(2) **Average of the flow of energy:** In order to reach the RTE, we had to assume that the magnitude of the Poynting vector within δV was approximately constant and equal to the volume averaged expression. This was needed to ensure that all particles *within* δV received the same amount power.

(3) **Average incident flow of energy much greater than local average scattered flow of energy:** This condition was necessary in order to obtain scattering as a separate contribution to the total energy flow. It implies that within δV the contribution of the scatterers *inside* δV is negligible compared to the overall contribution.

(4) **Incoherent scattering:** Starting with the definition of the specific intensity and the concept of the volume-averaged Poynting vector, we have always dealt with additive intensities instead of additive fields. This, of course, neglects all interference effects that obviously occur due to scattering between particles. This approximation, even though it seems quite drastic, is greatly alleviated (at least in tissue optics) due to the inherent movement of the scatterers which produce a 'self-averaging' effect as discussed in Chap. 2.

(5) **Statistically equivalent optical properties throughout the medium:** Another important approximation comes from the fact that we expect the optical properties within δV to be (at least statistically) equivalent to those *outside* δV. This was seen in the previous section, where we used $N/\delta V$ to defined the density, ρ, at $\mathbf{r}$. Even though we could assume a position-dependent ρ in Eq. (3.64), in order to reach the RTE one of the assumptions taken was that in the vicinity of $\mathbf{r}$ the optical properties should not change.

(6) **De-polarization neglected:** In order to use the scalar wave equation we neglected the term which couples Cartesian coordinates (it is the $\nabla(\nabla \cdot \mathbf{E})$ we encountered at the beginning of Chap. 2) which accounts for depolarization on interaction with the particles. This approximation is also greatly relieved by the fact that in tissues we have intrinsic movement.

These approximations should always be kept in mind when using the RTE, since more often than not we tend to assume that the RTE offers the exact solution to transport within scattering media. As can be seen from these approximations, clearly in cases where we have a very dilute concentration of particles the RTE should start to fail, and in the case of high concentration of particles as is well-known can fail in the presence of strong multiple scattering if coherent effects play a role. For example, as derived here the RTE fails to describe the enhanced backscattering peak that occurs when we illuminate a highly scattering sample and measure directly in the backscattering direction [Ogilvy (1991); Nieto-Vesperinas (2006); Roux *et al.* (2001)]. This effect is a fully coherent effect and therefore is neglected by our approximation on incoherent scattering.

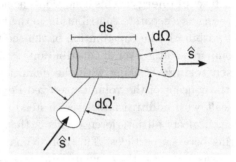

Fig. 3.10   Scattering of specific intensity incident upon the volume ds from the direction $\hat{s}'$ into the direction $\hat{s}$ (adapted from [Ishimaru (1978)]).

As has been seen throughout this chapter, the small differential volume δV has played a crucial role. One might tend to think that this δV is only needed for the definition of the volume-averaged Poynting vector, and that the specific intensity as presented in classical RTE books (see [Ishimaru (1978); Chandrasekhar (1960); Planck (1914)] for example) does not need an auxiliary differential volume for its definition. In order to relate more directly our volume-averaged expression of the Poynting vector to the specific intensity we can resolve directly to the definition of scattering of the specific intensity as shown in classical RTE works with the aid of Fig. 3.10. Following this figure, the decrease of specific intensity $dI(\mathbf{r}, \hat{s})$ for the volume ds can be expressed as:

$$dI(\mathbf{r}, \hat{s}) = -\rho ds(\sigma_{sc} + \sigma_a)I(\mathbf{r}, \hat{s}),$$

which results in the steady-state solution to the RTE when we neglect the contribution due to scattering, yielding $\hat{s} \cdot \nabla I(\mathbf{r}, \hat{s}) + \mu_t I(\mathbf{r}, \hat{s}) = 0$ (note that for the units of this equation to be correct ds cannot have units of volume but of length). As can be understood, this step requires a small volume (related to ds in Fig. 3.10) in order for ρ to be defined. This volume is equivalent to the δV we are currently discussing.

Therefore, it is now time to consider then: how small must δV be in order to be 'small enough'? As I will try to explain, this simple question has no direct answer as you might have already guessed. To being with, taking a look at the list of approximations we just went through we can see that approximations (1) and (4) imply that δV should be large so that the particles are far away from each other. On the other hand, approximations (2) and (3) suggest that that δV should be small so that the local contribution of scattering is small and the magnitude of the Poynting vector within δV is approximately constant. Additionally, these restrictions need to be balanced by the fact that the average optical properties within δV should be equivalent to those of the surrounding volume. The fact that all these apparent contradictions need to coexist for the RTE to be valid is one of the reasons that the RTE is sometimes considered an unphysical description of the problem of light propagation in particulate media [Mishchenko (2006)].

Taking the above into account, since there is no direct answer to the question of the actual size of δV, a good approximation could be that δV should be such that the average optical properties still hold, i.e. that it represents the same density of particles. Effectively, one could argue that the limit of the size of δV would be such that its volume is occupied by a single particle while maintaining the same particle density, i.e. $\delta V \simeq 1/\rho$, since the volume-averaged Poynting vector does not imply that several particles need to be enclosed in δV. Proof, however, of the validity of this approximation is currently not available, and in this limiting case the requirement that the value of the magnitude of the Poynting vector is approximately constant within δV probably would not hold.

3.7 Some Similarity Relations of the RTE

There are a few useful similarity relations that the RTE holds (for a complete discussion on this matter, see [Wyman *et al.* (1989)] and references therein). In particular, if $I(\mathbf{r}, \hat{\mathbf{s}}, t)$ is a solution to the RTE for a point source in space and time for a medium with transport coefficient $\mu_t = \mu_s + \mu_a$ and scattering diagram $p(\hat{\mathbf{s}}, \hat{\mathbf{s}}')$ then we can scale the specific intensity as:

$$\widetilde{I}(\widetilde{\mathbf{r}}, \hat{\mathbf{s}}, \widetilde{t}) = \left(\frac{\widetilde{\mu}_t}{\mu_t} \right)^3 I(\mathbf{r}, \hat{\mathbf{s}}, t), \qquad (3.72)$$

with $\widetilde{\mathbf{r}}$ and $\widetilde{t}$ defined as:

$$\widetilde{\mathbf{r}} = \frac{\mu_t}{\widetilde{\mu}_t}\mathbf{r}, \qquad \widetilde{t} = \frac{\mu_t}{\widetilde{\mu}_t}t, \qquad (3.73)$$

to a medium with total scattering coefficient $\widetilde{\mu}_t$ as long as the albedo $W = \mu_s/\mu_t$ and the $p(\hat{\mathbf{s}}, \hat{\mathbf{s}}')$ remain constant. An even more useful expression comes from the fact that if $I_0(\mathbf{r}, \hat{\mathbf{s}}, t|\mu_a = 0)$ represents the point-source solution for the non-absorbing medium ($\mu_a = 0$, $\mu_t = \mu_s$) then:

$$I_a(\mathbf{r}, \hat{\mathbf{s}}, t) = \exp(-\mu_a c_0 t) I_0(\mathbf{r}, \hat{\mathbf{s}}, t \,|\mu_a = 0), \qquad (3.74)$$

represents the solution for a medium with constant absorption μ_a. This similarity relation is exploited mostly when doing Monte Carlo simulations since only one solution per (μ_s, g) pair (where the previous similarity relation can be used) needs to be generated in order to simulate all μ_a dependence. We will come back to this last similarity relation when deriving the diffusion equation, since one of the approximations introduced invalidates Eq. (3.74).

3.8   The RTE and Monte Carlo

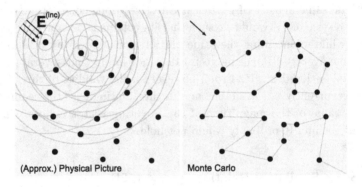

Fig. 3.11   Left: Idealized and very approximate physical picture of interaction of an electromagnetic wave with a random collection of particles. Scattering is assumed to be described by an out-going spherical wave (as in our far-field approximation). Note that only the interaction of a few scatterers with the incident wave is shown. Right: Solution to the transport of electromagnetic radiation provided by Monte Carlo.

During this chapter we have defined important quantities such as the volume-averaged Poynting vector, the specific intensity, and the equation that models light propagation in scattering media, the RTE. How this equation is solved in practice is a matter out of the scope of this book. There is, however, an approach very commonly used to model light transport in tissue based on Monte Carlo simulations: Monte Carlo has been used extensively to provide solutions to complex situations, and it is currently regarded as the most accurate solution to the problem of light transport in scattering media. There are very good reviews elsewhere that explain in detail how Monte Carlo can be applied to light transport in tissues, and therefore I will not include the details here (see [Wang *et al.* (1995); Jacques (2003); Wang and Wu (2007)] or [Martelli *et al.* (2010)] for examples). The important aspect of Monte Carlo that I would like to point out is that, at least in most of the cases where it is used to model transport in tissues, it *represents a solution of the RTE*. That is, the solution provided by Monte Carlo suffers from the same approximations as the RTE and should thus be regarded with great care. Only in those cases where Monte Carlo is employed to solve the complete set of Maxwell's equations can we be assured we are obtaining a statistical representation of the exact solution. Even though this approach is not the one commonly employed, so far the solutions provided by Monte Carlo to the problem of light transport in tissue have been of great help and as such Monte Carlo has proven to be an excellent tool.

Once this has been said, there is a very important issue regarding Monte Carlo which I believe has significantly altered our perception of how light propagates in tissues. Take for example Fig. 3.11, where a very simple interpretation of the physical picture is presented. On the left we have an approximated physical interpretation of what happens when an electromagnetic wave interacts with a collection of particles. A concept which would be easier for us grasp is that of a wave propagating in a pond and scattering from a random collection of pebbles within the pond: this would represent the scalar wave approximation, with the significant difference from water in a pond that our wave propagates at the speed of light. Picture now this in three dimensions but with polarization in mind and accounting for the electric and magnetic fields. This would be an interpretation of what Maxwell's equations formulates. As we have seen, when there is significant scattering and we can use certain approximations, the RTE may provide a good model of how energy propagates within this medium. One way to solve for this is using Monte Carlo, which uses the approach shown on the right of Fig. 3.11. In this case, in order to solve the RTE, we launch a ray of light in a given direction and assign a probability to scatter in a certain direction, following this ray until it is extinguished.

The problem arises, however, when this way of solving MC is understood as the physical picture: it is not accurate to assume that 'photons undergo collisions' or speak about the 'random walk of photons'. As Mishchenko mentions in [Mishchenko (2006)], the fact that photons are not localized makes the words like photon position, photon path or local flow of photons physically meaningless. Note that the picture of 'photons' as particles was discarded by Quantum Electrodynamics (QED) more than half a century ago, and currently the term 'photon' in QED refers to quantum of a single normal mode of the electromagnetic field [Mandel and Wolf (1995)] which is associated with plane waves of definite wave vector and definite polarization but infinite lateral extent.

3.8.1 *Photon Density*

Even though, and as has been mentioned in several instances of this book, I believe the use of the term 'photon' should be avoided whenever possible (and so far in all cases related to biomedical optics the need for Quantum Electrodynamics has not been necessary), since many research works make use of the concept of photons as particles I will here introduce the relationship between these photon-related quantities to the concepts discussed in this chapter regarding the average intensity, energy density and flux.

All quantities can be directly related to 'photons' by using the fact that the energy of a single photon of frequency ω is given by $E = \hbar\omega$, where $\hbar$ is the reduced Planck constant. Using this expression we obtain the following photon-dependent quantities:

- *Photon Density*

$$\phi(\mathbf{r}) = \frac{1}{\hbar\omega}u(\mathbf{r}) = \frac{1}{\hbar\omega c}U(\mathbf{r}), \qquad \left(Photons/cm^3\right) \qquad (3.75)$$

- *Photon Flux*

$$\mathbf{F}(\mathbf{r}) = \frac{1}{\hbar\omega}\mathbf{J}(\mathbf{r}), \qquad \left(Photons/cm^2 s\right) \qquad (3.76)$$

Note that the concept of photon density when the source is modulated at a certain frequency gave rise to what O'Leary, Boas, Chance and Yodh coined as 'Diffuse Photon Density Waves' (DPDWs) in their classic paper [O'Leary *et al.* (1992)], which opened a new line of research based on diffuse light imaging.

Key points

Before finishing this chapter, there are a few things that should be commented and remembered. First of all, the derivation of the RTE as presented here (and in greater detail in Appendix C) has been included with the sole purpose of offering a better understanding of the main approximations involved in the RTE, and thus in the diffusion approximation. It is quite common to assume that the solution to the RTE is the gold-standard solution. As can be understood from the approximations used to derive it, there are several limitations: for example, it does not have much physical sense finding the specific intensity at distances which are smaller than the discretized volume δV, i.e. the smallest volume which holds the same statistical properties as the whole medium. In the special case of the RTE in a non-scattering medium, this is reflected in the statistical properties of the absorbers. As scattering increases, δV will effectively decrease, and the RTE will be valid for smaller distances. However, if one wishes to recover the exact solution when dealing with small distances, I suggest using a full electromagnetic approach and not the RTE. Note, however, that as long as we are interested in the incoherent contribution of light the RTE has been shown to yield accurate results even in optically thin highly scattering media (see [Roux et al. (2001)] for a rigorous comparison between RTE and electromagnetism). Also, an important point to bear in mind is that neither the specific intensity nor the Poynting vector are observables. It is the flow of energy through an area which can be measured experimentally, and the average intensity might be approximated with the use of an integrating sphere, which is also based on measurements of flow of energy through an area.

Another important issue that should be remembered from this chapter is the physical picture one has in mind when working with light propagation in multiple scattering media such as tissues. We should try to avoid the concept of 'photon random walk' or 'probability of a photon being scattered in a certain direction' and use the more accurate representation given by scattering of electromagnetic waves.

From this chapter the following key points should be remembered:

- **Volume averaged flow of energy**: In order to reach the definition of the specific intensity, we need to measure the average flow of energy in a specific direction within a differential volume. The size of this differential volume will depend on the statistical optical properties of the medium, and its exact value is not well-defined. This imposes some restrictions on the validity of the RTE.
- **Multiple scattering neglected**: Even though the RTE includes a term which increases the flow of energy in a specific direction $\hat{s}_J$ due to scattered light, and effectively this term accounts for the contribution of practically the whole volume, this contribution only considers *single* scattering from each particle. As a result, we measure contributions

from all particles at a certain point **r**, but we do not take into account light that has scattered from more than one particle. This approximation, as we discussed in Chap. 2, is greatly alleviated by the fact that our media have in general self-averaging properties, in which case *on average* this multiple scattering contribution would effectively cancel out.

- **The Radiative Transfer Equation**: Once we have understood what approximations are needed in order to arrive to the RTE, we have a complete expression that will help us in modeling light transport in tissues. The RTE will be the starting point for the derivation of the diffusion equation.

- **The Equation of Flux Conservation**: A direct consequence of the RTE, it was obtained by integrating the RTE over all angles. It represents the conservation of energy flux and as such is independent of scattering. This equation, together with Fick's law, will yield the diffusion equation.

Further Reading

This book by no means intends to represent a treatise on the RTE, and therefore those interested in RTE should consult books centered on this subject. I would personally suggest reading the classic books on the RTE, such as the original book by Chandrasekhar (1960), the treatise on linear transport theory by Case and Zweifel (1967), the classic on radiative transport by Ishimaru (1978), and Martelli *et al.* (2010) for a detailed treatise on the RTE and solutions to it for the specific case of light transport in tissue. In all these cases, the radiative transport equation is derived in a phenomenological manner. However, a rigorous derivation of the RTE including polarization effects can be found for example in Mishchenko (2002). Additionally, I would recommend reading the papers by Mishchenko on Maxwell's equations and their relation with the RTE [Mishchenko (2006, 2010)]. I would also recommend the paper by the group of J. J. Greffet where the RTE is compared with rigorous numerical electromagnetic results in 2D [Roux *et al.* (2001)].

Regarding Monte Carlo, there is a great deal of bibliography on the subject. I personally recommend the papers by S. Jacques and L. Wang (see Wang *et al.* (1995) for example), or Jacques (2003) and Wang and

Wu (2007). A great source of information including codes for Monte Carlo simulation can be found in the webpage of S. Jacques and S. Prahl, http://omlc.ogi.edu/software/mc/ and in Martelli *et al.* (2010). Finally, for those interested in the deeper physical meaning of a photon in the Quantum Electrodynamic sense, I suggest reading Feynman (1998), Cohen-Tannoudji *et al.* (1997) and Mandel and Wolf (1995).

Chapter 4

Fick's Law and The Diffusion Approximation

Summary. In this chapter we shall go through the original derivation of Fick's law and its direct consequence, the diffusion approximation. The diffusion equation and the expression for the diffusion coefficient that will be used in the next chapters will be derived step by step from the Radiative Transfer Equation, discussing the implications of the approximations taken.

4.1 Historical Background

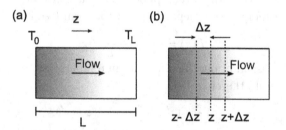

Fig. 4.1 Representation of the relation between flow and difference in concentration, (a) as presented by Fourier for an infinite slab at temperatures $T_0 > T_L$ and (b) in terms of particle diffusion flowing through z.

Almost two centuries ago, J. Fourier published in 1822 the *Analytical Theory of Heat* where he presented the equation of heat transfer (originally in French, see [Fourier and Freeman (1878)] for the English version). This work was an inspiration to G. Ohm, who applied it to charge conduction and presented what is now termed Ohm's law in 1827 (originally in German, later published in English [Ohm *et al.* (1891)]). Some years later, in 1855, it was applied by A. Fick to the diffusion of salts in a solvent[Fick

(1855)]. Both Fourier and Fick presented two equations that have since then been used with great success to predict transport of heat, particles or mass, charge, population dynamics, and our main concern, light in highly scattering media. The first equation related the flux with the change in concentration, while the second equation predicted the increase of concentration with time due to this flux. These are currently known as Fick's first law (or simply Fick's law) and Fick's second law, respectively. When dealing with heat transfer it is common to refer also to Fourier's Law, which relates flux of heat with temperature gradients through the diffusion coefficient.

In order to have a clear picture of what the diffusion approximation means, I believe deriving it in the same original context as used by both Fourier and Fick is very instructive. It is also important to understand that when both Fourier and Fick derived their equations several experiments pointed to the fact that the flux was proportional to the difference in temperature or salt concentration, respectively: what was missing was the theoretical derivation. Fourier postulated that, considering an infinite slab of width L, if we managed to keep one of the faces at a constant temperature T_0 while keeping the other face at $T_z < T_0$ there would be a linear dependence of temperature with distance. From the face of the slab at constant temperature T_0 to the opposite face at constant temperature T_z he expected a temperature dependence of the form (see Fig. 4.1(a)):

$$T(z) = T_0 + \frac{T_z - T_0}{L} z.$$

Following this assumption, the difference in temperature between two slices at z and z', would therefore be:

$$T(z) - T(z') = \frac{T_z - T_0}{L}(z - z').$$

This led him to postulate that for two different materials the ratio of their flux of heat F/F' considering the same distances $z - z'$ from two different points should be equivalent to the ratio of their difference in temperature (note that this quantity should always be positive and that we have set $T_z < T_0$):

$$\frac{F}{F'} = \frac{T_0 - T_z}{L} \frac{L'}{T_0' - T_z'},$$

Since this should hold for any material, Fourier established then that the flow of heat would be proportional to the gradient of temperatures:

$$F(z) = -K\frac{d}{dz}T(z),$$

where we have used the fact that $(T_0 - T_z)/L = -dT/dz$. This is what is sometimes termed *Fourier's law* and it is the formula Fick was based on when presenting his famous paper on liquid diffusion [Fick (1855)]. Note, however, that Fourier had to make use of the assumption that there was a linear dependence of the temperature with distance. It turns out that Fick's (or Fourier's) law does not need this assumption and it can be derived more rigorously. This rigorous derivation was originally presented in terms of particle diffusion, and it is the situation we will use now.

Let us begin assuming we have an infinite slab of width L, containing a homogeneous liquid with a certain number of molecules N. Let us assume that we somehow maintain different osmotic pressures at each side of this slab, and that we have reached equilibrium. In this case, defining the number of particles per unit volume as u, we will expect a change in particle density as we move from $z = 0$ to $z = L$, which will be constant for each z value, with lower values for u where the pressure is lower and vice-versa (note that we do not need to state that this dependence is linear). In order to reach this equilibrium there has to be a constant flow of particles throughout the volume, since we expect the molecules in the liquid to follow Brownian motion. In a particular slice z, the flux of particles F through z can be defined as the difference in number of particles at each side of z times the rate at which these particles traverse this virtual interface, divided by the area of the interface, A:

$$F(z) = [N(z - \Delta z) - N(z + \Delta z)] \times \frac{\text{Rate}}{A}, \qquad \left(particles/cm^2 s\right)$$

where $N(z + \Delta z)$ and $N(z - \Delta z)$ are the total number of particles in the differential slabs of width Δz immediately above and below z, respectively (see Fig. 4.1(b)). Assuming an average speed v_z of these particles in the z-direction and given the distance traveled by the particles in the z-direction to be Δz, we may assume the rate of particles traversing z as:

$$\text{Rate} = \frac{v_z}{\Delta z}, \qquad \left(s^{-1}\right)$$

in which case, the flux of particles through z becomes:

$$F(z) = -\frac{v_z}{A} \frac{1}{2} \frac{d}{dz} N(z),$$

where we have made use of $dN/2dz = (N(z + \Delta z) - N(z - \Delta z))/\Delta z$ when $\Delta z \to 0$. We can rewrite this in terms of the particle density u, since $N = uV$, being V the volume of the slice $V = A\Delta z$:

$$F(z) = -\frac{\Delta z^2}{2\Delta t} \frac{d}{dz} u(z), \qquad \left(particles/cm^2 s\right)$$

where we have re-written the average speed of the particles in the z-direction as $v_z = \Delta z / \Delta t$. Since this expression should not depend on the values we choose for Δz or Δt, we can group terms into a single constant coefficient, D, yielding:

$$F(z) = -D\frac{d}{dz}u(z), \qquad \left(particles/cm^2 s\right) \tag{4.1}$$

with

$$D = -\frac{\Delta z^2}{2\Delta t}, \qquad \left(cm^2/s\right) \tag{4.2}$$

the coefficient Δz representing the path-length described by the molecule on average during time Δt in the z-direction. The derivation just presented is very similar to the original one from Einstein in his famous paper on the theory of Brownian movement [Uhlenbeck and Ornstein (1930); Einstein (1956)], where he also found that the mean square value of the displacement of the particles was given by:

$$\Delta z^2 = 2D\Delta t = \frac{2k_B T}{f}\Delta t, \qquad \left(cm\right)$$

where k_B is Boltzmann's constant, T the temperature and f the frictional resistance on the particle defined as the friction force divided by the velocity. Substituting we obtain:

$$D = \frac{k_B T}{f}. \qquad \left(cm^2/s\right)$$

In the specific case of a solution sufficiently diluted we can use Stoke's law, which establishes the frictional resistance as $f = 6\pi\mu_B r_p$, yielding:

$$D = \frac{k_B T}{6\pi\mu_B r_p}, \qquad \left(cm^2/s\right) \tag{4.3}$$

where μ_B is the viscosity of the surrounding medium and r_p the particle radius. Eq. (4.3) is the Stokes-Einstein equation, commonly used for diffusion of spherical particles through liquids with low Reynolds number. As we shall soon see, when derived from the transport equation the diffusion coefficient depends on the scattering and absorption coefficients (and the speed of light in the medium, depending on the definition used).

All this derivation was presented for diffusion in one dimension. In three dimensions we can rewrite Fick's law as:

$$F(\mathbf{r}) = -D\nabla u(\mathbf{r}), \tag{4.4}$$

assuming particles diffuse isotropically in all directions[1].

---

[1]Note that this is not always the case. In liquid crystals, for example, light can have a different diffusion coefficient depending on the direction.

We have so far derived Fick's first law. Fick's second law, also present in Fourier's treatise on heat transport, can be derived following similar assumptions as before, considering the infinite slab of width L filled with a liquid and a certain number of particles N. Given the flux we have just derived, F, the increase in number of particles ΔN at a certain slice at z of width Δz for an interval of time Δt will be given by the sum of the inward fluxes going through z and $z + \Delta z$:

$$\Delta N(z) = [F(z) - F(z + \Delta z)] \Delta t A,$$

where the change of sign is due to the surface normals at z and $z + \Delta Z$ pointing in opposite directions. Rearranging and taking $\Delta z \to 0$ and $\Delta t \to 0$ we obtain:

$$\frac{\partial}{\partial t} N(z,t) = -\frac{\partial}{\partial z} F(z,t) \Delta z A,$$

which using $u = NV = NA\Delta z$ results in:

$$\frac{\partial}{\partial t} u(z,t) = -\frac{\partial}{\partial z} F(z,t),$$

which is usually referred to as Fick's second law, which represents the equation for conservation of mass. Introducing the expression for the flux F we finally obtain the *diffusion equation*, as originally formulated by Fourier and Fick:

$$\frac{\partial}{\partial t} u(z,t) = D \frac{\partial^2}{\partial z^2} u(z,t),$$

which in three dimensions can be expressed as:

$$\frac{\partial}{\partial t} u(z,t) = D \nabla^2 u(z,t). \tag{4.5}$$

Eq. (4.5) originally appeared in this form in Fourier's treatise on heat propagation with D replaced by $K/C\rho$ being K the conductance, C the specific heat and ρ the density, in which case it is usually termed the *heat equation*.

4.2 Diffuse Light

In the previous section we went over the derivation of the diffusion equation as presented by Fourier and Fick with regards to particle and heat diffusion, and presented the expression for the diffusion coefficient derived by Einstein. In the context of light, however, the concept of diffuse light took longer to develop, and the actual derivation of an expression for the diffusion coefficient was slightly more cumbersome. In particular, as we will soon see, depending on the approximations made the diffusion coefficient for light propagation may take different forms. To derive the diffusion equation for light propagation, we will approximately follow the steps of Ishimaru's derivation [Ishimaru (1978)], which is now quite standard in light propagation in turbid media. In order to derive the diffusion equation for light propagation, we need two main equations we have already encountered and that I include here again for convenience, namely, the radiative transfer equation:

$$\frac{1}{c_0}\frac{\partial}{\partial t}I(\mathbf{r},\hat{\mathbf{s}}) + \hat{\mathbf{s}}\cdot\nabla I(\mathbf{r},\hat{\mathbf{s}}) + \mu_t I(\mathbf{r},\hat{\mathbf{s}})$$

$$-\mu_t\int_{(4\pi)}I(\mathbf{r},\hat{\mathbf{s}}')p(\hat{\mathbf{s}},\hat{\mathbf{s}}')d\Omega' = \epsilon(\mathbf{r},\hat{\mathbf{s}}),\quad(4.6)$$

and the equation for energy conservation (see Eq. (3.70) from Chapter 3):

$$\frac{1}{c_0}\frac{\partial}{\partial t}U(\mathbf{r}) + \mu_a U(\mathbf{r}) + \nabla\cdot\mathbf{J}(\mathbf{r}) = S_0(\mathbf{r}),$$

where S_0 represented the average energy density of the source:

$$S_0(\mathbf{r}) = \int_{(4\pi)}\epsilon(\mathbf{r},\hat{\mathbf{s}})d\Omega.\quad\left(Watts/cm^3\right)$$

As we will soon see, in order to derive the diffusion equation we will not use the equation for energy conservation for the total average intensity U, but rather for the diffuse component of the average intensity.

4.2.1 *Reduced and Diffuse Intensity*

Before introducing the main approximation on the angular distribution of the specific intensity (or the volume averaged Poynting vector) which will yield the diffusion approximation, following Ishimaru's formalism (see, for example, Ch. 9 of [Ishimaru (1978)]) it is convenient to separate the contribution to the specific intensity into two components, the *reduced* or *collimated intensity*, and the *diffuse intensity*[2]:

$$I(\mathbf{r},\hat{\mathbf{s}}) = I_{ri}(\mathbf{r},\hat{\mathbf{s}}) + I_d(\mathbf{r},\hat{\mathbf{s}}).\quad(4.7)$$

The contribution of the reduced intensity, $I_{ri}(\mathbf{r},\hat{\mathbf{s}})$, will be the solution to the RTE which represents Beer's law:

$$\frac{1}{c_0}\frac{\partial}{\partial t}I_{ri}(\mathbf{r},\hat{\mathbf{s}}) + \hat{\mathbf{s}}\cdot\nabla I_{ri}(\mathbf{r},\hat{\mathbf{s}}) + \mu_{tr}I_{ri}(\mathbf{r},\hat{\mathbf{s}}) = \epsilon(\mathbf{r},\hat{\mathbf{s}}).\quad(4.8)$$

---

[2]The reason why we do this will become apparent in the next chapter where we will see that the point-source solution (the most common approximation to model light sources in diffuse light propagation) is a direct result of the contribution from the reduced intensity.

Note that I have purposely chosen μ_{tr} instead of μ_t in the above equation, anticipating the fact that the reduced intensity will lose less intensity the higher the anisotropy factor, g. In order for us to include μ_{tr}, however, we will need to add the $-g\mu_s$ contribution to the diffuse component (the definition of g and μ_s' will be specifically introduced in the derivation that follows[3]).

Introducing $I(\mathbf{r}, \hat{\mathbf{s}}) = I_{ri}(\mathbf{r}, \hat{\mathbf{s}}) + I_d(\mathbf{r}, \hat{\mathbf{s}})$ into Eq. (4.6) and making use of Eq. (4.8) we obtain the RTE for the diffuse component:

$$\frac{1}{c_0}\frac{\partial}{\partial t}I_d(\mathbf{r}, \hat{\mathbf{s}}) + \hat{\mathbf{s}} \cdot \nabla I_d(\mathbf{r}, \hat{\mathbf{s}}) + \mu_t I_d(\mathbf{r}, \hat{\mathbf{s}})$$

$$- \mu_t \int_{(4\pi)} I_d(\mathbf{r}, \hat{\mathbf{s}}')p(\hat{\mathbf{s}}, \hat{\mathbf{s}}')\mathrm{d}\Omega' = \epsilon_{ri}(\mathbf{r}, \hat{\mathbf{s}}), \quad (4.9)$$

where $\epsilon_{ri}(\mathbf{r}, \hat{\mathbf{s}})$ is the source energy density function which now takes into account the contribution of the reduced intensity[4]:

$$\epsilon_{ri}(\mathbf{r}, \hat{\mathbf{s}}) = \mu_t \int_{(4\pi)} I_{ri}(\mathbf{r}, \hat{\mathbf{s}}')p(\hat{\mathbf{s}}, \hat{\mathbf{s}}')\mathrm{d}\Omega' - g\mu_s I_{ri}(\mathbf{r}, \hat{\mathbf{s}}), \quad \left(Watts/cm^3 sr\right)$$

$$(4.10)$$

where once again the $g\mu_s I_{ri}(\mathbf{r}, \hat{\mathbf{s}})$ factor accounts for the fact that we are using μ_{tr} instead of μ_t in Eq. (4.8). We can now find the energy conservation equation for $I_d(\mathbf{r}, \hat{\mathbf{s}})$ by integrating Eq. (4.9) over (4π):

$$\frac{1}{c_0}\frac{\partial}{\partial t}U_d(\mathbf{r}) + \mu_a U_d(\mathbf{r}) + \nabla \cdot \mathbf{J}_d(\mathbf{r}) = S_{ri}(\mathbf{r}), \quad (4.11)$$

where $U_d(\mathbf{r})$ and $\mathbf{J}_d(\mathbf{r})$ are the average intensity and flux, respectively, of the diffuse component and S_{ri} represents the average energy density of the reduced intensity contribution of the source:

$$S_{ri}(\mathbf{r}) = \int_{(4\pi)} \epsilon_{ri}(\mathbf{r}, \hat{\mathbf{s}})\mathrm{d}\Omega,$$

which, introducing the expression for $\epsilon_{ri}(\mathbf{r}, \hat{\mathbf{s}})$ can be rewritten as:

$$S_{ri}(\mathbf{r}) = \mu_t \int_{(4\pi)} \int_{(4\pi)} I_{ri}(\mathbf{r}, \hat{\mathbf{s}})p(\hat{\mathbf{s}}, \hat{\mathbf{s}}')\mathrm{d}\Omega'\mathrm{d}\Omega - g\mu_s U_{ri}(\mathbf{r}).$$

---

[3]I am aware that it is at this point confusing, but hopefully it will make sense once we reach the expression for the diffusion equation. This step, by the way, is not usually encountered in the derivation of the diffusion equation found in the literature, where μ_t is traditionally used throughout.

[4]If you remember, $\epsilon(\mathbf{r}, \hat{\mathbf{s}})$ represents the amount of power per unit volume delivered to the system in direction $\hat{\mathbf{s}}$.

The first integral we had already encountered when deriving the energy conservation equation in the previous chapter (see Eq. (3.69)), yielding:

$$S_{ri}(\mathbf{r}) = \mu_s' U_{ri}(\mathbf{r}). \quad \left(Watts/cm^3 \right) \tag{4.12}$$

We can proceed in a similar way in the equation for $I_{ri}(\mathbf{r}, \hat{\mathbf{s}})$, Eq. (4.8), and obtain the following relationship for the reduced average intensity, $U_{ri}(\mathbf{r})$ and reduced intensity flux, $\mathbf{J}_{ri}(\mathbf{r})$:

$$\frac{1}{c_0}\frac{\partial}{\partial t}U_{ri}(\mathbf{r}) + \mu_{tr}U_{ri}(\mathbf{r}) + \nabla \cdot \mathbf{J}_{ri}(\mathbf{r}) = S_0(\mathbf{r}). \tag{4.13}$$

Of course, by adding Eq. (4.11) and Eq. (4.13) we recover the energy conservation equation derived in the previous chapter, where as expected from Eq. (4.7):

$$\mathbf{J}(\mathbf{r}) = \mathbf{J}_d(\mathbf{r}) + \mathbf{J}_{ri}(\mathbf{r}), \tag{4.14}$$

$$U(\mathbf{r}) = U_d(\mathbf{r}) + U_{ri}(\mathbf{r}). \tag{4.15}$$

We will now move on to introduce the main assumption of the diffusion approximation, namely the angular distribution of the diffuse component, $I_d(\mathbf{r}, \hat{\mathbf{s}})$.

4.2.2 Angular Distribution of Diffuse Light

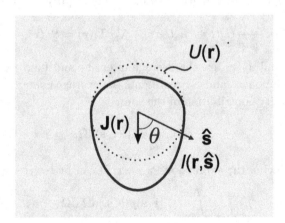

Fig. 4.2 Representation of the angular dependence of the intensity within the diffusion approximation.

It is clear that if at this point we simply introduce Fick's law as $\mathbf{J}_d(\mathbf{r}) = -D\nabla U_d(\mathbf{r})$ into the equation for energy conservation, Eq. (4.11), we would obtain the diffusion equation in a manner very similar to the approach used by Fourier and Fick, as was shown in the previous section. In doing so, however, we would have no information in terms of the approximations taken with regards to the flow of energy and, more importantly, what the diffusion coefficient actually depends on. Therefore, in order to derive the diffusion equation and obtain some information on the elements involved in the diffusion coefficient, we will first introduce an assumption on the light distribution, i.e. on a possible expression for $I_d(\mathbf{r}, \hat{\mathbf{s}})$. Given the fact that we expect the distribution of intensity inside a highly scattering medium to be quite uniform, and assuming the total flow of energy is in direction $\hat{\mathbf{s}}_J$ we can choose an approximation to $I(\mathbf{r}, \hat{\mathbf{s}})$ such that:

$$I_d(\mathbf{r}, \hat{\mathbf{s}}) \simeq f_0(\mathbf{r}) + f_1(\mathbf{r})\hat{\mathbf{s}}_J \cdot \hat{\mathbf{s}}, \tag{4.16}$$

where f_0 and f_1 are generic functions of $\mathbf{r}$ for which we now need to find an expression. Note that we cannot simply choose $I_d(\mathbf{r}, \hat{\mathbf{s}}) = f_0(\mathbf{r})$ and have no angular dependence on the specific intensity, since this would mean that there is no flow of energy[5]. Since $\hat{\mathbf{s}}_J \cdot \hat{\mathbf{s}} = \cos\theta$, Eq. (4.16) can be understood as the first order spherical harmonics expansion of the complete solution for the specific intensity, $I_d(\mathbf{r}, \hat{\mathbf{s}})$, and for that reason it is sometimes termed the P_1 approximation.

The next step is then to determine the expressions for the functions f_0 and f_1. The angular independent term can be obtained by introducing our approximation to $I_d(\mathbf{r}, \hat{\mathbf{s}})$ into the expression for $U_d(\mathbf{r})$, Eq. (3.8):

$$U_d(\mathbf{r}) = f_0(\mathbf{r}) \int_{(4\pi)} \mathrm{d}\Omega + f_1(\mathbf{r}) \int_{(4\pi)} \hat{\mathbf{s}}_J \cdot \hat{\mathbf{s}}\mathrm{d}\Omega.$$

Making use of $\int \mathrm{d}\Omega = 4\pi$ and the relation[6]:

$$\int_{(4\pi)} \hat{\mathbf{s}}_J \cdot \hat{\mathbf{s}}\mathrm{d}\Omega = 2\pi \int_{-1}^{+1} \cos\theta \mathrm{d}(\cos\theta) = 0, \tag{4.17}$$

we obtain $f_0(\mathbf{r}) = U_d(\mathbf{r})/4\pi$, as expected. Proceeding in a similar way, we may introduce the approximated specific intensity Eq. (4.16) into the expression for the flux $\mathbf{J}_d(\mathbf{r})$, Eq. (3.12). Since we assumed that the total flux pointed in direction $\hat{\mathbf{s}}_J$, we have that:

$$J_d(\mathbf{r}) = \mathbf{J}_d(\mathbf{r}) \cdot \hat{\mathbf{s}}_J = f_0(\mathbf{r}) \int_{(4\pi)} \hat{\mathbf{s}}_J \cdot \hat{\mathbf{s}}\mathrm{d}\Omega + f_1(\mathbf{r}) \int_{(4\pi)} (\hat{\mathbf{s}}_J \cdot \hat{\mathbf{s}})^2\mathrm{d}\Omega.$$

---

[5]Remember that $\mathbf{J}(\mathbf{r}) = \int I_d(\mathbf{r}, \hat{\mathbf{s}})\hat{\mathbf{s}}\mathrm{d}\Omega$ and therefore $\mathbf{J}(\mathbf{r})$ would be zero.

[6]As a reminder, $\mathrm{d}\Omega = \mathrm{d}\phi\mathrm{d}(\cos\theta) = 2\pi\mathrm{d}(\cos\theta)$ for this symmetry.

The solid angle integral can be easily solved:

$$\int_{(4\pi)} (\hat{\mathbf{s}}_J \cdot \hat{\mathbf{s}})^2 d\Omega = 2\pi \int_{-1}^{+1} \cos\theta^2 d\cos\theta = \frac{4\pi}{3}, \qquad (4.18)$$

in which case we obtain that $f_1(\mathbf{r}) = J_d(\mathbf{r})4\pi/3$. Within this approximation, the diffuse component of the specific intensity is therefore written as:

$$I_d(\mathbf{r}, \hat{\mathbf{s}}) \simeq \frac{U_d(\mathbf{r})}{4\pi} + \frac{3}{4\pi}\mathbf{J}_d(\mathbf{r}) \cdot \hat{\mathbf{s}}. \qquad (4.19)$$

 Building Block: **Specific Intensity for Diffuse Light**

This approximation to the specific intensity in highly scattering media will be the starting point to reach the diffusion equation and the expression for the diffusion coefficient.

4.3   Derivation of the Diffusion Equation

Once we have selected an approximated expression for the average flow of energy into a direction $\hat{\mathbf{s}}$, the next step is to introduce this approximation in the Radiative Transfer Equation for the diffuse component, Eq. (4.9). By doing so, we will retrieve a function equivalent to Fick's law for diffusion of light. Since this derivation has many terms and steps it is easy to get lost in the formulation. In order to avoid this as much as possible, I will start by individually introducing Eq. (4.19) into each of the terms of the RTE:

- The change of flow of energy with time:

$$\frac{1}{c_0}\frac{\partial}{\partial t}I_d(\mathbf{r}, \hat{\mathbf{s}}) = \frac{1}{4\pi c_0}\frac{\partial}{\partial t}U_d(\mathbf{r}) + \frac{3}{4\pi c_0}\frac{\partial}{\partial t}\mathbf{J}_d(\mathbf{r}) \cdot \hat{\mathbf{s}}.$$

- The ballistic contribution (Beer's law):

$$\hat{\mathbf{s}} \cdot \nabla I_d(\mathbf{r}, \hat{\mathbf{s}}) + \mu_t I_d(\mathbf{r}, \hat{\mathbf{s}}) =$$
$$\frac{1}{4\pi}\hat{\mathbf{s}} \cdot \nabla U_d(\mathbf{r}) + \frac{3}{4\pi}\hat{\mathbf{s}} \cdot \nabla\left(\mathbf{J}_d(\mathbf{r}) \cdot \hat{\mathbf{s}}\right) + \frac{\mu_t}{4\pi}\left(U_d(\mathbf{r}) + 3\mathbf{J}_d(\mathbf{r}) \cdot \hat{\mathbf{s}}\right).$$

which, using the fact that $\hat{\mathbf{s}} \cdot \nabla\left(\mathbf{J}_d(\mathbf{r}) \cdot \hat{\mathbf{s}}\right) = \nabla \cdot \mathbf{J}_d(\mathbf{r})$, becomes:

$$\hat{\mathbf{s}} \cdot \nabla I_d(\mathbf{r}, \hat{\mathbf{s}}) + \mu_t I_d(\mathbf{r}, \hat{\mathbf{s}}) =$$
$$\frac{1}{4\pi}\hat{\mathbf{s}} \cdot \nabla U_d(\mathbf{r}) + \frac{3}{4\pi}\nabla \cdot \mathbf{J}_d(\mathbf{r}) + \frac{\mu_t}{4\pi}\left(U_d(\mathbf{r}) + 3\mathbf{J}_d(\mathbf{r}) \cdot \hat{\mathbf{s}}\right).$$

- And, finally, the multiple scattering contribution:

$$\mu_t \int_{(4\pi)} I_d(\mathbf{r}, \hat{\mathbf{s}}')p(\hat{\mathbf{s}}, \hat{\mathbf{s}}')\mathrm{d}\Omega' = \frac{\mu_t}{4\pi}U_d(\mathbf{r}) \int_{(4\pi)} p(\hat{\mathbf{s}}, \hat{\mathbf{s}}')\mathrm{d}\Omega' +$$
$$\frac{3\mu_t}{4\pi} \int_{(4\pi)} p(\hat{\mathbf{s}}, \hat{\mathbf{s}}')\mathbf{J}_d(\mathbf{r}) \cdot \hat{\mathbf{s}}'\mathrm{d}\Omega'.$$

The contribution of this term is extremely important (it will be the one accounting for anisotropy in the diffusion coefficient), and since the steps that solve for it are usually missing in the derivations one can find in the literature I will include them here explicitly, at the expense of significantly extending the current derivation[7]. First of all, in order to proceed we need to introduce some assumption on the form of the phase function, $p(\hat{\mathbf{s}}, \hat{\mathbf{s}}')$ so that we can somehow obtain a solution to the solid angle integral. The simplest approach is to consider that the phase function depends only on the angle between $\hat{\mathbf{s}}$ and $\hat{\mathbf{s}}'$, which is equivalent to assuming that the scattering diagram of our particles is *on average* equivalent to that of spherical particles:

$$p(\hat{\mathbf{s}}, \hat{\mathbf{s}}') = p(\hat{\mathbf{s}} \cdot \hat{\mathbf{s}}').$$

We can now identify from Chapter 2 one of the properties of the phase function, namely that (see Eq. (2.45)):

$$\int p(\hat{\mathbf{s}} \cdot \hat{\mathbf{s}}')\mathrm{d}\Omega' = \frac{\mu_s}{\mu_t},$$

and also make use of the definition of the average of the scattering angle, $\hat{\mathbf{s}} \cdot \hat{\mathbf{s}}'$, which if you remember is the *anisotropy factor*, g, we derived in Chapter 2:

$$g = \frac{\int_{(4\pi)} p(\hat{\mathbf{s}} \cdot \hat{\mathbf{s}}')\hat{\mathbf{s}} \cdot \hat{\mathbf{s}}'\mathrm{d}\Omega'}{\int_{(4\pi)} p(\hat{\mathbf{s}} \cdot \hat{\mathbf{s}}')\mathrm{d}\Omega'}.$$

Including all these expressions in our multiple scattering contribution and rewriting $\mathbf{J}_d(\mathbf{r})$ as $\mathbf{J}_d(\mathbf{r}) = J_d(\mathbf{r})\hat{\mathbf{s}}_J$ we obtain:

$$\mu_t \int_{(4\pi)} I_d(\mathbf{r}, \hat{\mathbf{s}}')p(\hat{\mathbf{s}}, \hat{\mathbf{s}}')\mathrm{d}\Omega' = \frac{\mu_s}{4\pi}U_d(\mathbf{r}) + \frac{3\mu_t}{4\pi}J_d(\mathbf{r}) \int_{(4\pi)} p(\hat{\mathbf{s}} \cdot \hat{\mathbf{s}}')\hat{\mathbf{s}}_J \cdot \hat{\mathbf{s}}'\mathrm{d}\Omega',$$

where we now have to take great care with the term $\hat{\mathbf{s}}_J \cdot \hat{\mathbf{s}}'$ since it is not the $\hat{\mathbf{s}} \cdot \hat{\mathbf{s}}'$ we need. We can reach this expression, however, by

---

[7]I do believe, however, that understanding each and every one of the steps is highly necessary. I am quite certain that I am not the only one that has stared bewildered at a derivation which simply leapfrogged through steps that then took too long to decipher.

writing $\hat{s}_J \cdot \hat{s}'$ in terms of $\hat{s}_J \cdot \hat{s}$. Assuming the angle between $\hat{s}'$ and $\hat{s}$ is α, and the angle between $\hat{s}$ and $\hat{s}_J$ is β we can write $\hat{s}_J \cdot \hat{s}' = \cos(\alpha - \beta) = \cos(\alpha)\cos(\beta) + \sin(\alpha)\sin(\beta)$, or equivalently, $\hat{s}_J \cdot \hat{s}' = (\hat{s} \cdot \hat{s}')(\hat{s}_J \cdot \hat{s}) + (1 - (\hat{s} \cdot \hat{s}')^2)^{1/2}(1 - (\hat{s}_J \cdot \hat{s})^2)^{1/2}$, in which case the solid angle integral becomes:

$$\int_{(4\pi)} p(\hat{s} \cdot \hat{s}')\hat{s}_J \cdot \hat{s}'d\Omega' = (\hat{s}_J \cdot \hat{s})\int_{(4\pi)} p(\hat{s} \cdot \hat{s}')\hat{s} \cdot \hat{s}'d\Omega',$$

where the integral of $\sin(\alpha)$ over (4π) is zero[8]. Using the expression for g we now obtain:

$$\int_{(4\pi)} p(\hat{s} \cdot \hat{s}')\hat{s}_J \cdot \hat{s}'d\Omega' = (\hat{s}_J \cdot \hat{s})\frac{\mu_s}{\mu_t}g.$$

yielding the solution for the multiple scattering contribution as:

$$\mu_t \int_{(4\pi)} I_d(\mathbf{r}, \hat{s}')p(\hat{s}, \hat{s}')d\Omega' = \frac{\mu_s}{4\pi}U_d(\mathbf{r}) + \frac{3\mu_s g}{4\pi}\mathbf{J}_d(\mathbf{r}) \cdot \hat{s}.$$

We are finally in the position to regroup and rearrange all these terms, obtaining the RTE expression within the diffusion approximation as:

$$\frac{1}{c_0}\frac{\partial}{\partial t}U_d(\mathbf{r}) + \frac{3}{c_0}\frac{\partial}{\partial t}\mathbf{J}_d(\mathbf{r}) \cdot \hat{s} + \hat{s} \cdot \nabla U_d(\mathbf{r}) + 3\nabla \cdot \mathbf{J}_d(\mathbf{r}) +$$
$$\mu_a U_d(\mathbf{r}) + 3\left(\mu_a + (1 - g)\mu_s\right)\mathbf{J}_d(\mathbf{r}) \cdot \hat{s} = 4\pi\epsilon_{ri}(\mathbf{r}, \hat{s}). \quad (4.20)$$

At this point we can now distinguish the contribution of the scattering anisotropy on the overall flow of energy through the term $(1 - g)\mu_s$. This quantity, usually termed as the *reduced scattering coefficient*, was already introduced in Chapter 2, Section 2.5.1:

$$\mu'_s = (1 - g)\mu_s. \quad \left(cm^{-1}\right) \quad (4.21)$$

---

[8]Since $p(\hat{s} \cdot \hat{s}')$ is an even function and $\sin(\alpha)$ is odd, the result of $2\pi \int_{-1}^{1} \sqrt{1 - (\mu)^2}p(\mu)d\mu$, being $\mu = \hat{s} \cdot \hat{s}'$, will yield an even function which will thus cancel out for these integration limits. This integral can also be solved using the triple vector product relationship, which we will use to solve for the integral of $\epsilon_{ri}(\mathbf{r}, \hat{s})$ at the end of this section.

Building Block: **Reduced Scattering Coefficient**

The reduced scattering coefficient takes into account that forward scattering significantly diminishes the loss of the original direction of light propagation.

As mentioned previously, the reduced scattering coefficient accounts for the fact that particles might have high scattering coefficients and at the same time exhibit high forward scattering contributions. In this case the overall light intensity in the direction of propagation is lost at a slower pace, resulting in an effective scattering coefficient μ'_s which is lower than the original scattering coefficient μ_s. One can think of the reduced scattering coefficient as the absolute scattering coefficient that an equivalent collection of particles would have if their phase or scattering diagrams were isotropic. In analogy to the transport coefficient μ_t, it is also common to represent the reduced transport coefficient μ_{tr} as:

$$\mu_{\text{tr}} = \mu'_s + \mu_a = (1 - g)\mu_s + \mu_a. \qquad \left(cm^{-1}\right) \qquad (4.22)$$

Additionally, as we already saw in Chap. 2, in terms of μ_s and μ'_s it is also common to define the *scattering* and *reduced scattering mean free path* as l_{sc} and l^*_{sc}, respectively:

$$l_{\text{sc}} = \frac{1}{\mu_s}; \qquad l^*_{\text{sc}} = \frac{1}{\mu'_s} = \frac{l_{\text{sc}}}{1 - g}, \qquad \left(cm\right) \qquad (4.23)$$

and the *transport* and *reduced transport mean free paths* as l_{tr} and l^*_{tr} respectively:

$$l_{\text{tr}} = \frac{1}{\mu_s + \mu_a} = \frac{1}{\mu_t}; \qquad l^*_{\text{tr}} = \frac{1}{\mu'_s + \mu_a} = \frac{1}{\mu_{\text{tr}}}. \qquad \left(cm\right) \qquad (4.24)$$

We will return to this issue towards the end of this chapter, where we will discuss the deeper meaning of the transport and scattering mean free paths.

Once defined μ'_s, the final step in order to reach an expression equivalent to Fick's law consists of multiplying Eq. (4.20) by $\hat{s}$ and integrating over (4π) which gives:

$$\nabla U_d(\mathbf{r}) = -3(\mu'_s + \mu_a)\mathbf{J}_d(\mathbf{r}) - 3\frac{1}{c_0}\frac{\partial \mathbf{J}_d(\mathbf{r})}{\partial t} + 3\int_{(4\pi)} \epsilon_{ri}(\mathbf{r}, \hat{s})\hat{s}d\Omega, \qquad (4.25)$$

where we have made use of the following identities valid for any vector $\mathbf{A}$ (see Appendix B):

$$\int_{(4\pi)} \hat{\mathbf{s}} \cdot (\hat{\mathbf{s}} \cdot \mathbf{A}) d\Omega = \frac{4\pi}{3} \mathbf{A},$$

$$\int_{(4\pi)} \hat{\mathbf{s}} \cdot [\hat{\mathbf{s}} \cdot \nabla(\mathbf{A} \cdot \hat{\mathbf{s}})] \, d\Omega = \int_{(4\pi)} \hat{\mathbf{s}} \cdot [\nabla \cdot \mathbf{A}] \, d\Omega = 0,$$

$$\int_{(4\pi)} \frac{\partial A}{\partial t} \hat{\mathbf{s}} d\Omega = \frac{\partial A}{\partial t} \int_{(4\pi)} \hat{\mathbf{s}} d\Omega = 0.$$

Before continuing our derivation, in the above Eq. (4.25) we must distinguish the contribution of the flux due to the reduced intensity[9] (i.e. the flux contribution of the source that has not lost directionality due to scattering) which we will term $\mathbf{Q}_{ri}(\mathbf{r})$:

$$\mathbf{Q}_{ri}(\mathbf{r}) = \frac{1}{\mu_{\mathrm{tr}}} \int_{(4\pi)} \epsilon_{ri}(\mathbf{r}, \hat{\mathbf{s}}) \hat{\mathbf{s}} d\Omega. \qquad \left(Watts/cm^2 \right) \qquad (4.26)$$

As a quick reminder, $\epsilon_{ri}(\mathbf{r}, \hat{\mathbf{s}})$ represents the source energy density due to the reduced intensity contribution radiated into direction $\hat{\mathbf{s}}$, which when introduced into $\mathbf{Q}_{ri}(\mathbf{r})$ gives:

$$\mathbf{Q}_{ri}(\mathbf{r}) = \frac{\mu_t}{\mu_{\mathrm{tr}}} \int_{(4\pi)} I_{ri}(\mathbf{r}, \hat{\mathbf{s}}') \left(\int_{(4\pi)} p(\hat{\mathbf{s}} \cdot \hat{\mathbf{s}}') \hat{\mathbf{s}} d\Omega \right) d\Omega'$$

$$- \frac{g\mu_s}{\mu_{\mathrm{tr}}} \int_{(4\pi)} I_{ri}(\mathbf{r}, \hat{\mathbf{s}}) \hat{\mathbf{s}} d\Omega.$$

A more compact expression to the first integral can be reached by using the vector triple product rule[10], $\mathbf{a} \times (\mathbf{b} \times \mathbf{c}) = \mathbf{b}(\mathbf{a} \cdot \mathbf{c}) - \mathbf{c}(\mathbf{a} \cdot \mathbf{c})$ identifying $\mathbf{a} = \mathbf{c} = \hat{\mathbf{s}}'$ and $\mathbf{b} = \hat{\mathbf{s}}$ which gives:

$$\hat{\mathbf{s}} = \hat{\mathbf{s}}' \times (\hat{\mathbf{s}} \times \hat{\mathbf{s}}') + \hat{\mathbf{s}}'(\hat{\mathbf{s}}' \cdot \hat{\mathbf{s}}), \qquad (4.27)$$

in which case we may solve the integral over the phase function as:

$$\int_{(4\pi)} p(\hat{\mathbf{s}} \cdot \hat{\mathbf{s}}') \hat{\mathbf{s}} d\Omega = \frac{g\mu_s}{\mu_t} \hat{\mathbf{s}}' + \int_{(4\pi)} p(\hat{\mathbf{s}} \cdot \hat{\mathbf{s}}') \left(\hat{\mathbf{s}}' \times (\hat{\mathbf{s}} \times \hat{\mathbf{s}}') \right) d\Omega.$$

Since the vector product contribution is zero (for a detailed derivation see Appendix B[11]), the reduced intensity contribution to the diffuse flux

---

[9]Do not confuse it with the reduced intensity flux, $\mathbf{J}_{ri}(\mathbf{r})$.

[10]Note that this same expression could have been used to solve the integral over the phase function of $\hat{\mathbf{s}}_J \cdot \hat{\mathbf{s}}'$ encountered in this same section.

[11]It is clear from geometrical considerations that the resulting vector will be contained within the plane defined by $\hat{\mathbf{s}} \cdot \hat{\mathbf{s}}'$ and it will be perpendicular to $\hat{\mathbf{s}}'$ whatever the value of $\hat{\mathbf{s}}$, in which case the resulting integral is zero if we integrate over 2π.

becomes:

$$\mathbf{Q}_{ri}(\mathbf{r}) = \frac{g\mu_s}{\mu_{\text{tr}}} \int_{(4\pi)} I_{ri}(\mathbf{r},\hat{\mathbf{s}}')\hat{\mathbf{s}}'d\Omega' - \frac{g\mu_s}{\mu_{\text{tr}}} \int_{(4\pi)} I_{ri}(\mathbf{r},\hat{\mathbf{s}})\hat{\mathbf{s}}d\Omega = 0.$$

$$\left(Watts/cm^2 \right)$$

At this point it is important that we make a brief stop to understand why the reduced intensity contribution to the diffuse flux is zero. First of all, it is clear that if we had defined the equation that the reduced specific intensity has to follow in terms of μ_t instead of μ_{tr} (see Eq. (4.8)), the term $\mathbf{Q}_{ri}(\mathbf{r})$ would not be zero but equal to:

$$\mathbf{Q}_{ri}(\mathbf{r}) = \frac{g\mu_s}{\mu_s' + \mu_a}\mathbf{J}_{ri}(\mathbf{r}), \qquad \left(Watts/cm^2 \right)$$

where we have identified the reduced intensity flux, $\mathbf{J}_{ri}(\mathbf{r})$. In other words, by including the $-g\mu_s I_{ri}(\mathbf{r},\hat{\mathbf{s}})$ factor in the equation for the diffuse component we have already accounted for the anisotropy contribution of the reduced intensity, factor which otherwise would have been taken care of by the $g\mu_s/\mu_{\text{tr}}\mathbf{J}_{ri}(\mathbf{r})$ contribution. In this sense, by expressing the equation for $I_{ri}(\mathbf{r},\hat{\mathbf{s}})$ in terms of μ_{tr} we have also significantly simplified the derivation of the diffusion equation.

4.3.1 *The Diffusion Coefficient*

We can now move on to obtaining an equivalent to Fick's law in the context of light diffusion from Eq. (4.25), where we need to introduce one final assumption, namely that the contribution of the term $\partial\mathbf{J}_d(\mathbf{r})/\partial t$ is negligible. To justify this, we can first write Eq. (4.25) in a Fick's law form:

$$\mathbf{J}_d(\mathbf{r},t) = -\frac{1}{3(\mu_s' + \mu_a)}\nabla U_d(\mathbf{r},t) + \frac{l_{\text{tr}}}{c_0}\frac{\partial\mathbf{J}_d(\mathbf{r},t)}{\partial t}. \qquad (4.28)$$

Since J represents the rate at which energy traverses a unit area, what $\partial J/\partial t$ represents is the rate at which this rate changes (note that for a constant source the rate at which energy traverses a unit area is constant). We must compare this change of flux with time with $\tau = l_{\text{tr}}/c_0$, which can be understood as the time light takes to travel a distance of one transport mean free path (note that typical path-lengths are in the order of $1mm$). As long as we are not very close to a source[12], we can expect the rate of change in $\mathbf{J}_d(\mathbf{r})$ to be significantly smaller than $1/\tau$ or, equivalently:

$$\left| \frac{l_{\text{tr}}}{c_0}\frac{1}{J_d(\mathbf{r})}\frac{\partial J_d(\mathbf{r})}{\partial t} \right| \ll 1. \qquad (4.29)$$

---

[12]We could use our definition of δV from the previous chapter to define 'close'.

A few words must be said about this approximation before moving on. First of all, it is also possible not to include this approximation, in which case we could use Eq. (4.25) and the energy conservation Eq. (4.11) to generate the *telegrapher's equation*. It is through the introduction of this approximation that we reach the diffusion approximation instead. However, there is yet an additional implication of this approximation which has to do with the similarity relations of the RTE introduced in the previous chapter. By removing this temporal dependence of $\mathbf{J}_d(\mathbf{r})$ we are effectively invalidating the relationship which relates the specific intensity in the presence of absorption with the specific intensity for a point source in a non-absorbing medium (see Eq. (3.74)), in which case:

$$I_a(\mathbf{r},\hat{\mathbf{s}},t) \neq \exp(-\mu_a c_0 t) I_0(\mathbf{r},\hat{\mathbf{s}},t\,|\mu_a = 0),$$
$$U_a(\mathbf{r},t) \neq \exp(-\mu_a c_0 t) U_0(\mathbf{r},t\,|\mu_a = 0),$$
$$\mathbf{J}_a(\mathbf{r},t) \neq \exp(-\mu_a c_0 t) \mathbf{J}_0(\mathbf{r},t\,|\mu_a = 0).$$

Note that even though this relationship does not hold anymore, in the presence of low absorption it is still expected to yield very good results.

Making use of approximation Eq. (4.29) in Eq. (4.28) we finally obtain an expression for Fick's law in the context of light propagation:

$$\mathbf{J}_d(\mathbf{r}) = -D\nabla U_d(\mathbf{r}), \tag{4.30}$$

where we have introduced the *diffusion coefficient*, D as:

$$D = \frac{1}{3(\mu_s' + \mu_a)} = \frac{1}{3(\mu_s(1-g) + \mu_a)} = \frac{l_{tr}^*}{3}, \quad \left(cm\right) \tag{4.31}$$

 Building Block: **Diffusion Coefficient**

The diffusion coefficient will account for the spatial and temporal broadening of the average intensity in highly scattering and absorbing media.

Note that the diffusion coefficient defined in this manner has units of cm rather than cm^2/s as in Section 4.1 (see Eq. (4.3), for example), and it is quite common in current literature dealing with light diffusion to find it defined in this manner. An alternative (possibly with greater physical significance) definition is:

$$D_v = \frac{c_0}{3(\mu_s' + \mu_a)} = \frac{c_0}{3(\mu_s(1-g) - \mu_a)} = c_0 \frac{l_{tr}^*}{3}, \quad \left(cm^2/s\right) \tag{4.32}$$

in which case Fick's law is not defined in terms of the average intensity but in terms of the energy density, $u_d(\mathbf{r})$:

$$\mathbf{J}_d(\mathbf{r}) = -\frac{1}{c_0} D_v \nabla U_d(\mathbf{r}) = -D_v \nabla u_d(\mathbf{r}). \qquad (4.33)$$

Any choice of the expression for D is of course valid, as long as it is clear that one involves working with the average intensity U measured in $Watts/cm^2$, while the other involves the energy density u with units of J/cm^3. In the latter case, great care must be taken when dealing with index-mismatched interfaces since keeping track of the indexes is slightly more cumbersome.

4.3.2 The Diffusion Coefficient In Absorbing Media

In those cases where we are dealing with values of absorption comparable to the scattering coefficient we need to consider the derivation of the diffusion equation much more carefully. As has been suggested in the past, it might seem that high absorption precludes diffusion from happening. This, however, is not the case as has been shown both theoretically [Aronson and Corngold (1999); Elaloufi et al. (2003)], and experimentally [Ripoll et al. (2005)]. The derivation and proof of the validity of the diffusion approximation in highly absorbing media will not be presented here, but what is important is to understand that the diffusion approximation still holds as long as we define a diffusion coefficient as:

$$D_{abs} = \frac{1}{3(\mu_s' + a\mu_a)}, \qquad (4.34)$$

with a representing a factor which depends on the scattering, anisotropy and absorption coefficients which can be approximated to (see [Aronson and Corngold (1999); Ripoll et al. (2005)] for details on the derivation):

$$a = 1 - \frac{4}{5} \frac{\mu_s' + \mu_a}{\mu_s'(1 + g) + \mu_a}.$$

Typical values for a range from 0.2 to 0.6, being in the order of 0.5-0.55 in the case of tissue in the visible, assuming an anisotropy factor g of $g = 0.8$. Overall, $a = 0.2$ yields very good experimental results even for values of $\mu_a \simeq \mu_s'$ [Ripoll et al. (2005)]. Note that in the isotropic scattering case, $g = 0$, $a = 0.2$ is the value we recover since the equation above gives $a = 1/5 = 0.2$.

4.4 The Diffusion Equation

We are finally in the position to obtain the expression for the diffusion equation to light transport. For that, we simply need to introduce our approximated expression for the flux:

$$\mathbf{J}_d(\mathbf{r}) = -D\nabla U_d(\mathbf{r}),$$

into the energy conservation equation for the diffuse contribution:

$$\frac{1}{c_0}\frac{\partial}{\partial t}U_d(\mathbf{r}) + \mu_a U_d(\mathbf{r}) + \nabla \cdot \mathbf{J}_d(\mathbf{r}) = S_{ri}(\mathbf{r}),$$

which introducing the expression for S_{ri} in terms of $U_{ri}(\mathbf{r})$ yields a diffusion equation for the diffuse average intensity:

$$\frac{1}{c_0}\frac{\partial}{\partial t}U_d(\mathbf{r}) - \nabla \cdot \left(D\nabla U_d(\mathbf{r})\right) + \mu_a U_d(\mathbf{r}) = \mu'_s U_{ri}(\mathbf{r}). \tag{4.35}$$

 Building Block: **Diffusion Equation**

The diffusion equation will be from now on the governing equation to model light propagation in highly scattering media such as tissues.

Note that if we had defined the equation for $I_{ri}(\mathbf{r},\hat{\mathbf{s}})$ in terms of μ_t instead of μ_{tr}, we would have an extra contribution to the source term in the above equation which would account for the anisotropy of the reduced intensity, and would be given by $-g\mu_s/\mu_{\text{tr}}\nabla \cdot \mathbf{J}_{ri}(\mathbf{r})$.

In terms of the reduced intensity, S_{ri} is defined as:

$$S_{ri}(\mathbf{r}) = \mu'_s U_{ri}(\mathbf{r}) = \mu'_s \int_{(4\pi)} I_{ri}(\mathbf{r},\hat{\mathbf{s}})d\Omega, \tag{4.36}$$

where, allow me to insist, $I_{ri}(\mathbf{r},\hat{\mathbf{s}})$ is the solution to:

$$\frac{1}{c_0}\frac{\partial}{\partial t}I_{ri}(\mathbf{r},\hat{\mathbf{s}}) + \hat{\mathbf{s}} \cdot \nabla I_{ri}(\mathbf{r},\hat{\mathbf{s}}) + \mu_{\text{tr}}I_{ri}(\mathbf{r},\hat{\mathbf{s}}) = \epsilon(\mathbf{r},\hat{\mathbf{s}}). \tag{4.37}$$

Equations (4.35)-(4.37) represent the complete set of equations that determine the diffuse and reduced intensity components of the specific intensity, and from now on will be our starting set of equations dealing with specific expressions for $I_{ri}(\mathbf{r},\hat{\mathbf{s}})$ and thus solutions to $U_d(\mathbf{r})$ in the next chapter.

Regarding the diffusion equation Eq. (4.35), note that we have explicitly left D inside the brackets to account for a possible spatial dependence on μ'_s, in which case the divergence must be correctly applied (more on this in the next chapter). In the case we have a homogeneous medium Eq. (4.35) is reduced to the more common expression:

$$D\nabla^2 U(\mathbf{r}) - \mu_a U(\mathbf{r}) = \frac{1}{c_0}\frac{\partial}{\partial t}U(\mathbf{r}) - S_{ri}(\mathbf{r}). \qquad (4.38)$$

4.5 The Mean Free Path

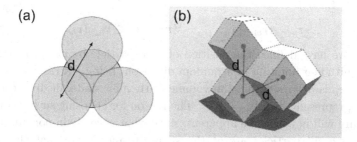

Fig. 4.3 (a) Hexagonal close packing (hpc) distribution and maximum volume which each unit can occupy, which corresponds to a rhombic dodecahedron (b).

In the previous sections we defined several mean free paths in terms of either scattering or transport coefficients which we had previously encountered in Chapter 2, namely the scattering and reduced scattering coefficients:

$$l_{\text{sc}} = \frac{1}{\mu_s}; \qquad l_{\text{sc}}^* = \frac{1}{\mu'_s},$$

and the transport and reduced transport mean free paths:

$$l_{\text{tr}} = \frac{1}{\mu_s + \mu_a} = \frac{1}{\mu_t}; \qquad l_{\text{tr}}^* = \frac{1}{\mu'_s + \mu_a} = \frac{1}{\mu_{\text{tr}}}.$$

Before continuing to solve the diffusion equation, I believe it is important to clarify the meaning of these quantities. Even though we briefly explained what each of these mean in Chapter 2, what is intended in this section is to clearly explain and prove one misconception very common in light propagation in multiple scattering media: contrary to what is commonly said, the scattering mean free path *does not* represent the average distance between scattering events. Since this erroneous concept has become mainstream, I believe a simple proof is necessary. There are, of course, several ways of proving it and what is shown here is only one of them.

Let us consider we have a collection of scatterers distributed so that they are arranged, on *average*, in a hexagonal close packing (hcp) distribution (see Fig. 4.3(a)). This distribution is the one you find in a stack of oranges, and is the most compact and dense distribution for spheres. In terms of scatterers, this means that each particle is located at the positions of the hcp, which we will define as $r_0 = r_{hcp}$, but is displaced by a *random number* so that it has an equal chance of occupying any point in space within the corresponding volume for the hcp, the rhombic dodecahedron[13] (see Fig. 4.3(b)):

$$r_p = r_{hcp} + r_\delta, \qquad \forall\, r_\delta \to r_p \in V_{hcp}(r_{hcp}) \qquad (4.39)$$

where we have defined the position of each particle by r_p. r_δ is a vector of magnitude δ pointing in a random direction, being δ a random number so that r_p is always within the volume of rhombic dodecahedron centered at r_{hcp}, represented as $V_{hcp}(r_{hcp})$. The reason for choosing such a complex structure will now become apparent, since what is really important is that *on average* all particles are at the same distance (note that the actual generation of an appropriate random δ requires that all volume V_{hcp} can be filled with equal probability) from their first neighbors, 12 in total. We will term this distance d (see Fig. 4.3(b)).

Once we have defined how the particles are arranged on average in space, we can assign some optical properties to them, such as a scattering cross-section σ_{sc}, an absorption cross-section σ_a, and an anisotropy g. The scattering and absorption coefficients will thus be defined by:

$$\mu_s = \rho\sigma_{sc}; \ \mu_s' = \rho(1-g)\sigma_{sc}; \ \mu_a = \rho\sigma_a,$$

where ρ is the density of the particles. Since we know the volume that one of these particles occupies, i.e. V_{hcp} (note that this is *not* the volume of the particle), the density will be given by:

$$\rho = \frac{1}{V_{hcp}}.$$

We now need to find the volume of the rhombic dodecahedron in terms of the distance between the average particle positions, r_{hcp}. It is known

---

[13]Note that you can choose your favorite geometry for this, as long as you can fill the full space with it (called space-filling tessellation) and you correctly account for it in the r_0 positions.

that the rhombic dodecahedron can contain an inscribed sphere tangent to all its faces of radius:

$$R = a\frac{2}{\sqrt{3}},$$

being a the edge length of the dodecahedron. The average distance between particles will therefore be $d = 2R$, in which case the edge length of the dodecahedron we need to fill the space is:

$$a = d\frac{\sqrt{3}}{4}.$$

The volume of the rhombic dodecahedron is defined in terms of its edge length as:

$$V_{hcp} = \frac{16}{9}\sqrt{3}a^3,$$

which in terms of d simply becomes:

$$V_{hcp} = d^3.$$

With our simple model we have then that the scattering and absorption coefficients are given by:

$$\mu_s = \frac{\sigma_{sc}}{d^3}, \; \mu_s = \frac{(1-g)\sigma_{sc}}{d^3}, \; \mu_a = \frac{\sigma_a}{d^3},$$

which gives the following scattering mean free path:

$$l_{sc} = \frac{d^3}{\sigma_{sc}}.$$

Clearly the scattering mean free path is *not* the average distance between particles, unless their scattering cross-section is, by chance[14], d^2. In a similar manner, we cannot interpret the reduced mean free path in terms of the distance between particles. These path-lengths are, of course, an indication of how dense and how scattering a medium is, but I do not think it is necessary to stretch their meaning further. Once we have understood that l_{sc} is not the average distance between particles any possible interpretation of its physical meaning could, in principle, be valid. From my point of view, these distances only reflect how light is attenuated on its propagation, nothing else. What is correct, however, is the interpretation that the transport mean free path l_{tr}^* is the mean distance light has to travel to become, on average, isotropic (i.e. to lose its original directionality) since this term accounts for the anisotropy of scattering and is reflected in the $U_{ri}(\mathbf{r})$ term of the diffusion equation.

---

[14]If you remember, we had discussed the relationship between the scattering cross-section and the geometrical cross-section in Chapter 2.

4.6   Limits of Validity of the Diffusion approximation

Before closing this chapter, a few words should be said regarding the limits of validity of the diffusion approximation. The study of how accurate the diffusion approximation is and what are its limits has been going on since its origins. Based on the assumptions and approximations we have introduced during its derivation it is quite common to state that, for example, the diffusion equation will break down very close to surfaces or sources (this would, in principle, invalidate the angular dependence we expected from $I_d(\mathbf{r}, \hat{\mathbf{s}})$, and possibly make the contribution of the term $\partial J/\partial t$ not negligible). Typically, these (flexible) limits are found by comparing the results of the diffusion approximation to the RTE, and not the complete solution to Maxwell's equations and usually neglect the contribution from the reduced intensity, $I_{ri}(\mathbf{r}, \hat{\mathbf{s}})$. In this way, going back to what we had already mentioned in Chap. 3, we are not doing a fair comparison: the RTE suffers from many approximations of its own which are obviously not being accounted for. If a rigorous comparison of theories is to be done, it should be with respect to the exact problem using Maxwell's equations, since the diffusion equation is not something unique to the RTE, it is an equation which can be also be derived directly from Maxwell[15].

On the other hand, when carefully applied (in the presence of high absorption for instance, or very close to sources) the diffusion approximation has yielded experimentally very good results, in most cases better than what was expected. This could be due to the fact that we are dealing with volume averaged quantities measured with real detectors. So the question still remains: what are the limits of validity of the diffusion approximation? Once again, in a way equivalent to what we discussed towards the end of Chap. 3, I believe there is no direct answer to this question. I would consider, however, that speaking of distances smaller or equivalent to the average distances between particles does not hold much significance neither in the context of the RTE nor in the context of the diffusion approximation. In a similar way, time-scales that confine light propagation to such small volumes should not be considered when dealing with diffuse light. Additionally, as happened with the RTE, results predicted by the diffusion ap-

---

[15]Note, however, that this derivation is significantly more complex than the one presented in this book which uses the RTE as a starting point. For those interested, the diffusion equation is found applying the Ward identity to an approximation of the Bethe-Salpeter equation. Since this more rigorous formalism is seldom used in light transport in tissues, I have opted for using the RTE as starting point.

proximation are expected to deviate significantly in cases where there might be a highly coherent contribution. A good measure of where it does not hold would be to analyze the angular dependence of the volume-averaged flow of energy and compare it with the approximation we introduced in Eq. (4.19), including the contribution of the reduced intensity.

In summary, what we can clearly determine is when the diffusion approximation is expected to hold (far away from sources and detectors, at long time-scales, and in the absence of coherent effects). Determining where it does not hold is a different matter, since there are no rigorously defined limits.

Key points

From this chapter the following important points should be remembered:

- **Angular dependence of the diffuse light**: In order to reach the diffusion equation, one of the main assumptions was that we can express the angular distribution energy flow as a sum of a monopolar (constant) and dipolar (cosine-dependent) term, what is sometimes also referred to as the P_1 approximation. This enabled us to obtain an expression for the flux which we could work with: Fick's law.
- **The diffusion coefficient**: When obtaining Fick's law for light propagation we obtained a definition for the diffusion coefficient. This expression, however, had to be re-visited when the contribution of absorption is comparable to the scattering coefficient and an expression which more properly accounts for absorption was presented.
- **The diffusion equation**: Introducing Fick's law into the equation for energy conservation gave us the diffusion equation, where we have a source term which accounts for the reduced intensity contribution. This will be our main equation for the remainder of the book.
- **The scattering mean free path**: We have briefly shown a 'back of the envelope' calculation that shows that the scattering mean free path *is not* the average distance between particles. The transport mean free path, on the other hand, represents the distance light has to travel to lose track of its initial direction.

Further Reading

Several works deal with the derivation of the diffusion approximation from the radiative transfer equation and present different steps. The one I have followed (not entirely, but mostly) is that of Ishimaru (1978). Other interesting derivations can be found in Furutsu [Furutsu (1980)], and Ito [Ito (1984)], which present a comparison of diffusion theories being a great source of reference. As briefly mentioned towards the end of this chapter, there are other ways of reaching the diffusion equation directly from Maxwell's equations. For those interested in this derivation, I recommend [Barabanenkov (1995, 1991)], and the excellent treatise by U. Frisch [Frisch (1965)]. On this subject I strongly recommend the book edited by Ping Sheng [Sheng (1990)] and his work on wave scattering, Sheng (2006). In these works the relationship between the Dyson and Bethe-Salpeter equations, together with the Ward identity and the diffusion coefficient are explained in great detail.

In this chapter we introduced an expression of the diffusion coefficient in highly absorbing media. The issue of the validity of the diffusion approximation in highly absorbing media had been a topic of debate for several years and is worth studying more extensively, in order to understand what were the points presented both in favor and against a dependence of the diffusion coefficient on absorption. Since there is extensive literature on the subject, I recommend reading the paper by Elaloufi *et al.* (2003) and following the references that can be found there, with special emphasis on Aronson and Corngold (1999).

Finally, what was presented dealt exclusively with the derivation of the diffusion equation. Many other approaches can be found in the literature such as the P_n approximation (if you remember, we truncated an expansion in spherical harmonics at the first order to obtain an expression for the diffuse specific intensity, $I_d(\mathbf{r}, \hat{\mathbf{s}})$, commonly referred to as the P_1 approximation), the telegrapher equation, Kubelka-Munk theory, and others. I refer the reader to books devoted to transport theory, such as Case and Zweifel (1967) and Ishimaru (1978) for further insight into this matter.

PART 2

Diffuse Light

Chapter 5

The Diffusion Equation

Summary. In this chapter we shall obtain the main solutions to the diffusion equation that was derived in the previous chapter, deriving the basic expressions that will be used during the remainder of this book. For that, we shall first solve for the homogeneous diffusion equation and obtain the time-dependent and constant illumination Green's functions. Finally, the expression for the diffusion equation in optically inhomogeneous media will be presented.

5.1 The Diffusion Equation in Infinite Homogeneous Media

The main goal of this chapter is to obtain the basic solutions to the diffusion equation we derived in the previous chapter. Even though all the information provided here can be found scattered (pun intended) around different books, I have opted for including in exhaustive detail all the derivations hoping it will help to understand what each expression means and where it came from. For that we will first obtain the expressions for Green's function both in the time-dependent and constant illumination cases, moving on to finding the solution to the more general problem using Green's theorem.

We shall start by presenting the diffusion equation in its simplest form, i.e. in an infinite and homogeneous medium. As we saw in the previous chapter, specifically when representing the time-dependence on the average intensity as $U(\mathbf{r}, t)$, by introducing Fick's law:

$$\mathbf{J}(\mathbf{r}, t) = -D\nabla U(\mathbf{r}, t), \tag{5.1}$$

into the equation for energy conservation:

$$\frac{1}{c_0}\frac{\partial}{\partial t}U(\mathbf{r}, t) + \mu_a U(\mathbf{r}, t) + \nabla \cdot \mathbf{J}(\mathbf{r}, t) = \mu'_s U_{ri}(\mathbf{r}, t), \tag{5.2}$$

we obtained the diffusion equation in an optically homogeneous medium:

$$\frac{1}{c_0}\frac{\partial}{\partial t}U(\mathbf{r},t) - D\nabla^2 U(\mathbf{r},t) + \mu_a U(\mathbf{r},t) = \mu_s' U_{ri}(\mathbf{r},t), \qquad (5.3)$$

with $U_{ri}(\mathbf{r},t)$ being:

$$U_{ri}(\mathbf{r},t) = \int_{(4\pi)} I_{ri}(\mathbf{r},\hat{\mathbf{s}},t)\mathrm{d}\Omega, \qquad (5.4)$$

and $I_{ri}(\mathbf{r},\hat{\mathbf{s}},t)$ representing the reduced intensity resulting from:

$$\frac{1}{c_0}\frac{\partial}{\partial t}I_{ri}(\mathbf{r},\hat{\mathbf{s}},t) + \hat{\mathbf{s}}\cdot\nabla I_{ri}(\mathbf{r},\hat{\mathbf{s}},t) + (\mu_a + \mu_s')I_{ri}(\mathbf{r},\hat{\mathbf{s}},t) = \epsilon(\mathbf{r},\hat{\mathbf{s}},t). \qquad (5.5)$$

Clearly, the solution to this set of equations depends on the source density, $\epsilon(\mathbf{r},\hat{\mathbf{s}},t)$, which represents the energy per unit volume injected into the system at point $\mathbf{r}$ into direction $\hat{\mathbf{s}}$. Note that we have dropped the 'd' subscript in all the above equations: unless specifically stated, we shall refer from now on to the diffuse component as U and J (instead of U_d and J_d) and to the reduced intensity components as U_{ri} and J_{ri}. We have now presented the set of equations we need to solve to obtain the solution for any specific expression of $\epsilon(\mathbf{r},\hat{\mathbf{s}},t)$. In order to obtain this more general solution we shall resort to the fundamental solution to Eq. (5.3): Green's function.

5.2 Green's Functions and Green's Theorem

Before attempting to obtain the expressions for the Green functions, I believe a brief introduction to Green's functions and Green's theorem is always handy. These were originally presented in a somewhat different form by George Green in 1828 in his famous work *An Essay on the Application of Mathematical Analysis to the Theories of Electricity and Magnetism*[1], [Green (1828)].

In order to understand the idea behind the use of Green's functions we can do some preliminary calculations that are not too rigorous in a mathematical sense, but very instructive. Let us assume we have a linear differential equation given by

$$\mathcal{L}\big[U(\mathbf{r},t)\big] = f(\mathbf{r},t), \qquad (5.6)$$

---

[1]As an interesting note, Green had no formal schooling (he was basically self-taught), and this essay was published at his own expense. Unfortunately it attracted little attention at the time and Green had to reluctantly return working at his father's mill, resuming his studies in 1830. For further insight in Green's life I suggest reading [Cannell (1999)].

where the linear operator $\mathcal{L}$ would be in our specific case given by $\mathcal{L} = [1/c_0 \partial/\partial t + \mu_a - D\nabla^2]$. The Green function should satisfy the equation:

$$\mathcal{L}[G(\mathbf{r}, \mathbf{r}', t, t')] = \delta(\mathbf{r} - \mathbf{r}')\delta(t - t'). \tag{5.7}$$

The reason for this choice and the proof that this equation will help us reach a solution to Eq. (5.6) can be found by using the properties of the delta function, in which case we may define $f(\mathbf{r}, t)$ as:

$$f(\mathbf{r}, t) = \int_0^\infty \int_V \delta(\mathbf{r} - \mathbf{r}')\delta(t - t')f(\mathbf{r}', t')d^3r'dt',$$

being V a volume large enough to enclose the delta function. We can now insert the equation for Green's function, Eq. (5.7), since it represents a delta:

$$f(\mathbf{r}, t) = \int_0^\infty \int_V \mathcal{L}[G(\mathbf{r}, \mathbf{r}', t, t')]f(\mathbf{r}', t')d^3r'dt',$$

where we can place the linear operator $\mathcal{L}$ outside of the integral, since it acts on $\mathbf{r}$ and t, not on $\mathbf{r}'$ and t':

$$f(\mathbf{r}, t) = \mathcal{L}\left[\int_0^\infty \int_V G(\mathbf{r}, \mathbf{r}', t, t')f(\mathbf{r}', t')d^3r'dt'\right].$$

Now we can identify further $f(\mathbf{r}, t)$ with U through Eq. (5.6):

$$\mathcal{L}[U(\mathbf{r}, t)] = \mathcal{L}\left[\int_0^\infty \int_V G(\mathbf{r}, \mathbf{r}', t, t')f(\mathbf{r}', t')d^3r'dt'\right],$$

which implies that:

$$U(\mathbf{r}, t) = \int_0^\infty \int_V G(\mathbf{r}, \mathbf{r}', t, t')f(\mathbf{r}', t')d^3r'dt',$$

should satisfy the differential equation Eq. (5.6). We can plug in this expression into Eq. (5.6) to verify this, which indeed will give us $f(\mathbf{r}, t)$ as a solution. Note that we can add the homogeneous solution (i.e. the solution to $\mathcal{L}[U(\mathbf{r}, t)] = 0$) to $U(\mathbf{r}, t)$ and still satisfy the differential equation, Eq. (5.6). In our particular case, for $\mathcal{L} = [\partial/c_0\partial t + \mu_a - D\nabla^2]$ this means that any solution of the form $U(\mathbf{r}, t) = \int \int G(\mathbf{r}, \mathbf{r}', t, t')f(\mathbf{r}', t')d^3r'dt' + A\exp(-\mu_a ct)$ with A being an arbitrary constant is also possible. The reason for this is that in this derivation we have not imposed any boundary conditions on G or U.

5.2.1　The Diffusion Equation and Green's Theorem

Once we have understood the general idea behind the Green functions, we shall now concentrate in obtaining the ones specific to our problem and derive rigorously the relationship between $U(\mathbf{r}, t)$ and $G(\mathbf{r}, \mathbf{r}', t, t')$. Since the Green function is required to satisfy the differential equation for a point source (see [Arfken and Weber (1995)] for a thorough study on Green's functions in several problems in physics, or the famous treatise by E. N. Economou [Economou (2006)]) we can express our diffusion equation in infinite homogeneous media as:

$$\frac{1}{c_0}\frac{\partial}{\partial t}G(\mathbf{r}, \mathbf{r}', t, t') - D\nabla^2 G(\mathbf{r}, \mathbf{r}', t, t')+$$

$$\mu_a G(\mathbf{r}, \mathbf{r}', t, t') = \delta(\mathbf{r} - \mathbf{r}')\delta(t - t'), \quad (5.8)$$

where G is the Green function and physically it represents the intensity at any point in space $\mathbf{r}$ and time t generated by an infinitely short pulse at time t' due to a point source at $\mathbf{r}'$. Note that the above equation is also sometimes presented with a 4π factor preceding the delta: if not included here, this factor will appear further down the derivation[2] and ultimately will yield the same solution for $U(\mathbf{r}, t)$. If we now multiply the above equation by the average intensity (i.e. the solution we seek), $U(\mathbf{r}, t)$, and subtract the equation we wish to solve, Eq. (5.3), multiplied by the Green function, $G(\mathbf{r}, \mathbf{r}', t, t')$ we obtain:

$$\left[\frac{U(\mathbf{r}, t)}{c_0}\frac{\partial}{\partial t}G(\mathbf{r}, \mathbf{r}', t, t') - \frac{G(\mathbf{r}, \mathbf{r}', t, t')}{c_0}\frac{\partial}{\partial t}U(\mathbf{r}, t)\right] -$$

$$D\left[U(\mathbf{r}, t)\nabla^2 G(\mathbf{r}, \mathbf{r}', t, t') - G(\mathbf{r}, \mathbf{r}', t, t')\nabla^2 U(\mathbf{r}, t)\right] =$$

$$\delta(\mathbf{r} - \mathbf{r}')\delta(t - t')U(\mathbf{r}, t) - \mu'_s U_{ri}(\mathbf{r}, t)G(\mathbf{r}, \mathbf{r}', t, t'), \quad (5.9)$$

where the μ_a contribution has canceled out. The next step is to integrate both in time and space the above equation in a volume V large enough to enclose the delta function. The first term of the equation vanishes due to the fact that:

$$\int_0^\infty \frac{\partial}{\partial t}G(\mathbf{r}, \mathbf{r}', t, t')\mathrm{d}t = G(\mathbf{r}, \mathbf{r}', t, t'), \quad \int_0^\infty \frac{\partial}{\partial t}U(\mathbf{r}, t)\mathrm{d}t = U(\mathbf{r}, t), \quad (5.10)$$

---

[2]It is also common to present the equation for the Green function with a negative sign for the deltas. All are perfectly valid, of course, as long as one maintains the same notation. If a 4π appears in the equation for G, then the integral equation for U in terms of G will appear divided by 4π.

while the second term requires attention. Considering only the spatial integral, this term results in:

$$\int_V \left[U(\mathbf{r},t)\nabla^2 G(\mathbf{r},\mathbf{r}',t,t') - G(\mathbf{r},\mathbf{r}',t,t')\nabla^2 U(\mathbf{r},t) \right] \mathrm{d}^3 r =$$

$$\int_V \nabla \cdot \left(U(\mathbf{r},t)\nabla G(\mathbf{r},\mathbf{r}',t,t') \right) \mathrm{d}^3 r -$$

$$\int_V \nabla \cdot \left(G(\mathbf{r},\mathbf{r}',t,t')\nabla U(\mathbf{r},t) \right) \mathrm{d}^3 r, \quad (5.11)$$

where the cross terms, $\nabla G \cdot \nabla U$, cancel out. We can now apply the divergence or Gauss' theorem which we already encountered in Chap. 2 when deriving the Poynting's vector in Sec. 2.2, yielding *Green's theorem*:

$$\int_V \left[U(\mathbf{r},t)\nabla^2 G(\mathbf{r},\mathbf{r}',t,t') - G(\mathbf{r},\mathbf{r}',t,t')\nabla^2 U(\mathbf{r},t) \right] \mathrm{d}^3 r =$$

$$\int_S \left[U(\mathbf{r},t)\hat{\mathbf{n}} \cdot \nabla G(\mathbf{r},\mathbf{r}',t,t') - G(\mathbf{r},\mathbf{r}',t,t')\hat{\mathbf{n}} \cdot \nabla U(\mathbf{r},t) \right] \mathrm{d}S, \quad (5.12)$$

being S the surface that encloses volume V, defined by its normal $\hat{\mathbf{n}}$. We can now go back to Eq. (5.9) and rewrite it as:

$$\int_0^\infty \int_V \delta(\mathbf{r} - \mathbf{r}')\delta(t - t')U(\mathbf{r},t)\mathrm{d}^3 r \mathrm{d}t =$$

$$\mu_s' \int_0^\infty \int_V U_{ri}(\mathbf{r},t)G(\mathbf{r},\mathbf{r}',t,t')\mathrm{d}^3 r \mathrm{d}t -$$

$$D \int_0^\infty \int_S \left[U(\mathbf{r},t)\hat{\mathbf{n}} \cdot \nabla G(\mathbf{r},\mathbf{r}',t,t') - \right.$$

$$\left. G(\mathbf{r},\mathbf{r}',t,t')\hat{\mathbf{n}} \cdot \nabla U(\mathbf{r},t) \right] \mathrm{d}S \mathrm{d}t. \quad (5.13)$$

This equation represents the complete solution for any general expression for $U_{ri}(\mathbf{r},t)$. In the case when we are dealing with infinite homogeneous optical properties, however, we may extend the volume to infinity in which case we can omit the surface integral since it will not contribute to the total intensity[3]. Introducing this assumption and integrating over the delta we obtain our general solution for an infinite medium:

$$U(\mathbf{r},t) = \mu_s' \int_0^\infty \int_V U_{ri}(\mathbf{r}',t')G(\mathbf{r},\mathbf{r}',t,t')\mathrm{d}^3 r' \mathrm{d}t', \quad (5.14)$$

---

[3] Rigorously speaking, we expect that very far away from the sources the intensity will behave as an outgoing spherical wave — just as we assumed when working in the far-field of the scatterers in Chap. 2 — and the surface integral would not contribute to the total intensity at $\mathbf{r}$. In the specific case of diffuse light, however, we also expect the intensity to be zero at infinity.

where we have interchanged $\mathbf{r}$ and $\mathbf{r}'$, and t and t' for convenience. For finite volumes (where we need to impose boundary conditions not yet established) the general solution for the finite homogeneous medium takes the form:

$$U(\mathbf{r},t) = \mu_s' \int_0^\infty \int_V U_{ri}(\mathbf{r}',t')G(\mathbf{r},\mathbf{r}',t,t')\mathrm{d}^3r'\mathrm{d}t' -$$

$$D \int_0^\infty \int_S \Big[U(\mathbf{r}',t')\hat{\mathbf{n}}' \cdot \nabla G(\mathbf{r},\mathbf{r}',t,t') -$$

$$G(\mathbf{r},\mathbf{r}',t,t')\hat{\mathbf{n}}' \cdot \nabla U(\mathbf{r}',t')\Big]\mathrm{d}S'\mathrm{d}t', \quad (5.15)$$

where it should be noted that U appears also within the surface integral which considerably complicates solving Eq. (5.15) and only has a closed-form analytical solution for simple geometries (as we shall see in the following chapters). For the time being, we can concentrate on our solution to the infinite homogeneous medium, Eq. (5.14), for which we shall now find the expressions for the Green function both for the time-dependent and constant illumination cases.

5.3 The Time-dependent Green's Function

Let us consider first the diffusion equation that the Green function must follow, Eq. (5.8), which we here rewrite for convenience:

$$\frac{1}{c_0}\frac{\partial}{\partial t}G(\mathbf{r},t) - D\nabla^2 G(\mathbf{r},t) + \mu_a G(\mathbf{r},t) = \delta(\mathbf{r}-\mathbf{r}')\delta(t-t'). \quad (5.16)$$

There are of course many ways of finding a solution to this equation. The one we shall follow here is one I consider very instructive, even though it is not the shortest. For the sake of clarity I will explain each step in detail, starting with the fact that great care must be taken with the c_0 factor. For that, two options are possible: we can solve the diffusion equation for the energy density $u = U/c_0$, in which case the c_0 needs only to be taken into account when describing intensity in terms of the average intensity U, or we multiply the complete equation by c_0, in which case this coefficient appears *also* in the right-hand side multiplying the delta[4]. Once this has

---

[4]The effects of not doing this are similar to what happens with the diffusion coefficient when we divide the complete equation by D. If this is done properly, then this term has to appear also in the right-hand side dividing the $\delta(\mathbf{r}-\mathbf{r}')\delta(t-t')$ factor. As we shall discuss later, neglecting this factor in the right-hand side is a common mistake that renders a Green function which is missing a D factor (or a c_0 factor, in the case we are discussing here).

been taken care of, the time-dependent Green's function can be found by first using separation of variables, $G(\mathbf{r}, t) = g(\mathbf{r})T(t)$, for the solution of the homogeneous equation:

$$\frac{\partial}{\partial t}G(\mathbf{r}, t) - Dc_0\nabla^2 G(\mathbf{r}, t) + \mu_a c_0 G(\mathbf{r}, t) = 0, \qquad (5.17)$$

in which case we obtain the following set of equations (remember that there should now be a c_0 multiplying the deltas when solving for the complete equation):

$$\frac{1}{T(t)}\frac{\partial T(t)}{\partial t} + \mu_a c_0 = \frac{Dc_0}{g(\mathbf{r})}\nabla^2 g(\mathbf{r}) = \lambda,$$

where since the left-hand-side has only a time-dependence and the right-hand-side only a spatial dependence, the only solution to both must be a constant which we have named λ (not to be confused with the wavelength). In this case, the solution for T will be given by:

$$T(t) = T_0 \exp(-\mu_a c_0 t)\exp(\lambda t),$$

where the constant T_0 can be obtained by considering the particular solution. Going back to Eq. (5.16) we can plug in the solution at time $t = t'$ for any general value of $\mathbf{r}$ which is given by (note that the time-derivative is zero in this case):

$$T(t')\Big(-Dc_0\nabla^2 g(\mathbf{r}) + \mu_a c_0 g(\mathbf{r})\Big) = c_0\delta(\mathbf{r} - \mathbf{r}').$$

Since $T(t')$ cannot have any spatial dependence, this means that the quantity in brackets must be equal to $\delta(\mathbf{r} - \mathbf{r}')$, and therefore[5]:

$$T(t') = T_0 \exp(-\mu_a c_0 t')\exp(\lambda t') = c_0,$$

which gives us the expression for T as:

$$T(t) = c_0 \exp\big[-\mu_a c_0(t - t')\big]\exp\big[\lambda(t - t')\big].$$

Note how the c_0 factor appears now in the expression for $T(t)$. We could use the same reasoning and include it in $g(\mathbf{r})$, but obviously not in both.

We can now solve for the spatial-dependent part of our homogeneous equation, Eq. (5.17), for which we have the following equation:

$$-Dc_0\nabla^2 g(\mathbf{r}) + \lambda g(\mathbf{r}) = 0,$$

which we can also solve by parts by writing $g(\mathbf{r}) = X(x)Y(y)Z(z)$ in which case:

$$g(\mathbf{r}) = A_{k_x}\exp(ik_x x)A_{k_z}\exp(ik_y y)A_{k_z}\exp(ik_z z),$$

---

[5]Note additionally that any arbitrary constant can be multiplied here, as long as we divide the spatial dependent part by the same constant, effectively canceling it out.

where the A's are constants to be determined and where k_x, k_y and k_z are the constants we have respectively chosen for X, Y and Z, which hold the following relationship:

$$k_x^2 + k_y^2 + k_z^2 = k^2 = -\frac{\lambda}{Dc_0},$$

which now gives us an expression for λ in terms of k. Note that in the expression for $g(\mathbf{r})$ above we have only considered the positive values of the k's. This will be justified further down since we will allow these constants to take any real value, both positive and negative. The actual physical meaning of k will be discussed in the next chapter (they refer to the components of the wave-vector). The values of the constants A_{kx}, A_{ky} and A_{kz} can be found once again by finding the particular solution for $\mathbf{r} = \mathbf{r}'$, for any general t in Eq. (5.8), in which case we obtain that:

$$g(\mathbf{r}')\Big(\frac{\partial T(t)}{\partial t} - c_0\mu_a T(t)\Big) = c_0\delta(t - t'),$$

which, following the same line of reasoning that we used for $T(t')$, since $g(\mathbf{r}')$ cannot have any time dependence we must have that $g(\mathbf{r}') = 1$ which results in:

$$A_{kx}A_{kz}A_{kz} = \exp(-ik_y x')\exp(-ik_x y')\exp(-ik_z z').$$

Since we can express $\lambda = c_0 D k^2$ we can now write the solution for the Green function in terms of a particular value of k as:

$$G_k(\mathbf{r} - \mathbf{r}', t - t') = c_0 \exp\big[-\mu_a c_0(t - t')\big] \exp\big[-Dc_0 k^2(t - t')\big] \times$$
$$\exp\big[ik_x(x - x') + ik_y(y - y') + ik_z(z - z')\big].$$

This expression, however, represents the solution for a certain value of k given by a set of k_x, k_y and k_z values. The complete solution will be given by the sum of all the possible values, each weighted by its own constant A_k:

$$G(\mathbf{r} - \mathbf{r}', t - t') = c_0 \exp\big[-\mu_a c_0(t - t')\big] \times$$
$$\int_{-\infty}^{\infty} A_{kx} \exp\big[-Dc_0 k_x^2(t - t') + ik_x x\big]\mathrm{d}k_x \times$$
$$\int_{-\infty}^{\infty} A_{ky} \exp\big[-Dc_0 k_y^2(t - t') + ik_y y\big]\mathrm{d}k_y \times$$
$$\int_{-\infty}^{\infty} A_{kz} \exp\big[-Dc_0 k_z^2(t - t') + ik_z z\big]\mathrm{d}k_z,$$

for which we have already identified the A constants above, with the only requirement that their values are *normalized* (remember we had g(r')=1 as a particular solution):

$$\int_{-\infty}^{\infty} \int_{-\infty}^{\infty} \int_{-\infty}^{\infty} A_{kx} A_{ky} A_{kz} \mathrm{d}k_x \mathrm{d}k_y \mathrm{d}k_z = 1.$$

Since $A_u = \exp(-ik_u u)$ and $\int \exp(-i(k_x x' + k_y y' + k_z z')) \mathrm{d}k_x \mathrm{d}k_y \mathrm{d}k_z = \delta(\mathbf{r}')(2\pi)^3$ (this integral is one of the expressions for a delta in three-dimensions, see Appendix A), we must normalize each dimension by 2π. Bearing this in mind, the complete Green function can be therefore expressed as:

$$G(\mathbf{r} - \mathbf{r}', t - t') = c_0 \exp\left[-\mu_a c_0 (t - t')\right] \mathcal{I}(x - x') \mathcal{I}(y - y') \mathcal{I}(z - z'), \quad (5.18)$$

where we have defined:

$$\mathcal{I}(u) = \frac{1}{2\pi} \int_{-\infty}^{\infty} \exp\left[-Dc_0 v^2 (t - t') + ivu\right] \mathrm{d}v, \quad (5.19)$$

as the last integral we need to solve in order to obtain our final solution. Before attempting this, however, there are several aspects of Eq. (5.18) which are quite remarkable and should be pointed out. First of all, the contribution of absorption is independent from the contribution of scattering (diffusion in our case)[6]. Second, and most important, diffusion in three dimensions is basically diffusion in one dimension (i.e. the solution to $\mathcal{I}$) to the power of 3. This can be generalized to N dimensions and holds true as long as we can consider the diffusion coefficient isotropic. In this case, extrapolating our solution to an N-dimensional system is therefore trivial.

We shall now solve Eq. (5.19) by rewriting it as a Gaussian integral for which we know the solution. As a quick reminder, the Gaussian integral, also known as the Euler-Poisson integral or Poisson integral is (see Appendix A):

$$\mathcal{I}(u) = \int_{-\infty}^{\infty} \exp(-v^2) \mathrm{d}v = \sqrt{\pi}. \quad (5.20)$$

In order to use the above expression, we simply have to rewrite the quantity in the exponent of our integral $\mathcal{I}(u)$ as $-(av + b)^2$, where a and b can be easily determined giving $a = \sqrt{Dc_0(t - t')}$ and $b = iu/2a$ in which case $\mathcal{I}(u)$ can be written as:

$$\mathcal{I}(u) = \exp(b^2) \int_{-\infty}^{\infty} \exp[-(av + b)^2] \mathrm{d}v,$$

---

[6]This is true as long as we do not consider that the diffusion coefficient has an absorption dependence. If we consider the diffusion equation simply in terms of μ_a and D the contribution of each is clearly separated.

which can be now directly solved using with the change of variable $v' = (av + b)$, $dv' = adv$, and Gauss' integral Eq. (5.20) resulting in:

$$\mathcal{I}(u) = \frac{1}{(4\pi Dc_0(t - t'))^{1/2}} \exp\left(\frac{-u^2}{4Dc_0(t - t')}\right).$$

We can finally now introduce this expression into Eq. (5.18) obtaining, at long last, the solution we had been searching for:

$$G(\mathbf{r}-\mathbf{r}', t-t') = c_0 \frac{\exp\left[-\mu_a c_0(t - t')\right]}{(4\pi Dc_0(t - t'))^{3/2}} \exp\left(\frac{-|\mathbf{r} - \mathbf{r}'|^2}{4Dc_0(t - t')}\right), \qquad \left(cm^{-2}s^{-1}\right)$$

$$(5.21)$$

where we have made use of the fact that $|\mathbf{r} - \mathbf{r}'| = (x - x')^2 + (y - y')^2 + (z - z')^2$.

 Building Block: **Time-dependent Green's Function**

The Time-dependent Green's function will be the basis to build the solution for an arbitrary distribution of sources with arbitrary temporal profiles.

Before moving on to the simpler case of constant illumination, it is important to underline that the derivation we just followed is one of many. The common approach to solve this equation is through a Laplace-Fourier transform (Laplace in time and Fourier in space), in which case by using tabulated integrals one reaches the solution in (almost) a straight-forward manner by writing the delta functions in terms of their Fourier components. This approach, however, does not show so clearly the relationship between the spatial dependence of the solution to the three-dimensional and that of the one dimensional situation. In this sense, I believe some insight into the properties of this particular Green function are lost.

Note: The Units of the Green Functions

It is a misconception that Green functions have no units. On the contrary, Green functions are required to have units in order for the fundamental integral that relates them with the solution to provide the appropriate units:

$$y(\mathbf{r}) = \int_V G(\mathbf{r} - \mathbf{r}')S(\mathbf{r}')\mathrm{d}^3 r'.$$

One possible reason for this is overlooking the fact that the delta functions have units also. In our case, for example, $\delta(\mathbf{r})$ has units of cm^{-3} and $\delta(t)$ units of s^{-1} which, given the structure of our diffusion equation, renders the time-dependent Green function units of $cm^{-2}s^{-1}$.

5.4 The Constant Illumination Green's Function

We shall now find the expression for the special case when we have constant illumination. As a first approach, we can use directly our result for the time dependent Green function (note that a similar approach can be found in [Mandelis (2011)]):

$$G_{cw}(\mathbf{r} - \mathbf{r}') = \int_{t'}^{\infty} G(\mathbf{r} - \mathbf{r}', t - t')\mathrm{d}t, \qquad (5.22)$$

where we consider that the Green function is zero at $t < t'$, since t' represents the time of the initial pulse. This integral can be solved directly as in [Mandelis (1995)] or in a more intuitive manner by performing the time integral before integrating in $\mathrm{d}k_x$, $\mathrm{d}k_y$ and $\mathrm{d}k_z$ in the previous section:

$$G_{cw}(\mathbf{r}-\mathbf{r}') = \frac{c_0}{(2\pi)^3} \int_{-\infty}^{\infty} \left(\int_{t'}^{\infty} \exp\left[-\mu_a c_0(t-t')\right] \exp\left[-Dc_0 k^2(t-t')\right] \mathrm{d}t \right) \times$$

$$\exp(i\mathbf{k} \cdot (\mathbf{r} - \mathbf{r}'))\mathrm{d}^3 k.$$

The integral over time can be directly solved in which case our expression becomes:

$$G_{cw}(\mathbf{r} - \mathbf{r}') = \frac{1}{(2\pi)^3} \int_{-\infty}^{\infty} \frac{1}{\mu_a + Dk^2} \exp(i\mathbf{k} \cdot (\mathbf{r} - \mathbf{r}'))\mathrm{d}^3 k. \qquad (5.23)$$

This expression appears very often in physics, in particular every time we deal with the wave equation. The solution to the above expression, however, is not that straightforward (alas!). Before attempting to solve it, lets go back to our original diffusion equation for the time-dependent average intensity, Eq. (5.3). Assuming we have a constant illumination, we will expect no temporal dependence on the average intensity U:

$$-D\nabla^2 U(\mathbf{r}) + \mu_a U(\mathbf{r}) = \mu_s' U_{ri}(\mathbf{r}), \qquad (5.24)$$

in which case the Green function should follow the following equation:

$$-D\nabla^2 G_{cw}(\mathbf{r} - \mathbf{r}') + \mu_a G_{cw}(\mathbf{r} - \mathbf{r}') = \delta(\mathbf{r} - \mathbf{r}'). \qquad (5.25)$$

This equation can clearly be converted to a Helmholtz equation by introducing the variable[7]:

$$\kappa_0 = \sqrt{-\frac{\mu_a}{D}}, \qquad (5.26)$$

in which case we can express the equation as:

$$\nabla^2 G_{cw}(\mathbf{r} - \mathbf{r}') + \kappa_0^2 G_{cw}(\mathbf{r} - \mathbf{r}') = -\frac{1}{D}\delta(\mathbf{r} - \mathbf{r}').$$

We can solve the above equation exactly like we did for the time-dependent Green function, by separation of variables. Identifying the constants for each Cartesian component as k_x, k_y and k_z as before, we obtain that the Green function must be of the form:

$$G_{cw}(\mathbf{r} - \mathbf{r}') = \frac{1}{(2\pi)^3} \int_{-\infty}^{\infty} A_{kx} A_{ky} A_{kz} \exp(i\mathbf{k} \cdot \mathbf{r}) \mathrm{d}^3 k.$$

Introducing this expression into Eq. (5.25) and using the expression for a delta in terms of k_x, k_y and k_z we can identify $A_{kx} A_{ky} A_{kz} = \exp(-i\mathbf{k} \cdot \mathbf{r}')/D(k^2 - \kappa_0^2)$ so that we obtain, as expected:

$$G_{cw}(\mathbf{r} - \mathbf{r}') = \frac{1}{D(2\pi)^3} \int_{-\infty}^{\infty} \frac{1}{k^2 - \kappa_0^2} \exp(i\mathbf{k} \cdot (\mathbf{r} - \mathbf{r}')) \mathrm{d}^3 k. \qquad (5.27)$$

which can be directly identified as Eq. (5.23). Our next task, therefore, is to solve this integral. As we have done so far, we shall proceed step by step, starting by converting the three-dimensional integral into an integral which depends only on the modulus of $\mathbf{k}$, $|\mathbf{k}| = k$. For that, we will write $\mathrm{d}^3 k = 2\pi k^2 \mathrm{d}k \mathrm{d}(\cos\theta)$, and take the angle θ with respect to $\mathbf{r} - \mathbf{r}'$ in which case $\mathbf{k} \cdot (\mathbf{r} - \mathbf{r}') = k|\mathbf{r} - \mathbf{r}'|\cos\theta$:

$$G_{cw}(\mathbf{r} - \mathbf{r}') = \frac{1}{D(2\pi)^2} \int_0^{\infty} \frac{1}{k^2 - \kappa_0^2} \left(\int_{-1}^{1} \exp(ik|\mathbf{r} - \mathbf{r}'|\cos\theta) \mathrm{d}(\cos\theta) \right) k^2 \mathrm{d}k.$$

The quantity within brackets looks uglier than it actually is, and it can be solved directly:

$$\int_{-1}^{1} \exp(ik|\mathbf{r} - \mathbf{r}'|v) \mathrm{d}v = \frac{1}{ik|\mathbf{r} - \mathbf{r}'|}.$$

---

[7]We shall deal with this very important quantity at the end of this chapter.

Introducing this result we obtain:

$$G_{cw}(\mathbf{r} - \mathbf{r}') = \frac{1}{iD(2\pi)^2 r} \int_0^\infty \frac{k}{k^2 - \kappa_0^2} \Big(\exp(ikr) - \exp(-ikr) \Big) dk,$$

where we have written $r = |\mathbf{r} - \mathbf{r}'|$ to simplify notation. We can now change the integration limits to account for both positive and negative contributions, yielding:

$$G_{cw}(\mathbf{r} - \mathbf{r}') = \frac{1}{iD(2\pi)^2 r} \int_{-\infty}^\infty \frac{k}{k^2 - \kappa_0^2} \exp(ikr) dk.$$

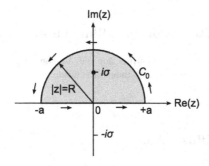

Fig. 5.1   Complex contour for solving the complex integral, Eq. (5.28).

Our next task, therefore, will be to solve this integral. The simplest way is to resort to complex variable integration. One can identify the above equation with a recurring formula in complex analysis, namely:

$$f(z) = \oint_C \frac{\exp(iRz)z dz}{z^2 + \sigma^2}. \tag{5.28}$$

This integral can be solved by using the complex contour shown in Fig. 5.1, by separating it into the real and complex contributions, and expanding the contour to infinity (i.e. $|z| \to \infty$ and $|a| \to \infty$):

$$f(z) = \lim_{|z| \to \infty} \int_{C_0} \frac{\exp(iRz)z dz}{z^2 + \sigma^2} + \int_{-\infty}^\infty \frac{\exp(ikR)k dk}{k^2 + \sigma^2} = 2\pi i \mathrm{Res}(i\sigma),$$

because the only pole inside the contour is $z = i\sigma$ (note that $z^2 + \sigma^2 = (z - i\sigma)(z + i\sigma)$). The integral over C_0 vanishes for $|z| \to \infty$ (the numerator grows as z while the denominator grows as z^2), in which case we obtain:

$$f(z) = \int_{-\infty}^\infty \frac{\exp(ikR)k dk}{k^2 + \sigma^2} = 2\pi i \left((z - i\sigma) \frac{\exp(iRz)z}{(z - i\sigma)(z + i\sigma)} \Big|_{z=i\sigma} \right),$$

resulting in:

$$\int_{-\infty}^{\infty} \frac{\exp(ikR)}{k^2 + \sigma^2} dk = \pi i \exp(-\sigma R),$$

which is the solution we were looking for. Identifying $r = |\mathbf{r} - \mathbf{r}'|$ as R, and $\sqrt{\mu_a/D}$ as σ we finally obtain the solution to our equation:

$$G_{cw}(\mathbf{r} - \mathbf{r}') = \frac{\exp(i\kappa_0 |\mathbf{r} - \mathbf{r}'|)}{4\pi D |\mathbf{r} - \mathbf{r}'|}. \tag{5.29}$$

 Building Block: **Constant Illumination Green's Function**

The constant illumination Green's function will be the basis to build the solution for an arbitrary distribution of time-independent sources.

5.5   Waves of Diffuse Light

A very interesting concept appears when we consider the time-dependent diffusion equation in the specific case where we modulate the source at a constant frequency ω. This approach is the one typically used for solving a time-dependent equation, by decomposing the temporal contribution into an infinite sum of frequencies. However, this approach is not only theoretical, but practical: as was shown by the group of A. Yodh in the early 90's, by modulating the intensity of the source one can create waves of diffuse light of a specific frequency ω, waves which they termed *diffuse photon density waves* or DPDWs [O'Leary *et al.* (1992); Boas *et al.* (1993, 1994)]. We already introduced them when discussing photon density in Chap. 3, and these waves can refract, scatter and interfere like any other wave, with the only specific property (and disadvantage) that they are highly damped. Since the frequencies used are in the MHz range, we can directly measure both the amplitude and the phase of these waves, offering an increased amount of information when compared to constant illumination.

The expression for DPDWs can be obtained by simply solving the diffusion equation in the Fourier domain, by assuming a temporal dependence of the source of the form:

$$\epsilon(\mathbf{r}, \hat{\mathbf{s}}, t) = \epsilon_{dc}(\mathbf{r}, \hat{\mathbf{s}}) + \epsilon_{\omega}(\mathbf{r}, \hat{\mathbf{s}}) \exp(-i\omega t), \tag{5.30}$$

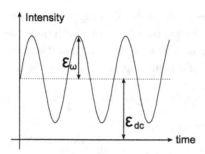

Fig. 5.2 Example of a source with its intensity modulated at frequency ω. Two components can be distinguished: a constant illumination component, ϵ_{dc}, and a modulated component ϵ_ω.

being ϵ_{dc} the continuous component (note that intensity cannot be negative, so in the best of cases the amplitude of $\epsilon_\omega(\mathbf{r}, \hat{\mathbf{s}})$ can be equal to $\epsilon_{dc}(\mathbf{r}, \hat{\mathbf{s}})$ but never higher – see Fig. 5.2). This modulated source would yield a reduced intensity given by:

$$\hat{\mathbf{s}} \cdot \nabla I_{ri}(\mathbf{r}, \hat{\mathbf{s}}, \omega) + \left(\mu_a + \mu'_s - i\frac{\omega}{c_0}\right) I_{ri}(\mathbf{r}, \hat{\mathbf{s}}, \omega) = \epsilon_\omega(\mathbf{r}, \hat{\mathbf{s}}), \quad (5.31)$$

$$\hat{\mathbf{s}} \cdot \nabla I_{ri}^{dc}(\mathbf{r}, \hat{\mathbf{s}}) + (\mu_a + \mu'_s) I_{ri}^{dc}(\mathbf{r}, \hat{\mathbf{s}}) = \epsilon_{dc}(\mathbf{r}, \hat{\mathbf{s}}), \quad (5.32)$$

being the time-dependent reduced intensity expressed as the sum of the continuous and modulated components, $I_{ri}(\mathbf{r}, \hat{\mathbf{s}}, t) = I_{ri}^{dc}(\mathbf{r}, \hat{\mathbf{s}}) + I_{ri}(\mathbf{r}, \hat{\mathbf{s}}, \omega) \exp(-i\omega t)$. This will give rise to a frequency-dependent average intensity of the form:

$$U(\mathbf{r}, t) = U_{dc}(\mathbf{r}) + U_\omega(\mathbf{r}) \exp(-i\omega t), \quad (5.33)$$

where now both $U_{dc}(\mathbf{r})$ and $U_\omega(\mathbf{r})$ will have units of $Watts/cm^2$. Introducing this expression into our time-dependent diffusion equation, Eq. (5.3), will give us for the modulated component:

$$-\frac{i\omega}{c_0} U_\omega(\mathbf{r}) - D\nabla^2 U_\omega(\mathbf{r}) + \mu_a U_\omega(\mathbf{r}) = \mu'_s U_{ri}(\mathbf{r}, \omega), \quad (5.34)$$

and for the continuous illumination component:

$$-D\nabla^2 U_{dc}(\mathbf{r}) + \mu_a U_{dc}(\mathbf{r}) = \mu'_s U_{ri}^{dc}(\mathbf{r}), \quad (5.35)$$

where now $U_{ri}(\mathbf{r}, \omega)$ and $U_{ri}^{dc}(\mathbf{r})$ are represented by:

$$U_{ri}(\mathbf{r}, \omega) = \int_{(4\pi)} I_{ri}(\mathbf{r}, \hat{\mathbf{s}}, \omega) d\Omega, \quad (5.36)$$

$$U_{ri}^{dc}(\mathbf{r}) = \int_{(4\pi)} I_{ri}^{dc}(\mathbf{r}, \hat{\mathbf{s}}) d\Omega. \quad (5.37)$$

We can solve the problem by finding the solution for a generic ω and then use the $\omega = 0$ solution to obtain the continuous (dc) components. Concentrating thus only on the ω dependent equations, we can easily regroup terms in Eq. (5.34), and represent this equation as a Helmholtz equation:

$$\nabla^2 U_\omega(\mathbf{r}) + \kappa_\omega^2 U_\omega(\mathbf{r}) = -\frac{1}{D}\mu_s' U_{ri}(\mathbf{r}, \omega),\qquad(5.38)$$

with the wavenumber now given by:

$$\kappa_\omega = \sqrt{-\frac{\mu_a}{D} + i\frac{\omega}{Dc_0}}.\qquad(5.39)$$

We have already obtained the solution to this equation, since it is identical to Eq. (5.25) but with a different κ_0, in which case we obtain that the Green function is given by:

$$G_\omega(\mathbf{r} - \mathbf{r}') = \frac{\exp(i\kappa_\omega|\mathbf{r} - \mathbf{r}'|)}{4\pi D|\mathbf{r} - \mathbf{r}'|},\qquad(5.40)$$

and the frequency-dependent average intensity will therefore be given by the solution to the general equation[8]:

$$U_\omega(\mathbf{r}) = \mu_s' \int_{-\infty}^{+\infty} U_{ri}(\mathbf{r}', \omega)G_\omega(\mathbf{r} - \mathbf{r}')\mathrm{d}^3r'.\qquad(5.41)$$

For a single modulation frequency ω, the temporal change in average intensity will be:

$$U(\mathbf{r}, t) = \mu_s' \int_{-\infty}^{\infty} U_{ri}^{dc}(\mathbf{r}')G_{cw}(\mathbf{r} - \mathbf{r}')\mathrm{d}^3r' +$$
$$\mu_s' \exp(-i\omega t) \int_{-\infty}^{\infty} U_{ri}(\mathbf{r}', \omega)G_\omega(\mathbf{r} - \mathbf{r}')\mathrm{d}^3r'.\qquad(5.42)$$

Once we have obtained the solution for ω, a general time-dependent solution can be also obtained by integrating all frequency components (note that in this case the $\omega = 0$ component is accounted for in the integral):

$$U(\mathbf{r}, t) = \mu_s' \int_0^\infty \int_{-\infty}^\infty U_{ri}(\mathbf{r}', \omega)G_\omega(\mathbf{r} - \mathbf{r}') \exp(-i\omega t)\mathrm{d}^3r'\mathrm{d}\omega.\qquad(5.43)$$

Since the constant illumination solution is a special case of the frequency-modulated illumination (i.e. for $\omega = 0$), we shall from now on drop the ω subscripts and refer to the average intensity as U, and the diffuse wavenumber as κ_0.

---

[8]Note that the $-1/D$ factor is already accounted for in the expression for $G_\omega(\mathbf{r} - \mathbf{r}')$, see Eq. (5.24) and Eq. (5.25)

5.6 The Diffusion Equation in Inhomogeneous Media

In the previous chapter we arrived to an expression for the diffusion equation through the use of Fick's law for light diffusion and the equation for energy balance. This has been the starting point in previous sections, and we have already obtained the expressions of the Green functions for the time-dependent, constant illumination and frequency-dependent cases. During this derivation, however, we considered the optical properties of the medium to be homogeneous. We shall now derive the complete diffusion equation for spatially dependent optical properties, $D(\mathbf{r})$ and $\mu_a(\mathbf{r})$, that is, the most general solution. For this purpose, we can describe the absorption coefficient as a sum of a constant value (could be the average, for example) and a spatially-dependent value:

$$\mu_a(\mathbf{r}) = \mu_{a0} + \overline{\mu}_a(\mathbf{r}), \qquad \left(cm^{-1} \right) \tag{5.44}$$

and obtain a similar expression for the scattering coefficient:

$$\mu_s(\mathbf{r}) = \mu_{s0} + \overline{\mu}_s(\mathbf{r}). \qquad \left(cm^{-1} \right) \tag{5.45}$$

We can also define a spatially dependent phase function, $p(\mathbf{r}, \hat{\mathbf{s}} \cdot \hat{\mathbf{s}}')$, which would give a spatially dependent anisotropy factor:

$$g(\mathbf{r}) = \frac{\int_{(4\pi)} p(\mathbf{r}, \hat{\mathbf{s}} \cdot \hat{\mathbf{s}}')\hat{\mathbf{s}} \cdot \hat{\mathbf{s}}' d\Omega'}{\int_{(4\pi)} p(\mathbf{r}, \hat{\mathbf{s}} \cdot \hat{\mathbf{s}}')d\Omega'},$$

where

$$\int_{(4\pi)} p(\mathbf{r}, \hat{\mathbf{s}} \cdot \hat{\mathbf{s}}')d\Omega' = \frac{\mu_s(\mathbf{r})}{\mu_a(\mathbf{r}) + \mu_s(\mathbf{r})}.$$

Using the spatially dependent anisotropy and the scattering coefficient we obtain the spatially dependent reduced scattering coefficient:

$$\mu_s'(\mathbf{r}) = \mu_{s0}' + \overline{\mu}_s'(\mathbf{r}), \qquad \left(cm^{-1} \right) \tag{5.46}$$

where $\overline{\mu}_s(\mathbf{r})$ is defined in terms of the spatially varying μ_s and $g(\mathbf{r})$ as:

$$\overline{\mu}_s'(\mathbf{r}) = \mu_{s0}' + (1 - g(\mathbf{r}))\,\overline{\mu}_s(\mathbf{r}).$$

Once we have defined the spatially dependent reduced scattering and absorption coefficients, the corresponding diffusion coefficient is defined as:

$$D(\mathbf{r}) = \frac{1}{D_0^{-1} + (\Delta D(\mathbf{r}))^{-1}} = D_0 + \overline{D}(\mathbf{r}), \tag{5.47}$$

where we have defined D_0 and $\overline{D}(\mathbf{r})$ as:

$$D_0 = \frac{1}{3(\mu'_{s0} + \mu'_{a0})}, \tag{5.48}$$

$$\overline{D}(\mathbf{r}) = -\frac{D_0^2}{D_0 + \Delta D(\mathbf{r})}, \tag{5.49}$$

being

$$\Delta D(\mathbf{r}) = \frac{1}{3\left(\overline{\mu}_s(\mathbf{r}) + \overline{\mu}_a(\mathbf{r})\right)}. \tag{5.50}$$

Using all the above expressions, we can now go back to our original derivation of the diffusion equation in the previous chapter and identically follow each step. Since the energy balance equation is defined locally (note that all our integrals are over the angular dependence of the specific intensity) most of the steps consist simply on substituting the homogeneous optical coefficients by the spatially-dependent ones. There are, however, a few critical steps that must be done with care. The first and most important is that in order to simplify things we want the reduced intensity to follow the same equation as before, with no spatially-dependent variables (compare with Eq. (4.8) in 4.2.1):

$$\frac{1}{c_0}\frac{\partial}{\partial t}I_{ri}(\mathbf{r},\hat{\mathbf{s}}) + \hat{\mathbf{s}} \cdot \nabla I_{ri}(\mathbf{r},\hat{\mathbf{s}}) + (\mu_{a0} + \mu'_{s0})\,I_{ri}(\mathbf{r},\hat{\mathbf{s}}) = \epsilon(\mathbf{r},\hat{\mathbf{s}}). \tag{5.51}$$

For this equation to hold true, we need to account for the spatially dependent μ_a and μ'_s in the source density term of the equation for the diffuse specific intensity:

$$\frac{1}{c_0}\frac{\partial}{\partial t}I_d(\mathbf{r},\hat{\mathbf{s}}) + \hat{\mathbf{s}} \cdot \nabla I_d(\mathbf{r},\hat{\mathbf{s}}) + [\mu_a(\mathbf{r}) + \mu'_s(\mathbf{r})]\,I_d(\mathbf{r},\hat{\mathbf{s}})$$
$$- [\mu_a(\mathbf{r}) + \mu'_s(\mathbf{r})]\int_{(4\pi)} I_d(\mathbf{r},\hat{\mathbf{s}}')p(\hat{\mathbf{s}},\hat{\mathbf{s}}')\mathrm{d}\Omega' = \epsilon_{ri}(\mathbf{r},\hat{\mathbf{s}}),$$

where now $\epsilon_{ri}(\mathbf{r},\hat{\mathbf{s}})$ is given by:

$$\epsilon_{ri}(\mathbf{r},\hat{\mathbf{s}}) = \mu_t(\mathbf{r})\int_{(4\pi)} I_{ri}(\mathbf{r},\hat{\mathbf{s}}')p(\hat{\mathbf{s}},\hat{\mathbf{s}}')\mathrm{d}\Omega' - g(\mathbf{r})\mu_s(\mathbf{r})I_{ri}(\mathbf{r},\hat{\mathbf{s}})$$
$$- \left(\overline{\mu}_a(\mathbf{r}) + \overline{\mu}'_s(\mathbf{r})\right)I_{ri}(\mathbf{r},\hat{\mathbf{s}}), \tag{5.52}$$

where now we have an extra term when compared to Eq. (4.10). This new term has a direct consequence in the contribution to the flux due to the reduced intensity (see Eq. (4.26)), $\mathbf{Q}_{ri}(\mathbf{r})$:

$$\mathbf{Q}_{ri}(\mathbf{r},t) = -\frac{\overline{\mu}'_s(\mathbf{r}) + \overline{\mu}_a(\mathbf{r})}{\mu'_s(\mathbf{r}) + \mu_a(\mathbf{r})}\int_{(4\pi)} I_{ri}(\mathbf{r},\hat{\mathbf{s}})\hat{\mathbf{s}}\mathrm{d}\Omega, \tag{5.53}$$

which is now non-zero. In terms of the flux from the reduced intensity contribution, $\mathbf{J}_{ri}(\mathbf{r})$, we may rewrite $\mathbf{Q}_{ri}(\mathbf{r})$ as:

$$\mathbf{Q}_{ri}(\mathbf{r}, t) = -\frac{\overline{\mu}'_s(\mathbf{r}) + \overline{\mu}_a(\mathbf{r})}{\mu'_s(\mathbf{r}) + \mu_a(\mathbf{r})} \mathbf{J}_{ri}(\mathbf{r}, t). \qquad (5.54)$$

This new contribution to the diffuse flux has to be accounted for in Fick's law, yielding:

$$\mathbf{J}(\mathbf{r}, t) = -D(\mathbf{r})\nabla U(\mathbf{r}, t) + \mathbf{Q}_{ri}(\mathbf{r}, t). \qquad (5.55)$$

The expression of the spatially-dependent energy conservation equation results in:

$$\frac{1}{c_0}\frac{\partial}{\partial t} U(\mathbf{r}, t) + \mu_a(\mathbf{r})U(\mathbf{r}, t) + \nabla \cdot \mathbf{J}(\mathbf{r}, t) = S_{ri}(\mathbf{r}, t),$$

where now S_{ri} has a new spatially-dependent contribution from Eq. (5.52):

$$S_{ri}(\mathbf{r}, t) = \mu'_s(\mathbf{r})U_{ri}(\mathbf{r}, t) - [\overline{\mu}'_s(\mathbf{r}) + \overline{\mu}_a(\mathbf{r})] U_{ri}(\mathbf{r}, t). \qquad (5.56)$$

Finally, introducing the expression for Fick's law, Eq. (5.55), into the energy conservation equation gives the general diffusion equation in inhomogeneous media:

$$\frac{1}{c_0}\frac{\partial}{\partial t} U(\mathbf{r}, t) - \nabla \cdot \left(D(\mathbf{r})\nabla U(\mathbf{r}, t) \right) + \mu_a(\mathbf{r})U(\mathbf{r}, t) =$$
$$[\mu'_{s0} - \overline{\mu}_a(\mathbf{r})] U_{ri}(\mathbf{r}, t) + \nabla \cdot \mathbf{Q}_{ri}(\mathbf{r}, t),$$

where it should be noted how the spatially-varying contribution of the reduced scattering coefficient to $U_{ri}(\mathbf{r})$ has canceled out. In order to obtain a simple expression for this equation, we can group all spatially-dependent coefficients in the right-hand side:

$$\frac{1}{c_0}\frac{\partial}{\partial t} U(\mathbf{r}, t) - D_0\nabla^2 U(\mathbf{r}, t) + \mu_{a0}U(\mathbf{r}, t) =$$
$$\mu'_{s0}U_{ri}(\mathbf{r}, t) + \mathcal{Q}_l(\mathbf{r}, t) + \mathcal{Q}_{nl}(\mathbf{r}, t; U), \qquad (5.57)$$

where we have written the linear and non-linear contributions as $\mathcal{Q}_l$ and $\mathcal{Q}_{nl}$, respectively:

$$\mathcal{Q}_l(\mathbf{r}, t) = -\overline{\mu}_a(\mathbf{r})U_{ri}(\mathbf{r}, t) + \nabla \cdot \left(\frac{\overline{\mu}'_s(\mathbf{r}) + \overline{\mu}_a(\mathbf{r})}{\mu'_s(\mathbf{r}) + \mu_a(\mathbf{r})} \mathbf{J}_{ri}(\mathbf{r}, t) \right), \qquad (5.58)$$
$$\mathcal{Q}_{nl}(\mathbf{r}, t; U) = -\overline{\mu}_a(\mathbf{r})U(\mathbf{r}, t) + \nabla \overline{D}(\mathbf{r}) \cdot \nabla U(\mathbf{r}, t). \qquad (5.59)$$

Solving this non-linear equation is a difficult task, and unfortunately obtaining a solution in closed form is only possible for simple geometries such as slabs, cylinders and spheres. Solving this equation for the general inhomogeneous case is however out of the scope of this book since one has to resort to numerical methods such as the Finite Element Method (FEM) or finite differences (FD). We will, however, obtain the analytical solutions to simple cases dealing with planar interfaces in Chap. 8.

5.7 Summary of Green's Functions

Before closing this chapter, it is convenient to present a summary of the expressions we derived since they might be very well lost and forgotten within the dense derivation we had to go through. Even though we have not derived these, I will also include the expression in one and two-dimensions, for the sake of completeness. The derivation of these expressions can be found in general mathematical methods books such as [Arfken and Weber (1995)]. Together with Green's functions, I will also include the expression of their gradient (we will need this expression — at least in 3D — in following chapters, specifically when expressing measurements at a detector in terms of Fick's law).

5.7.1 *1D Green's functions*

The expressions for Green's function and its gradient in homogeneous infinite media in 1D are given by:

- 1D time-dependent expressions:

$$G^{(1D)}(x - x', t - t') = c_0 \frac{\exp\left[-\mu_a c_0(t - t')\right]}{(4\pi D c_0(t - t'))^{1/2}} \times$$
$$\exp\left(\frac{-|x - x'|^2}{4D c_0(t - t')}\right), \quad (5.60)$$

$$\nabla G^{(1D)}(x - x', t - t') = -2\pi |x - x'| c_0 \frac{\exp\left[-\mu_a c_0(t - t')\right]}{(4\pi D c_0(t - t'))^{3/2}} \times$$
$$\exp\left(\frac{-|x - x'|^2}{4D c_0(t - t')}\right) \hat{\mathbf{u}}_{x-x'}, \quad (5.61)$$

- 1D frequency-dependent and constant illumination expressions:

$$G^{(1D)}(x - x') = \frac{i}{2D\kappa_0} \exp(i\kappa_0 |x - x'|), \quad (5.62)$$

$$\nabla G^{(1D)}(x - x') = -\frac{1}{2D} \exp(i\kappa_0 |x - x'|)\hat{\mathbf{u}}_{x-x'}, \quad (5.63)$$

with κ_0 given by:

$$\kappa_0 = \sqrt{-\frac{\mu_a}{D} + i\frac{\omega}{c_0}},$$

the *cw* being obtained for the $\omega = 0$ case.

5.7.2 2D Green's functions

The expressions for the Green function and its gradient in homogeneous infinite media in 2D are given by:

- 2D time-dependent expressions:

$$G^{(2D)}(\mathbf{R} - \mathbf{R}', t - t') = c_0 \frac{\exp\left[-\mu_a c_0(t - t')\right]}{4\pi D c_0(t - t')} \times$$
$$\exp\left(\frac{-|\mathbf{R} - \mathbf{R}'|^2}{4D c_0(t - t')}\right), \quad (5.64)$$

$$\nabla G^{(2D)}(\mathbf{R} - \mathbf{R}', t - t') = -2\pi|\mathbf{R} - \mathbf{R}'|c_0 \frac{\exp\left[-\mu_a c_0(t - t')\right]}{(4\pi D c_0(t - t'))^2} \times$$
$$\exp\left(\frac{-|\mathbf{R} - \mathbf{R}'|^2}{4D c_0(t - t')}\right)\hat{\mathbf{u}}_{\mathbf{R}-\mathbf{R}'}, \quad (5.65)$$

- 2D frequency-dependent and constant illumination expressions:

$$G^{(2D)}(\mathbf{R} - \mathbf{R}') = \frac{i}{4D} H_0^{(1)}(\kappa_0 |\mathbf{R} - \mathbf{R}'|), \quad (5.66)$$

$$\nabla G^{(2D)}(\mathbf{R} - \mathbf{R}') = \frac{i\kappa_0}{4D} H_1^{(1)}(\kappa_0 |\mathbf{R} - \mathbf{R}'|)\hat{\mathbf{u}}_{\mathbf{R}-\mathbf{R}'}, \quad (5.67)$$

where $H_0^{(1)}$ and $H_0^{(1)}$ represent the zeroth and first order Hankel functions of the first kind.

5.7.3   3D Green's functions

The expressions for the Green function and its gradient in homogeneous infinite media in 3D are given by:

- 3D time-dependent expressions:

$$G^{(3D)}(\mathbf{r} - \mathbf{r}', t - t') = c_0 \frac{\exp\left[-\mu_a c_0 (t - t')\right]}{(4\pi D c_0 (t - t'))^{3/2}} \times$$

$$\exp\left(\frac{-|\mathbf{r} - \mathbf{r}'|^2}{4 D c_0 (t - t')}\right), \quad (5.68)$$

$$\nabla G^{(3D)}(\mathbf{r} - \mathbf{r}', t - t') = -2\pi |\mathbf{r} - \mathbf{r}'| c_0 \frac{\exp\left[-\mu_a c_0 (t - t')\right]}{(4\pi D c_0 (t - t'))^{5/2}} \times$$

$$\exp\left(\frac{-|\mathbf{r} - \mathbf{r}'|^2}{4 D c_0 (t - t')}\right) \hat{\mathbf{u}}_{\mathbf{r}-\mathbf{r}'}, \quad (5.69)$$

- 3D frequency-dependent and constant illumination expressions:

$$G^{(3D)}(\mathbf{r} - \mathbf{r}') = \frac{\exp(i\kappa_0 |\mathbf{r} - \mathbf{r}'|)}{4\pi D |\mathbf{r} - \mathbf{r}'|}, \quad (5.70)$$

$$\nabla G^{(3D)}(\mathbf{r} - \mathbf{r}') = \frac{\exp(i\kappa_0 |\mathbf{r} - \mathbf{r}'|)}{4\pi D |\mathbf{r} - \mathbf{r}'|} \left(i\kappa_0 - \frac{1}{|\mathbf{r} - \mathbf{r}'|}\right) \hat{\mathbf{u}}_{\mathbf{r}-\mathbf{r}'}.$$

$$(5.71)$$

As a reminder, in all the cases presented in this section $\hat{\mathbf{u}}_{\mathbf{a}-\mathbf{b}}$ represents a unitary vector defined as:

$$\hat{\mathbf{u}}_{\mathbf{a}-\mathbf{b}} = \frac{\mathbf{a} - \mathbf{b}}{|\mathbf{a} - \mathbf{b}|}$$

Key points

From this chapter the following important points should be remembered:

- **Time-dependent behavior**: As we saw when deriving the time-dependent expression for the Green function, the spatial dependence can be easily generalized to N-dimensions. This is due to the fact that our equation can be solved using separation of variables. This can be seen directly in the summary of the time-dependent expressions presented in the previous section.

- **Dependence of the Green function on the diffusion coefficient**: A common mistake is to neglect the diffusion coefficient which divides the expression for the Green function in the frequency and constant illumination domains:

$$G(r) = \frac{\exp(i\kappa_0 r)}{4\pi D r}$$

This represents a problem specifically when dealing with several media with different diffusion coefficients, in which case the average intensity and flux obtained from these Green functions will not adhere to the proper boundary conditions, and jumps can be observed (expressions for the Green function at an interface will be discussed in detail in the chapter which deals with diffuse light at interfaces, Chap. 8).

- **4π factors**: It is all too easy to get confused with the large number of 4π factors that we need to keep track of. Additionally, depending on the equation one uses for the definition of the Green function, this factor might appear (or not) in the expression for the Green function, sometimes preceded by a negative sign. It is crucial that these expressions are used consistently, in which case they will always give the same final result for the average intensity. A similar problem occurs when working with Fourier transforms, since there is quite some freedom in their definition, specifically regarding the factor that precedes the integral.

- **The reduced intensity**: All expressions have been derived based on the reduced intensity, $U_{ri}(\mathbf{r})$, generated inside the scattering medium which if you recall represents the intensity that has not lost its original direction and thus follows Beer's law. Once $U_{ri}(\mathbf{r})$ is determined, we can find the average diffuse intensity by making use of the Green functions derived in this chapter. The actual expression of $U_{ri}(\mathbf{r})$ will depend on the angular and spatial distribution of our source densities and we will soon investigate some examples which will give us specific expressions for the total average intensity.

Further Reading

Most of what has been presented can be found in several books that deal either with numerical methods or the diffusion equation. In particular, the core of this chapter has been the use of Green functions and their derivation within the context of light diffusion. To gain further insight into Green's functions I recommend Morse and Feshbach (1953) and Arfken and Weber (1995) as general mathematical methods textbooks, and Economou (2006) for their use in greater depths. More specific to diffusion, the book by Mandelis (2011) includes a detailed derivation of many solutions to the diffusion equation in the context mostly of heat transport, but it also contains a minor chapter on diffuse photon density waves. In this chapter we did not attempt to solve the diffusion equation but rather to derive the Green functions that will enable us to solve analytically simple geometries in the following chapters; these references will be of great help in those cases also. With regards to solving integrals through complex variable, even though we just skimmed over it, they are of great importance specially for those aiming at obtaining analytical solutions to cumbersome equations. For those interested in learning greater insight in this powerful tool I recommend the same mathematical methods books, both Morse and Feshbach (1953) and Arfken and Weber (1995). With regards to Diffuse Photon Density Waves, the original experiments of the groups of A. G. Yodh and E. Gratton created pretty much a field on its own: I recommend reading the original papers from the early 90s such as [O'Leary *et al.* (1992); Boas *et al.* (1994)] and [Fishkin and Gratton (1993)]. Finally, for those interested in the human aspect of how these approaches came to be, and in this way gain a much better understanding of the development of the mathematical methods in physics, I would recommend reading further on the lives of Green, Gauss, Fourier and Cauchy, to name but a few.

Chapter 6

Propagation and Spatial Resolution of Diffuse Light

Summary. Once we have determined the basic equations governing light transport in diffusive media, in this chapter we will explore how diffuse light propagates within highly scattering and absorbing media and how this propagation affects both the resolving power of diffuse light and its intensity. In order to provide the necessary tools to analyze light propagation, we will introduce the angular spectrum representation and the transfer function. This representation will be the basis for accounting for boundary conditions and complex source distributions in the following chapters.

6.1 Propagation of Diffuse Light

In the previous chapter we derived the expression for the diffusion equation in an infinite homogeneous medium and found Green's functions that will help us solve any general distribution of sources within this highly scattering medium. We will now introduce the angular spectrum representation of the diffuse light propagation. As we will see in the following chapters of this book, this representation is extremely handy to account for complex boundary conditions and source distributions, and gives a great deal of insight into the basis behind diffuse light propagation: why diffuse light is 'blurry' and what relationship does this have with resolution. This will help us answer ultimate questions such as what is the resolution I might expect when recovering in 3D the concentration of a fluorophore, and it is directly linked to how certain we can be that we have recovered the original image (i.e. how ill-posed the problem is).

Let us consider the infinite homogeneous medium we introduced in Sec. 5.1 of Chap. 5, where we have a diffusive medium with optical properties defined by its diffusion coefficient D, absorption coefficient μ_a, and index of refraction n_0 yielding a speed of light given by c_0. We will decompose the average intensity both into its modulation frequency components (i.e. the temporal Fourier transform) and into its spatial frequency components. For that, we will use the homogeneous solution to the frequency-dependent expression of the diffusion equation (see Eq. (5.38)), in which case the average intensity U within this medium will be given by:

$$-D\nabla^2 U(\mathbf{r}) + \left(\mu_a - i\frac{\omega}{c_0}\right) U(\mathbf{r}) = 0, \qquad (6.1)$$

where ω is the intensity modulation frequency. We have already obtained a solution to this equation in Chap. 5, by describing the average intensity U as a sum of plane waves of wavevector $\mathbf{k}$ and amplitude $\widetilde{U}(\mathbf{k})$:

$$U(\mathbf{r}) = \int_{-\infty}^{\infty} \widetilde{U}(\mathbf{k}) \exp(i\mathbf{k} \cdot \mathbf{r}) \mathrm{d}^3 k.$$

Introducing this expression into the diffusion equation and writing $\mathbf{k}$ in its Cartesian components, $\mathbf{k} = (K_x, K_y, K_z)$ gives:

$$\int_{-\infty}^{\infty} \left[-D\left(-K_x^2 - K_y^2 - K_z^2\right) + \mu_a\right] \widetilde{U}(\mathbf{k}) \exp(i\mathbf{k} \cdot \mathbf{r}) \mathrm{d}^3 k = 0.$$

which, making use of the fact that:

$$|\mathbf{k}|^2 = K_x^2 + K_y^2 + K_z^2, \qquad \left(cm^{-2}\right) \qquad (6.2)$$

gives us a solution which can only exist if:

$$\left[Dk^2 + \mu_a\right] \widetilde{U}(\mathbf{k}) = 0,$$

and therefore:

$$Dk^2 + \mu_a + i\frac{\omega}{c_0} = 0. \qquad \left(cm^{-1}\right) \qquad (6.3)$$

Expressing the modulus of the wavevector $|\mathbf{k}|$ as $|\mathbf{k}| = \kappa_0$ gives us the diffusion wavenumber κ_0 we encountered in the previous chapter (see Eq. (5.39)):

$$\kappa_0 = \sqrt{-\frac{\mu_a}{D} + \frac{i\omega}{Dc_0}}. \qquad (6.4)$$

 Building Block: **The Diffusion Wavenumber**

This complex quantity represents the wavenumber of a diffuse wave generated by modulating the intensity of the source and it dictates how light propagates in a diffusive medium.

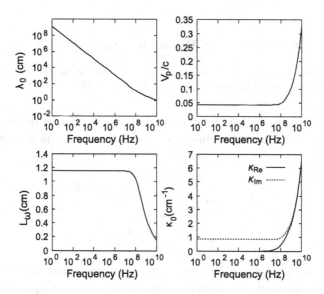

Fig. 6.1 Values of λ_0, V_p, L_d, $\kappa_{\Re e}$, and $\kappa_{\Im m}$ versus the modulation frequency f, where $f = \omega/2\pi$. In all cases $\mu_a = 0.025cm^{-1}$, $\mu_s' = 10cm^{-1}$ and $n = 1.333$, corresponding to values similar to those of tissue at the near-infrared. Note that V_p is presented in units of the speed of light c.

6.1.1 The Diffusion Wavenumber

Before moving on to finding the angular spectrum, let us analyze in greater detail the wavenumber κ_0 we just obtained, since all the properties of propagation and resolution are determined by it. The first important quantity one can obtain from the wavenumber is in the constant illumination (cw) case where we may express it as:

$$\kappa_0 = i\sqrt{\frac{\mu_a}{D}},$$

for which the solution to Eq. (6.1) becomes:

$$U(\mathbf{r}) = \frac{\exp(-\sqrt{\mu_a/D}\,r)}{4\pi D r}.$$

In this case, $\sqrt{D/\mu_a}$ can be understood as a decay length, which represents the distance that light has to travel to reduce its average intensity by e:

$$L_d = \sqrt{\frac{D}{\mu_a}}. \quad \left(cm\right) \tag{6.5}$$

We will term this distance the *diffusion length* and it represents a fundamental quantity when dealing with diffuse light propagation.

Building Block: **The Diffusion Length**

This quantity represents the distance a plane wave of diffuse light has to travel in order for its average intensity to decrease by a factor of e.

Once we have defined the diffusion length, let us now decompose the frequency-dependent wavenumber in its real and imaginary parts, $\kappa_0 = \kappa_{\Re e} + i\kappa_{\Im m}$:

$$\kappa_{\Re e} = \frac{1}{\sqrt{2}} \sqrt{\left(\frac{\omega^2}{D^2 c_0^2} + \frac{\mu_a^2}{D^2} \right)^{\frac{1}{2}} - \frac{\mu_a}{D}}, \tag{6.6}$$

$$\kappa_{\Im m} = \frac{1}{\sqrt{2}} \sqrt{\left(\frac{\omega^2}{D^2 c_0^2} + \frac{\mu_a^2}{D^2} \right)^{\frac{1}{2}} + \frac{\mu_a}{D}}. \tag{6.7}$$

Eqs. (6.6)–(6.7) show that for low values of ω we have that $\kappa_{\Re e} \sim 0$, $\kappa_{\Im m} \sim L_d$, whereas for high values of ω we have that $\kappa_{\Re e} \simeq \kappa_{\Im m}$. In terms of $\kappa_{\Re e}$ and $\kappa_{\Im m}$, we can define a *wavelength* associated with the intensity modulation — i.e. the wavelength of the diffuse density wave — as λ_0, a *phase velocity* V_p, and a frequency dependent *decay length*, L_ω as:

$$\lambda_0 = \frac{2\pi}{\kappa_{\Re e}}, \qquad \left(cm \right) \tag{6.8}$$

$$V_p = \frac{\omega}{\kappa_{\Re e}}, \qquad \left(cm/s \right) \tag{6.9}$$

$$L_\omega = \frac{1}{\kappa_{\Im m}}. \qquad \left(cm \right) \tag{6.10}$$

An example of values for λ_0, V_p, L_d, $\kappa_{\Re e}$, and $\kappa_{\Im m}$ for different values of the modulation frequency f, being $f = \omega/2\pi$, is shown in Fig. 6.1. As seen in this figure $\kappa_{\Re e}$, and $\kappa_{\Im m}$ are constant practically up to $f \sim 10^8$ Hz since for those cases $\omega/c \ll \mu_a$. Therefore, for values $f < 100$MHz, the following approximations are very accurate:

$$\kappa_{\Re e} \simeq \frac{\omega}{2 L_d} \frac{1}{\mu_a c_0}, \qquad\qquad \kappa_{\Im m} \simeq \frac{1}{L_d}, \tag{6.11}$$

$$\lambda_0 \simeq \frac{2 L_d}{f} \mu_a c_0, \qquad\qquad V_p \simeq 2 L_d \mu_a c_0,$$

and of course $L_\omega \simeq L_d$.

If we now consider in Fig. 6.1 the results for a modulation frequency of $f = 200$MHz, which would correspond to a frequency commonly used in experiments, we see that it would give us $\lambda_0 \sim 10cm$, and $L_d \sim 1cm$. This points to the fact that these waves are greatly attenuated before completing one cycle, decreasing by a factor of $\sim \exp(-10)$ after traveling one single diffuse wavelength[1]. In fact, $L_d \ll \lambda_0$ is a general characteristic of diffuse waves, and as a direct consequence the interference effects between them are very low. Note, however, that diffuse waves still interfere: their destructive interference has been used to detect small inhomogeneities within diffusive media (see [Intes *et al.* (2001); Chen *et al.* (2002); Intes *et al.* (2002)], for example). Also, as shown in Fig. 6.1, the phase velocity is of the order of $c/20$, thus implying that although these waves are greatly attenuated, they propagate at high speeds. A thorough study of the effect of high frequency modulation can be found in the original works of Fishkin and Gratton ([Fishkin *et al.* (1996); Fishkin and Gratton (1993)]).

6.2   The Angular Spectrum Representation

After this brief hiatus to examine the properties of the diffuse wavenumber, we will now go back to our original objective: the angular spectrum representation of diffuse light. From Eq. (6.3) we can obtain the fundamental relationship in Fourier's space which will dictate diffuse light propagation:

$$K_z(\mathbf{K}) = \pm q(\mathbf{K}) = \pm\sqrt{\mathbf{K}^2 - \kappa_0^2}, \qquad (6.12)$$

where $\mathbf{K} = (K_x, K_y)$. Note that we have selected K_z for convenience, but we could have selected any of the other Cartesian coordinates with the same result (we have not yet set any boundaries on our medium). Since κ_0 is a complex number (as we have discussed in the previous section, in the case $\omega = 0$ it is a pure imaginary number), $q(\mathbf{K}) = q_{\Re e} + iq_{\Im m}$ is always complex due to the fact that $q_{\Im m} \neq 0$ for all possible values of ω (as can be understood from Eq. (6.7), $\kappa_{\Im m} \neq 0$ no matter what value we choose). Making use of Eq. (6.12), clearly our solution to Eq. (6.1) will depend only on $\mathbf{K}$ and will have the form:

$$U(\mathbf{r}) = \int_{-\infty}^{\infty} \mathcal{A}(\mathbf{K}) \exp(iq(\mathbf{K})z) \exp(i\mathbf{K} \cdot \mathbf{R}) \mathrm{d}^2 K \quad z \geq 0, \qquad (6.13)$$

$$U(\mathbf{r}) = \int_{-\infty}^{\infty} \mathcal{B}(\mathbf{K}) \exp(-iq(\mathbf{K})z) \exp(i\mathbf{K} \cdot \mathbf{R}) \mathrm{d}^2 K \quad z \leq 0, \qquad (6.14)$$

---

[1] Just in case there is any trace of doubt here, the actual wavelength of the light used will give us the values of the optical coefficients (μ_a, D, and n_0). The wavelength we are referring to here is the wavelength of the diffuse wave generated by modulating the intensity of our source.

where we have chosen $q_{\Re e} > 0$ and $q_{\Im m} > 0$ so that the average intensity satisfies the *Sommerfeld's radiation condition* at infinity:

$$\lim_{r \to \infty} r \left(\frac{\partial U(\mathbf{r})}{\partial r} - i\kappa_0 U(\mathbf{r}) \right) = 0, \tag{6.15}$$

which simply requires the average intensity U to behaves as an outgoing spherical wave[2] at $r = |\mathbf{r}| \to \infty$. $q_{\Im m} > 0$ implies that the wave is damped on propagation, and $q_{\Re e} > 0$ represents an outgoing wave. Going back to Eqs. (6.13)–(6.14), these can be seen as the solutions for sources or objects at $z < 0$ and $z > 0$, respectively. The most convenient form of ensuring $iq(\mathbf{K})z \leq 0$ is by expressing the z-dependence is through $|z|$ yielding:

$$U(\mathbf{R}, z) = \int_{-\infty}^{\infty} \mathcal{A}(\mathbf{K}) \exp(iq(\mathbf{K})|z|) \exp(i\mathbf{K} \cdot \mathbf{R}) \mathrm{d}^2 K, \tag{6.16}$$

where $\mathcal{A}$ is usually termed *the angular spectrum* of $U(\mathbf{r})$ at $z = 0$, its origin coming from the fact that in the far-field of an electromagnetic wave — not a diffuse wave, by the way; with a diffuse wave we would need to resort to complex angles — we can write $K_x = \kappa_0 \cos\theta$ and $K_y = \kappa_0 \sin\theta$, thus the term 'angular'. It is also convenient to group $\mathcal{A}$ and the $q(\mathbf{K})z$ dependence into one function $\widetilde{U}(\mathbf{K}, z)$ in which case Eq. (6.16) would represent its 2D Fourier transform at z:

$$U(\mathbf{R}, z) = \int_{-\infty}^{\infty} \widetilde{U}(\mathbf{K}, z) \exp(i\mathbf{K} \cdot \mathbf{R}) \mathrm{d}^2 K, \tag{6.17}$$

with:

$$\widetilde{U}(\mathbf{K}, z) = \mathcal{A}(\mathbf{K}) \exp(iq(\mathbf{K})|z|). \tag{6.18}$$

This equation gives us the value of the average intensity at a plane z that was originated by an angular spectrum $\mathcal{A}$ at $z = 0$. Evidently, using this formula we can relate the average intensity at a plane z to the average intensity at a plane $z_0 < z$ as:

$$\widetilde{U}(\mathbf{K}, z) = \widetilde{U}(\mathbf{K}, z_0) \exp(iq(\mathbf{K})(z - z_0)).$$

As we will see in Sec. 6.5, by inverting the effect that propagation has on the original signal we are capable of *partially*[3] recovering the intensity at any plane $z < z_0$, a technique termed *back-propagation*.

---

[2] You might remember that this same condition appeared when making use of Green's theorem in Sec. 5.2.1 of Chap. 5, even though we did not write it explicitly.

[3] I wish to over-emphasize that the complete recovery of the angular spectrum is not physically possible, there will always be some limitation on our sensitivity to high frequencies plus the inherent noise present in any measurement.

To find the expression for the diffuse flux, $J_n(\mathbf{r})$, we can proceed in a similar way to what was just done to obtain the average intensity . Since the total flux density that traverses any plane $z = z_0$ is given by Fick's law, in terms of the angular spectrum representation it may be written as:

$$\widetilde{J}_n(\mathbf{K}, z_0) = -D \left. \frac{\partial \widetilde{U}(\mathbf{K}, z)}{\partial z} \right|_{z=z_0} = -iDq(\mathbf{K})\widetilde{U}(\mathbf{K}, z_0), \qquad (6.19)$$

where the surface normal $\hat{\mathbf{n}}$ at $z = z_0$ is pointing in the $+z$ direction, i.e. $\hat{\mathbf{n}} = (0, 0, 1)$. Note that if we consider it pointing inward, we must change the sign on the above equation.

It is important to remember that all the above expressions result from the fact that we chose to express the gradient in Eq. (6.1) in Cartesian coordinates. We will now proceed to obtain the complete solution to Eq. (6.1). For that, we will first find the Green function's expression in K-space.

6.2.1 *Angular spectrum of a point source: The Green Function in K-space*

We have already derived the Green function which represents the solution to Eq. (6.1) in Chap. 5 where we had obtained:

$$g(\mathbf{r} - \mathbf{r}') = \frac{\exp\left(i\kappa_0 |\mathbf{r} - \mathbf{r}'|\right)}{4\pi D |\mathbf{r} - \mathbf{r}'|}, \qquad (6.20)$$

which represents the solution to a point source. Our current goal, therefore, is to obtain the angular spectrum of Eq. (6.20) or, equivalently, to obtain the 2D Fourier transform of the Green function at the plane $z = z'$, which we may express as (I have included a list of transforms in Appendix A, for reference):

$$\widetilde{g}(\mathbf{K}, z = z') = \frac{1}{4\pi^2} \int_{-\infty}^{\infty} \frac{\exp\left(i\kappa_0 R\right)}{4\pi D R} \exp(-i\mathbf{K} \cdot \mathbf{R}) \mathrm{d}^2 R. \qquad (6.21)$$

In order to differentiate the angular spectrum of a point source from the more general $\widetilde{g}(\mathbf{K}, z)$ we will express it as $\widetilde{g}(\mathbf{K}, z = z') = \mathcal{A}_\delta(\mathbf{K})$. Since we are in an infinite homogeneous medium, in the previous equation we can introduce $\mathrm{d}^2 R = R dR d\theta$ resulting in:

$$\mathcal{A}_\delta(\mathbf{K}) = \frac{1}{4\pi D} \frac{1}{4\pi^2} \int_0^\infty \frac{\exp\left(i\kappa_0 R\right)}{R} \left[\int_0^{2\pi} \exp(-iKR\cos\theta)d\theta \right] R dR. \qquad (6.22)$$

In the above equation we can identify the expression for the Bessel function of the first kind and zero order:

$$J_0(KR) = \frac{1}{2\pi} \int_0^{2\pi} \exp(-iKR\cos\theta)d\theta. \tag{6.23}$$

Using this expression we obtain that the result for a delta source is:

$$\mathcal{A}_\delta(\mathbf{K}) = \frac{1}{4\pi D}\frac{1}{2\pi} \int_0^\infty \frac{\exp(i\kappa_0 R)}{R} J_0(KR)Rd^2Rd\theta, \tag{6.24}$$

where we can identify the *Hankel transform*:

$$\mathcal{H}\{f(R)\} = \int_0^\infty f(R)J_0(KR)RdR, \tag{6.25}$$

$$\mathcal{H}^{-1}\{\widehat{f}(K)\} = \int_0^\infty \widehat{f}(K)J_0(KR)KdK, \tag{6.26}$$

which is very useful since we can calculate 2D transforms with 1D integrals, as long as there is a cylindrical symmetry (as in our case). In terms of the Hankel transform, the solution for a point source becomes:

$$\mathcal{A}_\delta(\mathbf{K}) = \frac{1}{4\pi D}\frac{1}{2\pi}\mathcal{H}\left\{\frac{\exp(i\kappa_0 R)}{R}\right\},$$

which, given the fact that the Hankel transform for this function is well-known:

$$\mathcal{H}\left\{\frac{\exp(i\kappa_0 R)}{R}\right\} = \frac{i}{\sqrt{\kappa_0^2 - K^2}},$$

gives us the final result for a point source:

$$\mathcal{A}_\delta(\mathbf{K}) = \frac{1}{4\pi D}\frac{i}{2\pi q(\mathbf{K})}. \tag{6.27}$$

 Building Block: **Angular Spectrum of a Point Source**

The angular spectrum of a point source and thus the Green function in K-space will be fundamental for solving complex source distributions and accounting, rigorously, for the boundary conditions.

Note that one of the properties of the Hankel transform is that we may write $f(K) = f(\mathbf{K})$. Eq. (6.27) will be fundamental for our derivation, since we can now write the Green function as:

$$\widetilde{g}(\mathbf{K}, z - z') = \mathcal{A}_\delta(\mathbf{K})\exp(iq(\mathbf{K})|z - z'|),$$

or equivalently:

$$\tilde{g}(\mathbf{K}, z - z') = \frac{1}{4\pi D} \frac{i}{2\pi q(\mathbf{K})} \exp(iq(\mathbf{K})|z - z'|). \qquad (6.28)$$

In terms of the K-space Green function we can express the real space Green function as:

$$g(\mathbf{r} - \mathbf{r}') = \frac{1}{4\pi D} \int_{-\infty}^{\infty} \frac{i}{2\pi q(\mathbf{K})} \exp(i\mathbf{K} \cdot (\mathbf{R} - \mathbf{R}')) \exp(iq(\mathbf{K})|z - z'|) \mathrm{d}^2 K,$$

$$(6.29)$$

which is the Weyl representation of a diverging spherical wave.

We may now express the complete solution accounting for the energy introduced into the scattering medium through the reduced intensity $U_{ri}(\mathbf{r})$ (see Eq. (5.38) from Chap. 5). In terms of $\tilde{g}(\mathbf{K}, z)$ the complete solution can be written as:

$$U(\mathbf{r}) = \int_{-\infty}^{\infty} \left[\int_V \mu_s' U_{ri}(\mathbf{R}', z') \tilde{g}(\mathbf{K}, z - z') \times \right.$$

$$\left. \exp(i\mathbf{K} \cdot (\mathbf{R} - \mathbf{R}')) \mathrm{d}^3 r' \right] \mathrm{d}^2 K,$$

or equivalently:

$$U(\mathbf{r}) = \frac{1}{4\pi D} \int_{-\infty}^{\infty} \left[\int_V \mu_s' U_{ri}(\mathbf{R}', z') \frac{i}{2\pi q(\mathbf{K})} \times \right.$$

$$\left. \exp(i\mathbf{K} \cdot (\mathbf{R} - \mathbf{R}')) \exp(iq(\mathbf{K})|z - z'|) \mathrm{d}^3 r' \right] \mathrm{d}^2 K. \quad (6.30)$$

Since the volume V extends to infinity — remember we are dealing with an infinite homogeneous medium —, by writing $\mathrm{d}^3 r' = \mathrm{d}^2 R' \mathrm{d}z'$ we can identify terms within the volume integral as the Fourier transform of the reduced intensity:

$$\tilde{U}_{ri}(\mathbf{K}, z) = \frac{1}{4\pi^2} \int_{-\infty}^{\infty} U_{ri}(\mathbf{R}', z) \exp(-i\mathbf{K} \cdot \mathbf{R}') \mathrm{d}^2 R', \qquad (6.31)$$

in which case we can express Eq. (6.30) as:

$$U(\mathbf{r}) = \frac{\mu_s'}{4\pi D} \int_{-\infty}^{\infty} \frac{i}{2\pi q(\mathbf{K})} \left[\int_{-\infty}^{\infty} 4\pi^2 \tilde{U}_{ri}(\mathbf{K}, z') \exp(iq(\mathbf{K})|z - z'|) \mathrm{d}z' \right] \times$$

$$\exp(i\mathbf{K} \cdot \mathbf{R}) \mathrm{d}^2 K. \quad (6.32)$$

This equation, as expected, once again gives the Green function if we consider the contribution of the reduced intensity to be a point source, i.e. $\mu'_s U_{ri}(\mathbf{R}, z) = \delta(\mathbf{R} - \mathbf{R}')\delta(z - z')$, in which case $\mu'_s \widetilde{U}_{ri}(\mathbf{K}, z) = \exp(-i\mathbf{K} \cdot \mathbf{R}')\delta(z-z')/4\pi^2$ (see Appendix A for some properties of the delta function).

We have now found the relationship between the intensity in any plane z and the angular spectrum. The function that relates both is the subject of our next section: the *transfer function* and its expression in real-space, the *impulse response*.

6.3 Spatial Transfer Function and Impulse Response

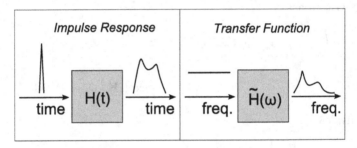

Fig. 6.2   Original meanings of the impulse response and the transfer function in electrical engineering.

Traditionally, in electrical engineering the *impulse response* has been defined as the response of a system (could be a circuit, for example) to a delta impulse in time, hence the name. The idea behind this is shown schematically in Fig. 6.6: in order to understand the effect of a system, we introduce a delta perturbation and measure the output. The function that reflects this change is termed the impulse response. Similarly, we can study the effect of a system on the possible frequencies we might input, by introducing a number of frequencies of known intensity and measuring the output. This second function is usually termed the *transfer function* and represents the temporal Fourier transform (or Laplace transform, depending on how it is defined) of the impulse response. In our diffusive medium we may define something equivalent to these functions, namely a function which reflects the frequency dependent change on propagation, and a function reflecting the temporal spread on propagation. These we have actually already found in Chap. 5, since they are represented by the frequency-dependent and the time-dependent Green functions, respectively. As a reminder, the time-dependent Green function was given by:

$$G(\mathbf{r}-\mathbf{r}',t-t') = c_0 \frac{\exp\left[-\mu_a c_0(t-t')\right]}{(4\pi D c_0(t-t'))^{3/2}} \exp\left(\frac{-|\mathbf{r}-\mathbf{r}'|^2}{4 D c_0(t-t')}\right). \qquad \left(cm^{-2}s^{-1}\right)$$

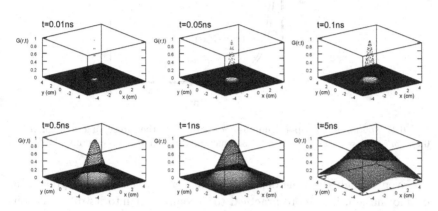

Fig. 6.3 Spatial and temporal impulse response, i.e. the Green function, for $\mu_a = 0.025 cm^{-1}$, $\mu_s' = 10 cm^{-1}$, and $n = 1.333$ measured at different time points and normalized to its maximum value at each time point.

In Fig. 6.3 we show the spatio-temporal impulse response (i.e. the time-dependent Green function) as derived in Chap. 5 for different time points normalized so that we can follow how the function widens in space as time progresses. Note how the relevant time-scale is in the nanosecond range and how the impulse response spreads out as time goes by. In other words, as light diffuses from the object or sources there is a spatial spread of the total energy density: if we wait long enough the spatial distribution of intensity will be approximately constant throughout the volume until it is all absorbed. In order to have a clearer idea of how the impulse response behaves in time, Fig. 6.4 shows the temporal profile for several distances.

In terms of the spatio-temporal impulse response, we had found in Chap. 5, Eq. (5.14) that the average intensity U was related to an original source temporal and spatial distribution, $S(\mathbf{r},t)$, as:

$$U(\mathbf{r},t) = \int_t^{\infty} \int_V S(\mathbf{r}',t')G(\mathbf{r},\mathbf{r}',t,t')\mathrm{d}^3r'\mathrm{d}t',$$

which in the specific case we solved in Eq. (5.14) was given by $S(\mathbf{r},t) = \mu_s' U_{ri}(\mathbf{r},t)$.

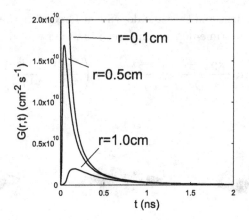

Fig. 6.4 Spatial and temporal impulse response, i.e. the Green function, for $\mu_a = 0.025cm^{-1}$, $\mu_s' = 10cm^{-1}$, and $n = 1.333$ measured at different time points.

Even though the time-dependent transfer function offers a great deal of information, it is more practical to decompose it into its different contributions, namely, its effect as a spatial filter (in the spatial frequency domain), and its effect as a temporal filter (in the modulation frequency, or frequency-dependent domain). Analyzing these different contributions will be the main goal of this section.

6.3.1 *Spatial Transfer Function and Impulse Response*

Following what we derived in terms of the angular spectrum representation, as long as the sources[4] are at $z_s < z$, and we are in an infinite homogeneous medium the Fourier transform at any plane $z' > z$ can be related to the Fourier transform at z as (see Eq. (6.18)):

$$\widetilde{U}(\mathbf{K}, z') = \widetilde{U}(\mathbf{K}, z) \exp(iq(\mathbf{K})|z - z'|), \qquad (6.33)$$

where $\exp(iq(\mathbf{K})|z - z'|)$ is usually termed the *transfer function*:

$$\widetilde{H}(\mathbf{K}, z - z') = \exp(iq(\mathbf{K})|z - z'|). \qquad (6.34)$$

---

[4]Note that here we use the term sources to refer to both actual emitting sources — could be fluorescence, for example — and scattering or absorbing objects.

Building Block: **Spatial Transfer Function**

The spatial transfer function represents a spatial frequency filter which relates the average intensity propagating in a specific direction to the average intensity at a larger distance of propagation.

The transfer function accounts for the spatial frequency decay of the diffuse wavefront on propagation (note that since $q_{\Im m} > 0$ for all values of $\mathbf{K}$ we expect *all* spatial frequencies to be damped by this transfer function). Eq. (6.33) represents the deeper meaning of the transfer function, since as long as the sources are at $z_s < z$ the average intensity at any plane $z' > z$ will be given by:

$$U(\mathbf{R}, z') = \int_{-\infty}^{\infty} \widetilde{U}(\mathbf{K}, z)\widetilde{H}(\mathbf{K}, z - z')\exp(i\mathbf{K} \cdot \mathbf{R})\mathrm{d}^2 K. \qquad (6.35)$$

Our goal now is to obtain an expression of this function in real-space. For that, we can now take the expression in Eq. (6.28) further if we realize that:

$$\frac{\partial \widetilde{g}(\mathbf{K}, z - z')}{\partial z'} = \frac{1}{4\pi D}\frac{1}{2\pi}\exp[iq(\mathbf{K})(z - z')], \qquad (6.36)$$

or equivalently:

$$\widetilde{H}(\mathbf{K}, z - z') = 8\pi^2 D\frac{\partial \widetilde{g}(\mathbf{K}, z - z')}{\partial z'}. \qquad (6.37)$$

Additionally, the Green function has also the following property:

$$\frac{\partial \widetilde{g}(\mathbf{K}, z - z')}{\partial z'} = -iq(\mathbf{K})\widetilde{g}(\mathbf{K}, z - z'). \qquad (6.38)$$

With the help of Eq. (6.37), Eq. (6.35) can be also rewritten as:

$$U(\mathbf{R}, z') = 8\pi^2 D \int_{-\infty}^{\infty} \widetilde{U}(\mathbf{K}, z)\frac{\partial \widetilde{g}(\mathbf{K}, z - z')}{\partial z'}\exp(i\mathbf{K} \cdot \mathbf{R})d\mathbf{K}. \qquad (6.39)$$

If we now introduce in Eq. (6.39) the expression for $\widetilde{U}(\mathbf{K}, z)$ in terms of its inverse Fourier transform the above equation becomes:

$$U(\mathbf{R}, z') = 8\pi^2 D \int_{-\infty}^{\infty} \left[\frac{1}{4\pi^2} \int_{-\infty}^{\infty} U(\mathbf{R}', z)\exp(-i\mathbf{K} \cdot \mathbf{R}')\mathrm{d}^2 R' \right] \times$$

$$\frac{\partial \widetilde{g}(\mathbf{K}, z - z')}{\partial z'}\exp(i\mathbf{K} \cdot \mathbf{R})\mathrm{d}^2 K,$$

equation which can be reorganized and the inverse Fourier transform of $\partial \widetilde{g}(\mathbf{K}, z)/\partial z$ identified:

$$U(\mathbf{R}, z') = 2D \int_{-\infty}^{\infty} U(\mathbf{R}', z) \frac{\partial g(\mathbf{R} - \mathbf{R}', z - z')}{\partial z'} \mathrm{d}^2 R', \qquad (6.40)$$

Eq. (6.40) is equivalent to the *first Rayleigh-Sommerfeld integral formula* defined in electromagnetism. Note that if we introduce Eq. (6.36) into Eq. (6.39) in the $z = z'$ case we recover the expression for a delta:

$$\int_{-\infty}^{\infty} \left. \frac{\partial \widetilde{g}(\mathbf{K}, z - z')}{\partial z'} \right|_{z'=z} \exp(i\mathbf{K} \cdot (\mathbf{R} - \mathbf{R}')) \mathrm{d}^2 K = \frac{4\pi^2}{4\pi D} \frac{1}{2\pi} \delta(\mathbf{R} - \mathbf{R}'),$$

which when introduced back into Eq. (6.40), yields, as should be if we did all substitutions correctly, $U(\mathbf{R}, z')|_{z'=z} = U(\mathbf{R}, z)$.

With what we have just derived we have enough material to recover the expression for the spatial impulse response H, which can be obtained by comparing Eq. (6.40) to Eq. (6.39) and Eq. (6.35), yielding:

$$H(\mathbf{R} - \mathbf{R}', z - z') = \int_{-\infty}^{\infty} \widetilde{H}(\mathbf{K}, z - z') \exp(i\mathbf{K} \cdot (\mathbf{R} - \mathbf{R}')) \mathrm{d}^2 K, \quad (6.41)$$

and in terms of the Green function:

$$H(\mathbf{R} - \mathbf{R}', z - z') = 2D \frac{\partial g(\mathbf{R} - \mathbf{R}', z - z')}{\partial z'}. \qquad (6.42)$$

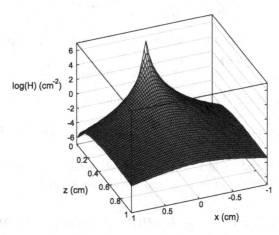

Fig. 6.5   Impulse response for the $\omega = 0$ case for $\mu_a = 0.025 cm^{-1}$, $\mu_s' = 10 cm^{-1}$ and $n = 1.333$, corresponding to values similar to that of tissue at the near-infrared (compare with Fig. 6.1)

Introducing the expression for g we derived in the previous section into the above equation gives us the explicit expression of the spatial impulse response for an infinite homogeneous medium:

$$H(\mathbf{R} - \mathbf{R}', z - z') = \frac{\exp(i\kappa_0|\mathbf{r} - \mathbf{r}'|)}{2\pi|\mathbf{r} - \mathbf{r}'|} \times$$
$$\left(i\kappa_0 - \frac{1}{|\mathbf{r} - \mathbf{r}'|}\right)\frac{-(z - z')}{|\mathbf{r} - \mathbf{r}'|}. \qquad \left(cm^{-2}\right) \quad (6.43)$$

 Building Block: **Spatial Impulse Response**

Since the intensity on propagation is a convolution between the spatial impulse response and the original wavefront, the spatial impulse response is the function which will dictate both the amount of blurring present — and thus resolution — in the plane of measurement and the decay in intensity.

In terms of the spatial impulse response, assuming we know the average intensity at $z = 0$ — for example, making use of Eq. (6.42) — the average intensity at any point in space $z > 0$ can be written as:

$$U(\mathbf{R}, z) = \int_{-\infty}^{\infty} H(\mathbf{R} - \mathbf{R}', z)U(\mathbf{R}', z = 0)\mathrm{d}^2R', \qquad (6.44)$$

which means that the total average intensity measured at a certain distance z from the object or collection of sources is the convolution of the impulse response with the total average intensity that emerges from the object or collection of sources at $z = 0$:

$$U(\mathbf{R}, z) = H(\mathbf{R} - \mathbf{R}', z) * U(\mathbf{R}', z = 0).$$

As can be understood, the closer our impulse response is to a delta, the more information (i.e. the larger the amount of spatial frequencies) will reach any given plane z. An example of how the impulse response looks like for the $\omega = 0$ case is shown in Fig. 6.5. This figure shows how on propagation the function not only widens — which would considerably blur the original wavefront at $z = 0$ — but also decreases drastically in intensity.

Once we have derived the spatial transfer function and impulse response we can proceed to study the effect propagation in diffusive media has on resolution since we can now obtain directly the z-dependent resolution of diffuse waves.

Principles of Diffuse Light Propagation

6.4   Spatial Resolution

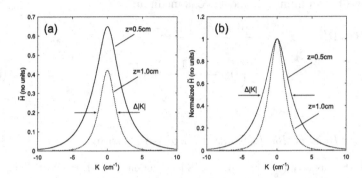

Fig. 6.6   Transfer function at $z = 0.5$ and $z = 1$cm for $\mu_a = 0.025 cm^{-1}$, $\mu'_s = 10 cm^{-1}$ and $n = 1.333$ in the $\omega = 0$ case (compare with Fig. 6.1 and Fig. 6.5).

Based on the expression for the diffuse wavenumber we have derived the functions that account for the changes in intensity of all the spatial frequencies that compose the diffuse wave-front. The transfer function specifically tells us how much each frequency K is damped on propagation, while the impulse response relates the average intensity at a plane z with the source of intensity at z' through a convolution. From these two quantities we can also obtain information on what would be the resolution at a certain plane of measurement z. But, before going on, to avoid misconceptions it is convenient to define the term *spatial resolution*. This definition should not be confused with the resolution limit in an inverse problem, which represents the limit of recovery of the optical parameters of an object and their distribution in space by solving the inverse problem (this will be the focus of Chap. 10). The definition of spatial resolution or *resolving power* that we will use here is the standard meaning used in optics: through the Oxford dictionary definition of resolution as 'the smallest interval measurable by a telescope or other scientific instrument' or as stated in Born and Wolf (1999) 'resolving power is a measure of the ability to separate images of two neighboring object points'. In other words, the resolving power represents the ability to separate two point objects or fine details through measurements taken at a certain distance from the scattering object or emitting source.

The criterion we will use to define the limit of spatial resolution is the well-known *Rayleigh criterion*, which sets the resolution limit as half of the full width at half maximum of the measured intensity. As with any criterion, we must not forget that it does not measure the absolute limit of resolution, but rather gives a measure to predict whether or not two objects — be these scattering objects or sources — will be resolved when measured at a certain distance. So as to obtain the resolution at a certain plane z, let us assume we have a point source at $z = 0$. In terms of our previous derivation, the spatial resolution limit is given by the full width at half maximum of the transfer function $\widetilde{H}$ (see Eq. (6.34)):

$$\widetilde{H}(\mathbf{K}, z) = \exp(iq(\mathbf{K})z). \tag{6.45}$$

An example of what this function looks like is shown in Fig. 6.6, where we see how this function not only decreases the overall intensity for all spatial frequencies (Fig. 6.6(a)), but considerably narrows on propagation (Fig. 6.6(b)). This figure shows how the full width at half maximum ($\Delta|\mathbf{K}|$) can be used to approximate this function to a pill-box or band-pass filter of the same width by assuming that any spatial frequency higher than $\Delta|\mathbf{K}|$ will have a very small contribution at z. Finding the analytical expression for the width of this spatial filter will be the purpose of this section.

6.4.1 *Resolution of Propagating Scalar Waves*

Before proceeding with recovering the resolution limit of diffuse waves, it is useful to first derive the simpler case of propagating scalar waves, i.e. waves with a wavelength equivalent to the diffuse waves we are studying but that do not suffer strong attenuation on propagation. In other words, they have the same $\kappa_{\Re e}$ but their $\kappa_{\Im m}$ values are zero. Although these waves do not necessarily exist, we can compare them to solutions of the scalar wave equation if we re-scale the wavelength. For these propagating waves the full width at half maximum, which we have define as $\Delta|\mathbf{K}|$, of the transfer function is obtained from the condition:

$$|\widetilde{H}(\mathbf{K}, z)| = \exp(-\sqrt{|\mathbf{K}|^2 - \kappa_0^2}\, z) = \frac{1}{2}.$$

If we take logarithms on both sides we obtain:

$$|\mathbf{K}|^2 - \kappa_0^2 = \left(\frac{\ln 2}{z}\right)^2,$$

and therefore the value of $|\mathbf{K}|$ that yields the value $\widetilde{H} = 1/2$ is simply:

$$|\mathbf{K}| = \sqrt{\kappa_0^2 + \left(\frac{\ln 2}{z}\right)^2}.$$

The full width half maximum of $\widetilde{H}$ will be twice this quantity, in which case the limit of spatial frequency for propagating waves ($\kappa_{\Im m} = 0$) is given by:

$$\Delta|\mathbf{K}| = 2\sqrt{\kappa_0^2 + \left(\frac{\ln 2}{z}\right)^2}. \tag{6.46}$$

Since the spatial frequency expressed by two objects separated a distance Δd can be represented by $2\pi/\Delta d$, we can find the minimum separation between two objects from the maximum spatial frequency we can resolve, $\Delta|\mathbf{K}|$:

$$\Delta d = \frac{2\pi}{\Delta|\mathbf{K}|}, \tag{6.47}$$

Δd being the resolution limit given by the full width at half maximum of H. Making use of Eq. (6.46) we obtain:

$$\frac{\Delta d}{\lambda_0} = \frac{1}{2}\left[1 + \left(\frac{\ln 2}{2\pi z/\lambda_0}\right)^2\right]^{-1/2}, \tag{6.48}$$

where $\Delta d/\lambda_0$ is the spatial resolution limit in units of the wavelength. When z increases, we see in Eq. (6.48) that the spatial resolution limit tends to:

$$\lim_{z \gg \lambda_0} \Delta d = \frac{\lambda_0}{2}, \tag{6.49}$$

namely, we retrieve the well-known *Rayleigh limit* of optical imaging valid for $z \gg \lambda_0$. In frequency space, Eq. (6.46) shows that as z increases,

$$\lim_{z \gg \lambda_0} \Delta|\mathbf{K}| = 2\kappa_0.$$

Once we have obtained the resolution limit for propagating waves, we can now follow the same approach to obtain that of diffuse waves. As we will see, this latter derivation is slightly more cumbersome due to the contribution of $\kappa_{\Im m}$ to the diffuse wavenumber κ_0.

6.4.2   *Resolution of Diffuse Waves*

In order to retrieve $\Delta|\mathbf{K}|$ for diffuse waves, we must consider that the expression for q is:

$$q(\mathbf{K}) = \sqrt{\kappa_0^2 - |\mathbf{K}|^2} = \sqrt{\kappa_{\Re e}^2 - \kappa_{\Im m}^2 - |\mathbf{K}|^2 + 2i\kappa_{\Re e}\kappa_{\Im m}},$$

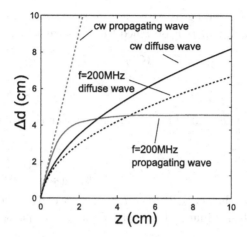

Fig. 6.7 Spatial resolution limit Δd in cm as z increases, for $\mu_a = 0.025cm^{-1}$, $\mu'_s = 10cm^{-1}$ and $n = 1.333$ in the cw and $f = 200$ MHz cases (see Fig. 6.6 for the transfer function of these optical parameters). The corresponding case for propagating waves (i.e. $\kappa_{\Im m} = 0$ with the same $\kappa_{\Re e}$) is also shown.

where as a reminder $\kappa_0 = \kappa_{\Re e} + i\kappa_{\Im m}$. After some manipulation, we can rewrite q as $q(\mathbf{K}) = q_{\Re e} + iq_{\Im m}$ being the real component given by:

$$q_{\Re e}(\mathbf{K}) = \frac{1}{\sqrt{2}}\left(\sqrt{|\kappa_0|^4 - 2(\kappa_{\Re e}^2 - \kappa_{\Im m}^2)|\mathbf{K}|^2 + |\mathbf{K}|^4} + \right.$$
$$\left. \kappa_{\Re e}^2 - \kappa_{\Im m}^2 - |\mathbf{K}|^2 \right)^{1/2}, \quad (6.50)$$

and the imaginary component:

$$q_{\Im m}(\mathbf{K}) = \frac{1}{\sqrt{2}}\left(\sqrt{|\kappa_0|^4 - 2(\kappa_{\Re e}^2 - \kappa_{\Im m}^2)|\mathbf{K}|^2 + |\mathbf{K}|^4} - \right.$$
$$\left. \kappa_{\Re e}^2 + \kappa_{\Im m}^2 + |\mathbf{K}|^2 \right)^{1/2}. \quad (6.51)$$

As expected, when $|\mathbf{K}| = 0$ we recover $q(\mathbf{K} = 0) = \kappa_0$. Since the maximum of $\widetilde{H}$ decays as z increases (see Fig. 6.6), we must define $\Delta|\mathbf{K}|$ normalizing $\widetilde{H}$ to its maximum, i.e. its value at $\mathbf{K} = 0$. Note that this was not necessary when dealing with propagating scalar waves in the previous section, since the intensity at $\mathbf{K} = 0$ is constant, whatever the distance. In the diffuse wave case, the value of $|\mathbf{K}|$ that brings the modulus of the *normalized* transfer function to its half maximum is given by:

$$\frac{|\widetilde{H}(\mathbf{K}, z)|}{|\widetilde{H}(\mathbf{K} = 0, z)|} = \exp\left(-(q_{\Im m}(\mathbf{K}) - \kappa_{\Im m})z\right) = \frac{1}{2},$$

in which case, proceeding as with the propagating scalar waves, the full width at half maximum $\widetilde{H}$ is obtained from the equation:

$$\left(|\kappa_0|^4 - 2(\kappa_{\Re e}^2 - \kappa_{\Im m}^2)|\mathbf{K}|^2 + |\mathbf{K}|^4\right)^{1/2}$$
$$- \kappa_{\Re e}^2 + \kappa_{\Im m}^2 + |\mathbf{K}|^2 = 2\left(\kappa_{\Im m} + \frac{\ln 2}{z}\right)^2.$$

After some tedious algebra, we finally recover the value for $\Delta|\mathbf{K}|$ we were looking for:

$$\Delta|\mathbf{K}| = 2\sqrt{\left(\kappa_{\Im m} + \frac{\ln 2}{z}\right)^2 + \kappa_{\Re e}^2 - \kappa_{\Im m}^2 - \frac{\kappa_{\Im m}^2 \kappa_{\Re e}^2}{(\kappa_{\Im m} + \ln 2/z)^2}}. \quad (6.52)$$

Using the fact that $\kappa_0 = 2\pi/\lambda_0 + i/L_\omega$ (see Sec. 6.1), the corresponding full width at half maximum of the spatial impulse response H is:

$$\frac{\Delta d}{\lambda_0} = \frac{1}{2}\left[\left(\frac{\lambda_0}{2\pi L_\omega} + \frac{\ln 2}{2\pi z/\lambda_0}\right)^2 - \left(\frac{\lambda_0}{2\pi L_\omega}\right)^2 + \right.$$
$$\left. 1 - \left(1 + \frac{\ln 2}{z/L_\omega}\right)^{-2}\right]^{-1/2}. \quad (6.53)$$

As can be seen from the expression for $\Delta|\mathbf{K}|$ Eq. (6.52), as z increases, $z \gg \lambda_0$, we see that:

$$\lim_{z \gg \lambda_0} \Delta|\mathbf{K}| = 0.$$

Additionally, the expression for the resolution limit of diffuse waves, Eq. (6.53), shows that $\Delta|\mathbf{K}|/\lambda_0$ has no upper limit:

$$\lim_{z \gg \lambda_0} \Delta d = \pi\sqrt{\frac{zL_\omega}{2\ln 2}}, \quad (6.54)$$

and tends to infinity as $z^{1/2}$, monotonically worsening the resolution (see Fig. 6.7). This is in great contrast with propagating scalar waves where frequencies in the range $|\mathbf{K}| \leq \kappa_{\Re e}$ propagate to infinity, and hence the resolution quickly settles at the expected $\lambda_0/2$ value (see the *cw* case for 200MHz shown in Fig. 6.7). In the case of diffuse waves, we see that the further away we measure, the closest our wavefront resembles that of a plane wave with a quickly diminishing intensity.

In the particular case where we have no intensity modulation, i.e $\omega = 0$, we reach from Eq. (6.52) that the spatial resolution of the diffuse waves is given by:

$$\Delta d_{cw} = \lim_{\lambda_0 \to \infty} \Delta d = \pi\left[\left(\frac{1}{L_d} + \frac{\ln 2}{z}\right)^2 - \left(\frac{1}{L_d}\right)^2\right]^{-1/2}, \quad (6.55)$$

where $L_d = \sqrt{D/\mu_a}$ is the diffusion length. The resolution when working in *cw* is lower than when modulating the intensity (see Fig. 6.7); however, it is important to remember that the transfer function *decays at a faster rate* when modulating the intensity or, in other words, the imaginary part of the wave number, $\kappa_{\Im m}$, increases with the frequency. In practical terms, this means that even though intensity modulation might increase your resolution and overall information content, by working in the *cw* regime we are optimizing the penetration and sensitivity[5]. Proceeding in a similar fashion with the propagating waves presented earlier, we obtain that:

$$\Delta d_{cw} = \lim_{\lambda_0 \to \infty} \Delta d = \frac{\pi}{\ln 2} z. \qquad (6.56)$$

This regime is commonly termed the *electrostatic limit* in electromagnetism and corresponds to the near-field regime where we are at sub-wavelength distances from the object (considered either as a primary source or as a scattering or absorbing object). In this range, if all distances involved are much smaller than the wavelength, retardation effects can be neglected (the near field and all related effects are thoroughly studied in Nieto-Vesperinas (2006)). An example of how a propagating wave would behave in the electrostatic limit is shown in Fig. 6.7, where it is important to underline the fact that in the case of propagating waves the behavior of the spatial resolution with respect to z no longer depends on the optical properties of the medium at this limit. This is in contrast with diffuse waves, which always depend on the optical properties of the medium where they propagate (see Eq. (6.55)).

6.5 Backpropagation of Diffuse Light

As we discussed in Sec. 6.2, the relationship between the angular spectrum that leaves the object, $\mathcal{A}$, and our measurement at a plane z was given by (see Eq. (6.17)):

$$\widetilde{U}(\mathbf{K}, z) = \mathcal{A}(\mathbf{K}) \exp(iq(\mathbf{K})|z|), \qquad (6.57)$$

---

[5]A proper choice of modulation frequency will ultimately depend on the information you intend to extract from the system. Note that resolution as defined here is the resolving power at a plane after propagation, and not the resolution of the image obtained after solving the inverse problem. Even though they are related, the ultimate resolution of your reconstruction will have a higher dependence on the signal to noise ratio of your signal that on the resolving power at that specific modulation frequency.

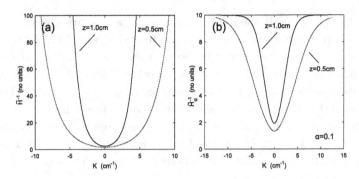

Fig. 6.8   Inverse of Transfer function, $\widetilde{H}^{-1}$ (a) and regularized inverse of transfer function, $\widetilde{H}_\alpha^{-1}$ (b) at $z = 0.5$ and $z = 1$cm for $\mu_a = 0.025 cm^{-1}$, $\mu_s' = 10 cm^{-1}$ and $n = 1.333$ in the $\omega = 0$ case (compare with Fig. 6.6). The regularization parameter used for $\widetilde{H}_\alpha^{-1}$ is $\alpha = 0.1$.

and we saw how using this formula we can relate the average intensity at a plane z to the average intensity at a plane $z < z_0$ as:

$$\widetilde{U}(\mathbf{K}, z) = \widetilde{U}(\mathbf{K}, z_0) \exp(iq(\mathbf{K})(z - z_0)).$$

Since this equation relates the intensity at the measurement plane with the intensity that left the object, we could invert the effect of propagation from z to z_0:

$$\widetilde{U}(\mathbf{K}, z_0) \simeq \widetilde{U}(\mathbf{K}, z) \exp(-iq(\mathbf{K})|z - z_0|).$$

In this respect Eq. (6.57) is a key equation when dealing with diffuse light propagation and represents our ability to recover information on an object or fluorescent probe we might want to image. In fact, we can use directly this equation to attempt to recover the angular spectrum of the object:

$$\mathcal{A}(\mathbf{K}) \simeq \widetilde{U}(\mathbf{K}, z) \exp(-iq(\mathbf{K})|z|), \qquad (6.58)$$

a technique which is termed *back-propagation*. As we have seen in the previous section, the function $\exp(iq(\mathbf{K})|z|)$ decays very fast for high spatial frequencies. This means that applying Eq. (6.58) imposes severe problems (see Fig. 6.8). In particular, we need to regularize the inverse of this function in order to remove high frequency noise, for example:

$$\mathcal{A}(\mathbf{K}) \simeq \widetilde{U}(\mathbf{K}, z) \frac{1}{\alpha + \exp(iq(\mathbf{K})|z|)}, \qquad (6.59)$$

where α is the regularization parameter. Equivalently, we may write the regularized back-propagation as:

$$\mathcal{A}(\mathbf{K}) \simeq \widetilde{U}(\mathbf{K}, z) \widetilde{H}_\alpha^{-1}(\mathbf{K}),$$

with:

$$\widetilde{H}_\alpha^{-1}(\mathbf{K}) = \frac{1}{\alpha + \exp(iq(\mathbf{K})|z|)}.$$

Note that this expression for the regularized back-propagation function is by no means unique. There are many other approaches to inverting $\widetilde{H}$ by applying different filters (Hann, Hamming, Gaussian, etc.) to the product $\widetilde{U}(\mathbf{K}, z) \exp(-iq(\mathbf{K})|z|)$. Whatever the approach used, they all try to attain the same goal: to constrain the function so as to recover as much information as possible contained in the lower frequency part of the spectrum. As can be clearly seen in Fig. 6.8 the regularization parameter will need to be more severe the larger the value of $|z|$ (i.e the longer the propagation distance). This result showcases how information is lost on propagation, no matter how sensitive our measurement is: as the information that left our object (i.e. the angular spectrum, $\mathcal{A}$) travels through the medium, high frequencies are very quickly damped by the transfer function. Based on the fact that there is no such thing as a noise-free measurement, when trying to extract the original angular spectrum from a measurement taken at a distance we will never be able to recover all the spatial frequencies that composed the original angular spectrum.

Key points

From this chapter the following important points should be remembered:

- **Diffuse wavenumber**: The wavenumber is the fundamental quantity which will dictate wave propagation. In the case of diffuse waves, we have that this wavenumber is either a pure imaginary (in the cw case), or a complex number with a positive imaginary component. This fact is responsible for the fast degradation of resolution as intensity propagates further into a scattering medium.
- **Angular spectrum**: We have found that a convenient way to represent the intensity that leaves an object (or is emitted by a source) is the angular spectrum representation, which accounts for the different spatial frequencies and the contribution of each one at the plane closest to the object.
- **Spatial Transfer function**: The spatial transfer function represents how an original spatial frequency distribution of intensity, i.e. the angular spectrum, propagates within the medium. We have found that

this function holds all the information regarding the resolving power we may have at a detector plane.

- **Loss of Information**: Due to the fact that high frequencies are highly damped on propagation, there is an inevitable loss of information. This means that the complete information that left the object (i.e. its angular spectrum) can never be fully recovered. The wider the spatial transfer function the more information can be retrieved (i.e. the less higher spatial frequencies are damped). In a similar way, we can say that the narrower the spatial impulse response, the higher the amount of information measured. This is directly related to back-propagation and how it can be used to attempt retrieve the original angular spectrum.

Further Reading

The core of this chapter has been the angular spectrum representation. To gain further insight into the angular spectrum I recommend Born and Wolf (1999), Mandel and Wolf (1995) and Nieto-Vesperinas (2006). I also recommend Nieto-Vesperinas (2006) on the issue of near-field optics and the electrostatic regime mentioned in this section. Most of what has been presented in this chapter with regards to diffuse waves can be also found in [Ripoll *et al.* (1999)] where the issue of resolution, back-propagation and the effect on noise on back-propagation is presented in greater detail following a similar notation. Finally, the topic of back-propagation is a very interesting and potentially very useful one that I have barely covered here. For further insight on this approach I suggest reading the original works by Matson, [Matson *et al.* (1997); Matson (1997)] and [Matson and Liu (2000)]. Experimentally, the group of Arjun Yodh applied algorithms based on back-propagation in [Durduran *et al.* (1999)] and [Li *et al.* (2000)], where a detailed derivation of both theory and experiment can be found serving as excellent starting points. The effect of back-propagation is also considered in a more advanced fashion by Schotland and Markel who they take this effect to the next level by deriving a direct inversion formula in [Markel and Schotland (2001)][6], and have recently applied this to experiments in [Konecky *et al.* (2008)].

---

[6]This approach is equivalent to back-propagating instead of from a plane to another plane, from the measurement surface to the complete volume simultaneously.

Chapter 7

The Point Source Approximation

Summary. In this chapter we present the reason behind the most commonly used approximation in diffuse light propagation, namely the assumption that collimated light can be approximated to an isotropic point source located at one transport mean free path l_{tr} inside the diffusive medium. We will derive this approximation and present the necessary expressions to account for more general source and detector profiles.

7.1 General Solution

Since the earliest works on light diffusion and its application in biomedicine, the point source approximation has been the most commonly used expression for modeling how diffuse light is generated within a highly scattering medium (see for example the classical paper of M. Patterson, B. Chance and B. Wilson [Patterson *et al.* (1989)]). Even though it is an approximation that has been thoroughly validated both experimentally and with Monte-Carlo, its formal derivation is not that well-known. In this first section we will derive this approximation from first principles making use of what we have derived so far, in particular with regards to the angular spectrum representation we discussed in Chap. 6. We will complete this derivation for the general case of modulated intensity, since all cases (both *cw* and time-domain results) can be obtained from it.

As we saw in Chap. 5, the general solution for modulated intensity in an infinite homogeneous medium was given by (see Eq. (5.38)):

$$\nabla^2 U_\omega(\mathbf{r}) + \kappa_\omega^2 U_\omega(\mathbf{r}) = \frac{-1}{D} S_\omega(\mathbf{r}), \qquad (7.1)$$

with the wavenumber given by:

$$\kappa_\omega = \sqrt{-\frac{\mu_a}{D} + i\frac{\omega}{Dc_0}}. \tag{7.2}$$

We also saw that for this case the Green function is given by:

$$G_\omega(\mathbf{r} - \mathbf{r}') = \frac{\exp(i\kappa_\omega|\mathbf{r} - \mathbf{r}'|)}{4\pi D|\mathbf{r} - \mathbf{r}'|}, \tag{7.3}$$

and the frequency-dependent average intensity is represented by the solution to the general equation:

$$U_\omega(\mathbf{r}) = \int_V S_\omega(\mathbf{r}')G_\omega(\mathbf{r} - \mathbf{r}')\mathrm{d}^3r'. \tag{7.4}$$

In Chap. 5 we saw that the complete solution which accounts for the reduced intensity contribution was found for $S_\omega(\mathbf{r}) = \mu_s' U_{ri}(\mathbf{r})$ (see Eq. (5.38)). Temporarily, however, we will assume a general distribution of sources modulated at frequency ω, $S_\omega(\mathbf{r})$, and will come back to the complete solution when dealing with a collimated light source.

In order to solve Eq. (7.4) we will express the above equations in terms of their angular spectrum. For that will make use of the expression for the Green function in K-space (see Eq. (6.28) in Chap. 6):

$$G_{cw}(\mathbf{r} - \mathbf{r}') = \int_{-\infty}^{\infty} \frac{1}{4\pi D} \frac{i}{2\pi q(\mathbf{K})} \times$$
$$\exp(i\mathbf{K}(\mathbf{R} - \mathbf{R}'))\exp(iq(\mathbf{K})|z - z'|)\mathrm{d}^2K, \tag{7.5}$$

with $q(\mathbf{K})$ given by:

$$q(\mathbf{K}) = \sqrt{\kappa_0^2 - \mathbf{K}^2}. \tag{7.6}$$

Introducing this expression for the Green function in Eq. (7.4) we obtain:

$$U_\omega(\mathbf{r}) = \frac{1}{4\pi D} \int_{-\infty}^{\infty} \left[\int_V S_\omega(\mathbf{R}', z') \frac{i}{2\pi q(\mathbf{K})} \times \right.$$
$$\left. \exp(i\mathbf{K}(\mathbf{R} - \mathbf{R}'))\exp(iq(\mathbf{K})|z - z'|)\mathrm{d}^3r' \right] \mathrm{d}^2K, \tag{7.7}$$

Since we have an infinite diffusive medium we can extend the limits of the volume integral to infinity and identify the Fourier transform of the source S_ω:

$$\widetilde{S}_\omega(\mathbf{K}, z) = \frac{1}{4\pi^2} \int_{-\infty}^{\infty} S_\omega(\mathbf{R}', z)\exp(-i\mathbf{K} \cdot \mathbf{R}')\mathrm{d}^2K, \tag{7.8}$$

in which case we can express Eq. (7.7) as (compare also with Eq. (6.32) of Chap. 6):

$$U_\omega(\mathbf{r}) = \frac{1}{4\pi D} \int_{-\infty}^{\infty} \frac{i}{2\pi q(\mathbf{K})} \left[\int_{-\infty}^{\infty} 4\pi^2 \widetilde{S}_\omega(\mathbf{K}, z') \times \right.$$
$$\left. \exp(iq(\mathbf{K})|z - z'|) dz' \right] \exp(i\mathbf{K} \cdot \mathbf{R}) d^2 K, \quad (7.9)$$

or directly in the angular spectrum representation:

$$\widetilde{U}_\omega(\mathbf{K}, z) = \frac{1}{4\pi D} \frac{i}{2\pi q(\mathbf{K})} \int_{-\infty}^{\infty} 4\pi^2 \widetilde{S}_\omega(\mathbf{K}, z') \exp(iq(\mathbf{K})|z - z'|) dz'. \quad (7.10)$$

This equation represents the complete solution in an infinite homogeneous medium for a modulated source of arbitrary distribution $S_\omega(\mathbf{r})$. We will now study under which conditions Eq. (7.10) can be approximated to a point source, but before that, let us derive the expression for a point source as a reference.

7.1.1 *Solution for a point source*

Let us consider a point source as the source distribution in an otherwise infinite and homogeneous medium. The expression for a point source is given by:

$$S_\omega(\mathbf{R}, z_s) = S_0 \delta(\mathbf{R} - \mathbf{R}_s) \delta(z - z_s), \quad \left(Watts/cm^3 \right) \quad (7.11)$$

with S_0 having units of *Watts* and representing the total power emitted by the source (don't forget that the delta functions have units too). Making use of Eq. (7.8) this expression becomes (note the $4\pi^2$ factor needed to normalize the delta function — see Appendix A):

$$S_\omega(\mathbf{K}, z_s) = S_0 \frac{\exp(-i\mathbf{K} \cdot \mathbf{R}_s)}{4\pi^2} \delta(z - z_s), \quad (7.12)$$

which gives the following average intensity in K-space:

$$\widetilde{U}_\omega(\mathbf{K}, z) = \frac{S_0}{4\pi D} \frac{i}{2\pi q(\mathbf{K})} \exp(-i\mathbf{K} \cdot \mathbf{R}_s) \exp(iq(\mathbf{K})|z - z_s|) \quad (7.13)$$

or, in terms of the angular spectrum of a point source $\mathcal{A}_\delta$:

$$\widetilde{U}_\omega(\mathbf{K}, z) = S_0 \mathcal{A}_\delta(\mathbf{K}) \exp(-i\mathbf{K} \cdot \mathbf{R}_s) \exp(iq(\mathbf{K})|z - z_s|).$$

In the case where we have a collection of N point sources with intensities $S_0^{(i)}$ located at positions $(\mathbf{R}_i, z_i)$ the total average intensity U would result in:

$$\widetilde{U}_\omega(\mathbf{K}, z) = \mathcal{A}_\delta(\mathbf{K}) \sum_{i=0}^{N} S_0^{(i)} \exp(-i\mathbf{K} \cdot \mathbf{R}_i) \exp(iq(\mathbf{K})|z - z_i|). \quad (7.14)$$

7.2 Solution for a collimated source

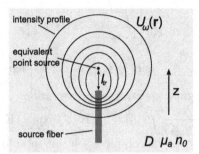

Fig. 7.1 Representation of a source fiber emitting a collimated beam inside an infinite homogeneous diffusive medium of optical properties defined by D, μ_a and n_0. Note how the intensity profile for distances far from the source is equivalent to that of a point source.

As we saw in Chap. 5 and mentioned above in Eq. (7.4), the general solution accounting for the reduced intensity contribution was represented by:

$$S_\omega(\mathbf{r}) = \mu'_s U_{ri}(\mathbf{r}), \tag{7.15}$$

with $U_{ri}(\mathbf{r})$ being given by (see Chap. 4, Eq. (4.36)):

$$U_{ri}(\mathbf{r}) = \int_{(4\pi)} I_{ri}(\mathbf{r}, \hat{\mathbf{s}}) d\Omega, \tag{7.16}$$

and $I_{ri}(\mathbf{r}, \hat{\mathbf{s}})$ being the the solution to:

$$\frac{1}{c_0}\frac{\partial}{\partial t} I_{ri}(\mathbf{r}, \hat{\mathbf{s}}) + \hat{\mathbf{s}} \cdot \nabla I_{ri}(\mathbf{r}, \hat{\mathbf{s}}) + \mu_{\text{tr}} I_{ri}(\mathbf{r}, \hat{\mathbf{s}}) = \epsilon(\mathbf{r}, \hat{\mathbf{s}}). \tag{7.17}$$

Let us now consider that in our infinite homogeneous medium we have a collimated source, due to, for example, a fiber inside the medium pointing in the +z direction, $\mathbf{u}_z$, as show in Fig. 7.1:

$$\epsilon(\mathbf{r}, \hat{\mathbf{s}}) = S_0 \delta(\mathbf{r} - \mathbf{r}_s)\delta(\hat{\mathbf{s}} - \mathbf{u}_z). \quad \left(Watts/cm^3 sr\right)$$

Neglecting the effect that the fiber can have on light propagation, the solution to Eq. (7.17) would be :

$$I_{ri}(\mathbf{r}, \hat{\mathbf{s}}) = S_0 \delta(\mathbf{R} - \mathbf{R}_s) \exp(-\mu_{\text{tr}} z)\delta(\hat{\mathbf{s}} - \mathbf{u}_z), \ \forall z \geq 0. \quad \left(Watts/cm^2 sr\right) \tag{7.18}$$

In this case the expression for the energy density of the source, Eq. (7.15), results in:

$$S_\omega(\mathbf{R}, z) = S_0 \mu'_s \delta(\mathbf{R} - \mathbf{R}_s) \exp(-\mu_{\mathrm{tr}}z), \quad \forall z \geq 0, \qquad (7.19)$$

and its expression in K-space becomes (see Eq. (7.8)):

$$\widetilde{S}_\omega(\mathbf{K}, z) = \frac{S_0}{4\pi^2} \mu'_s \exp(-\mu_{\mathrm{tr}}z) \exp(-i\mathbf{K} \cdot \mathbf{R}_s), \quad \forall z \geq 0, \qquad (7.20)$$

where the $4\pi^2$ factor appears due to the Fourier transform of the point source (see Appendix A).

Introducing Eq. (7.20) into Eq. (7.10) we obtain:

$$\widetilde{U}_\omega(\mathbf{K}) = \frac{S_0 \mu'_s}{4\pi D} \frac{i}{2\pi q(\mathbf{K})} \exp(-i\mathbf{K} \cdot \mathbf{R}_s) \times$$

$$\int_0^\infty \exp(-\mu_{\mathrm{tr}}z) \exp(iq(\mathbf{K})|z - z'|) \mathrm{d}z'. \qquad (7.21)$$

The above equation represents the complete solution for a collimated source accounting for the reduced intensity contribution and simply represents a sum of point sources with intensity decaying as we move further into z; for this reason it is also commonly referred to as an exponentially decaying line of sources. Finding its solution amounts to solving the following integral (assuming always a $\mathbf{K}$ dependence on q):

$$f(\mathbf{K}, z) = \int_0^\infty \exp(-\mu_{\mathrm{tr}}z') \exp(iq|z - z'|) \mathrm{d}z'.$$

Note that great care needs to be taken in order to solve this equation due to the $|z - z'|$ factor. The correct way to solve for this equation is by dividing space into the $z > z'$ and $z < z'$ contributions:

$$f(\mathbf{K}, z) = \int_0^z \exp(-\mu_{\mathrm{tr}}z') \exp(iq(z - z')) \mathrm{d}z' +$$

$$\int_z^\infty \exp(-\mu_{\mathrm{tr}}z') \exp(-iq(z - z')) \mathrm{d}z',$$

which, grouping terms and distinguishing the cases between $\Re\{-iq\} > \mu_{\mathrm{tr}}$ and $\Re\{-iq\} < \mu_{\mathrm{tr}}$ (the effect $|z - z'|$ needs to be still accounted for) gives:

$$f(\mathbf{K}, z) = \exp(iqz) \int_0^z \exp(-(\mu_{\mathrm{tr}} + iq)z') \mathrm{d}z' +$$

$$\exp(-iqz) \int_z^\infty \exp(-(\mu_{\mathrm{tr}} - iq)z') \mathrm{d}z'.$$

This can be solved, taking great care with the fact that q is a complex number, yielding:

- For $\Re\{-iq(\mathbf{K})\} \leq \mu_{\mathrm{tr}}$:

$$f(\mathbf{K}, z) = \frac{1}{\mu_{\mathrm{tr}}^2 + q^2} \left((\mu_{\mathrm{tr}} - iq) \exp(iqz) + 2iq \exp(-\mu_{\mathrm{tr}}z) \right).$$

- For $\Re\{-iq(\mathbf{K})\} > \mu_{\mathrm{tr}}$:

$$f(\mathbf{K}, z) = \frac{-1}{\mu_{\mathrm{tr}}^2 + q^2} \left((\mu_{\mathrm{tr}} - iq) \exp(iqz) - \right.$$

$$\left. \exp(-\mu_{\mathrm{tr}}z) \left[(\mu_{\mathrm{tr}} - iq) \exp(2(iq + \mu_{\mathrm{tr}})z) + (\mu_{\mathrm{tr}} - iq) \right] + 2iq \exp(-\mu_{\mathrm{tr}}z) \right).$$

We are now in the position to obtain a solution for the total average intensity $\widetilde{U}_\omega$ for a collimated source (or exponentially decaying line of sources):

- For $\Re\{-iq(\mathbf{K})\} \leq \mu_{\mathrm{tr}}$:

$$\widetilde{U}_\omega(\mathbf{K}, z) = S_0 \mu_s' \mathcal{A}_\delta(\mathbf{K}) \exp(-i\mathbf{K} \cdot \mathbf{R}_s) \frac{1}{\mu_{\mathrm{tr}}^2 + q(\mathbf{K})^2} \times$$

$$\left[(\mu_{\mathrm{tr}} - iq(\mathbf{K})) \exp(iq(\mathbf{K})z) + 2iq(\mathbf{K}) \exp(-\mu_{\mathrm{tr}}z) \right]. \quad (7.22)$$

- For $\Re\{-iq(\mathbf{K})\} > \mu_{\mathrm{tr}}$:

$$\widetilde{U}_\omega(\mathbf{K}, z) = S_0 \mu_s' \mathcal{A}_\delta(\mathbf{K}) \exp(-i\mathbf{K} \cdot \mathbf{R}_s) \frac{-1}{\mu_{\mathrm{tr}}^2 + q(\mathbf{K})^2} \times$$

$$\left[(\mu_{\mathrm{tr}} - iq(\mathbf{K})) \exp(iq(\mathbf{K})z) - \exp(-\mu_{\mathrm{tr}}z) \times \right.$$

$$\left. \left((\mu_{\mathrm{tr}} - iq(\mathbf{K})) \exp\left(2(iq(\mathbf{K}) + \mu_{\mathrm{tr}})z\right) + (\mu_{\mathrm{tr}} + iq(\mathbf{K})) \right) \right]. \quad (7.23)$$

7.3 Point Source Approximation to a collimated source

Eqs. (7.22)–(7.23) are the complete and formal derivation of a pencil beam of collimated light in an infinite diffusive medium. Since $\mathcal{A}_\delta$ acts as a low pass filter, one approximation we may take is to assume we are always in the case $\Re\{-iq(\mathbf{K})\} < \mu_{\mathrm{tr}}$ (this holds true as long as we are in the low spatial frequency regime, i.e. far away from sources). In this case we may approximate $\mu_{\text{tr}}^2 + q^2 \simeq \mu_{\text{tr}}^2$ and the $\exp(iqz)$ will dominate over $\exp(-\mu_{\text{tr}}z)$ yielding:

$$\widetilde{U}_\omega(\mathbf{K}, z) \simeq S_0 \frac{\mu_s'}{\mu_{\text{tr}}} \mathcal{A}_\delta(\mathbf{K}) \exp(-i\mathbf{K} \cdot \mathbf{R}_s) \left[\left(1 - \frac{iq(\mathbf{K})}{\mu_{\text{tr}}} \right) \exp(iq(\mathbf{K})z) \right].$$

Now, as long as $-iq(\mathbf{K}) < \mu_{\text{tr}}$, neglecting higher order terms we can introduce the following approximation:

$$\left(1 - \frac{iq(\mathbf{K})}{\mu_{\text{tr}}} \right) \simeq \exp \left(-\frac{iq(\mathbf{K})}{\mu_{\text{tr}}} \right), \tag{7.24}$$

in which case our expression becomes identical to that of a point source:

$$\widetilde{U}_\omega(\mathbf{K}, z) \simeq S_0 \frac{\mu_s'}{\mu_{\text{tr}}} \mathcal{A}_\delta(\mathbf{K}) \exp(-i\mathbf{K} \cdot \mathbf{R}_s) \exp \left(iq(\mathbf{K})(z - 1/\mu_{\text{tr}}) \right). \tag{7.25}$$

It is Eq. (7.24) which represents the core of the point source approximation. This expression explains the origin of the $z_s = 1/\mu_{\text{tr}} = l_{\text{tr}}$ traditionally used to describe the incident source as a point source, since this expression clearly represents the solution to a point source at $\mathbf{r}_s \equiv (\mathbf{R}_s, 1/\mu_{\text{tr}})$ (see Fig. 7.1). Comparing this expression with Eq. (7.13), the total average intensity due to a collimated source can be approximated to (see Chap. 5, Sec. 5.7.3):

$$U_\omega(\mathbf{r}) \simeq S_0 \frac{\exp(i\kappa_0 |\mathbf{r} - \mathbf{r}_s^*|)}{4\pi D |\mathbf{r} - \mathbf{r}_s^*|}, \qquad \left(Watts/cm^2 \right) \tag{7.26}$$

with $\mathbf{r}_s^*$ given by:

$$\mathbf{r}_s^* = \mathbf{r}_s + \mathbf{u}_{col} \frac{1}{\mu_{\text{tr}}}, \tag{7.27}$$

being $\mathbf{r}_s$ the position of the collimated source and $\mathbf{u}_{col}$ the direction that this collimated source points into (in the derivation we just completed we had that $\mathbf{r}_s = (\mathbf{R}_s, z = 0)$ and $\mathbf{u}_{col} = (0, 0, 1)$). Note that we have also used the approximation $\mu_s'/\mu_{\text{tr}} \simeq 1$, which will be justified in the next subsection.

 Building Block: **Point Source Approximation**

The point source approximation to a collimated source will be extremely handy for simplifying the expressions used to model light propagation in diffuse media, to solve both the forward and inverse problems.

7.3.1   Limits of Validity

Whenever introducing an approximation it is always necessary to understand the limits for which we can assume this approximation gives accurate results. Based on our previous derivation, the point source approximation is only valid as long as:

$$-iq(\mathbf{K}) < \mu_{\text{tr}},$$

$$\sqrt{\mu_a/D + |\mathbf{K}|^2} < \mu_{\text{tr}},$$

$$|\mathbf{K}|^2 < \mu_{\text{tr}}^2 - \mu_a/D,$$

$$|\mathbf{K}| < \mu_{\text{tr}}\sqrt{1 - \mu_a/D\mu_{\text{tr}}^2}.$$

If we assume D to be $D \simeq 1/3\mu_{\text{tr}}$ we obtain that the conditions so that the point source approximation are accurate are:

$$|\mathbf{K}| < \mu_{\text{tr}}\sqrt{1 - 3\mu_a/\mu_{\text{tr}}},$$

$$z > 1/\mu_{\text{tr}},$$

Note that since $\mu_{\text{tr}} = \mu_a + \mu_s'$, the first condition is true as long as $\mu_s' > 2\mu_a$. This also implies that the higher the μ_s'/μ_a ratio, the more relaxed the condition on $\mathbf{K}$, which is expected since the higher the scattering, the more pronounced the damping of high frequencies. Additionally, this justifies the approximation we introduced in Eq. (7.26) for $\mu_s'/\mu_{\text{tr}} \simeq 1$. As an important note, it is quite common to assume that when the point-source approximation breaks down so does the diffusion approximation. As we have seen this is clearly not the case and it points to the fact that if the limits of the diffusion approximation need to be established they should be tested against the complete solution for a collimated source, Eq. (7.21).

7.4   Accounting for the Source Profile

In the preceding sections we derived the expression for either a point source or a pencil beam of excitation. In both these cases, we assumed a delta function for the width of the collection of sources. Using the same formulation used in this chapter it is straight forward to introduce any source profile in order to more correctly model our sources. In particular, it is quite realistic to assume our source (a laser, for example) has a Guassian profile and to define our spot size in terms of its half-width.

Assuming, therefore, that we can model our source as a Gaussian beam of half-width w_g, we will assume we have our collimated source impinging at $z = 0$ and pointing into $+z$ in an otherwise infinite diffusive medium[1]. In this case its expression would be given by (compare with Eq. (7.19)):

$$S_\omega(\mathbf{R}, z) = \frac{S_0 \mu'_s}{\pi w_g^2} \exp\left(-(\mathbf{R} - \mathbf{R}_s)^2 / w_g^2\right) \exp(-\mu_{\text{tr}} z). \tag{7.28}$$

being its Fourier transform expressed as:

$$\widetilde{S}_\omega(\mathbf{K}) = \frac{1}{4\pi^2} \frac{S_0 \mu'_s}{\pi w_g^2} \exp(-\mu_{\text{tr}} z) \int_{-\infty}^{\infty} \exp(-\mathbf{R}^2 / w_g^2) \exp(-i\mathbf{K} \cdot \mathbf{R}) \mathrm{d}^2 R.$$

Since the Fourier transform of a Gaussian is another Gaussian of half-width $\widetilde{w}_g = 2/w_g$ (see Appendix A), the above equation becomes:

$$\widetilde{S}_\omega(\mathbf{K}) = \frac{1}{4\pi^2} S_0 \mu'_s \exp(-\mu_{\text{tr}} z) \exp(-\mathbf{K}^2 / \widetilde{w}_g^2) \exp(-i\mathbf{K} \cdot \mathbf{R}_s). \tag{7.29}$$

Clearly, following the same steps used to derive Eq. (7.21) we obtain:

$$\widetilde{U}_\omega(\mathbf{K}, z) = \widetilde{U}_\omega^{col}(\mathbf{K}, z) \exp(-\mathbf{K}^2 / \widetilde{w}_g^2), \tag{7.30}$$

where $\widetilde{U}_\omega^{col}$ represents the solution for collimated source with a delta profile, i.e Eq. (7.21). If we use the delta source approximation derived in Sec. 7.3 the above equation reduces to:

$$\widetilde{U}_\omega(\mathbf{K}, z) \simeq \frac{S_0}{4\pi D} \exp(-\mathbf{K}^2 / \widetilde{w}_g^2) \exp(-i\mathbf{K} \cdot \mathbf{R}_s) \times$$

$$\frac{i}{2\pi q(\mathbf{K})} \exp(iq(\mathbf{K})|z - z_s|), \tag{7.31}$$

with z_s given by $z_s = 1/\mu_{\text{tr}}$. In terms of the Green function the above becomes:

$$\widetilde{U}_\omega(\mathbf{K}, z) \simeq S_0 \exp(-\mathbf{K}^2 / \widetilde{w}_g^2) \exp(-i\mathbf{K} \cdot \mathbf{R}_s) \widetilde{g}(\mathbf{K}, |z - z_s|). \tag{7.32}$$

Note how, as expected, the expression for $U_\omega(\mathbf{r})$ in real-space amounts to a convolution of a point source with a Gaussian of half-width w_g.

Clearly any generic profile, $\widetilde{f}_{src}(\mathbf{K})$, for an isotropic source would yield an average intensity profile of the form:

$$\widetilde{U}_\omega(\mathbf{K}, z) \simeq S_0 \widetilde{f}_{src}(\mathbf{K}) \exp(-i\mathbf{K} \cdot \mathbf{R}_s) \widetilde{g}(\mathbf{K}, |z - z_s|), \tag{7.33}$$

and in the case of a collimated beam in the $+z$ direction:

$$\widetilde{U}_\omega(\mathbf{K}, z) = \widetilde{U}_\omega^{col}(\mathbf{K}, z) \widetilde{f}_{src}(\mathbf{K}),$$

with $\widetilde{U}_\omega^{col}$ given by the solution to Eq. (7.21), given by Eqs. (7.22)–(7.23).

---

[1] How the beam got there will be covered in the next section, once we take into account the boundary conditions.

Key points

From this chapter the following important points should be remembered:

- **Point source approximation:** The point source approximation is one of the approximations in diffuse light propagation most commonly used. Its limits of validity have been extensively studied both theoretically and experimentally and seems to break down only under extreme conditions. However, it is very important to understand where it comes from and what it represents: an approximation to the reduced intensity created by a collimated source. We saw how the intensity of this reduced intensity decays exponentially as it travels deeper into the medium. The fact that the point-source approximation fails, however, is not indicative of the diffusion approximation breaking down and must not be confused.

- **Source Profile:** We saw that it is straightforward to account in the Fourier domain for the source profile in any of our expressions for the average intensity, be it the point-source approximation or the solution to a collimated source.

Further Reading

This short chapter was centered on the derivation of the point source approximation, since this derivation is currently lacking in the literature. However, even though a thorough theoretical derivation is currently not easy to find, there are a great deal of research papers which put to test this approximation either with Monte Carlo results or with experiments. In particular, I suggest reading [Spott and Svaasand (2000)] from the group of B. Tromberg with respect to collimated sources and to follow the references therein. Additionally the solution for a collimated source this is represented as an integral in real-space in Farrell *et al.* (1992) which is a paper I strongly recommend generally to get acquainted with light reflectance measurements and more specifically to gain insight in this matter. With respect to what happens with the diffusion equation as we approach the sources, I suggest reading the works of the group of K. Rinzema, in particular [Graaff and Rinzema (2001)].

Chapter 8

Diffuse Light at Interfaces

Summary. In this chapter we present the boundary conditions for diffuse light both at a diffusive/diffusive interface (D-D) and a diffusive/non-diffusive interface (D-N), together with the expression of their reflection and transmission coefficients. These can be used to find the solution for any combination of diffuse slabs without any approximation, the basis of this derivation being the angular spectrum representation presented in Chap. 6. We will also consider non-contact geometries for sources and detectors.

8.1 Diffusive/Diffusive (D-D) Interfaces

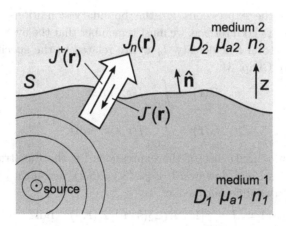

Fig. 8.1 Two semi-infinite diffusive media separated by an arbitrary surface S defined by its normal $\hat{n}$, with the flux contribution in upward, $J^+(\mathbf{r})$, and downward, $J^-(\mathbf{r})$, directions contributing to the total flux $J_n(\mathbf{r})$.

In order to model how diffuse light interacts in the presence of a boundary, we will first derive the boundary conditions and reflection and transmission coefficients for diffuse light at interfaces. For that, we will base ourself on the derivation of reflectivity and transmissivity of the electromagnetic field intensity, which we derived in the context of the radiative transfer equation in Sec. 3.5 of Chap. 3. In particular, assuming we have an arbitrary interface, S, defined by a unit normal $\hat{n}$ between two diffusive media (see Fig. 8.1) we will make use of Eqs. (3.52)–(3.53):

$$J^+(\mathbf{r}) = \int_{(2\pi)^+} [1 - R_{1\to 2}(\hat{\mathbf{s}} \cdot \hat{\mathbf{n}})]\, I(\mathbf{r}, \hat{\mathbf{s}})\hat{\mathbf{s}} \cdot \hat{\mathbf{n}}d\Omega, \tag{8.1}$$

$$J^-(\mathbf{r}) = \int_{(2\pi)^-} [1 - R_{2\to 1}(-\hat{\mathbf{s}} \cdot \hat{\mathbf{n}})]\, I(\mathbf{r}, \hat{\mathbf{s}})\hat{\mathbf{s}} \cdot (-\hat{\mathbf{n}})d\Omega, \tag{8.2}$$

where $R_{i\to j}$ represented the reflectivity while traveling from medium i to medium j through direction $\hat{\mathbf{s}}$ (see Eqs. (3.46)–(3.47)), in which case the angle of incidence is defined through $\hat{\mathbf{n}} \cdot \hat{\mathbf{s}} = \cos\theta$ and $(-\hat{\mathbf{n}}) \cdot \hat{\mathbf{s}} = -\cos\theta$, respectively. Note that this expression is defined locally at the interface and thus $\mathbf{r} \in S$, and we must assume a spatial dependence on the surface normal, $\hat{\mathbf{n}}(\mathbf{r})$. As we saw, $J^+(\mathbf{r})$ and $J^-(\mathbf{r})$ hold the following relationship (see Eq. (3.20) from Chap. 3):

$$J_n(\mathbf{r}) = J^+(\mathbf{r}) - J^-(\mathbf{r}). \quad \left(Watts/cm^2\right) \tag{8.3}$$

8.1.1 *D-D Boundary Conditions*

To arrive to the expressions for the boundary conditions for a diffusive/diffusive (D-D) interface, we must remember that the average intensity $U(\mathbf{r})$ and the total flux density $J_n(\mathbf{r})$ are related to the specific intensity $I(\mathbf{r}, \hat{\mathbf{s}})$ by (see Chap. 3):

$$U(\mathbf{r}) = \int_{(4\pi)} I(\mathbf{r}, \hat{\mathbf{s}})d\Omega,$$

$$J_n(\mathbf{r}) = \int_{(4\pi)} I(\mathbf{r}, \hat{\mathbf{s}})\hat{\mathbf{s}} \cdot \hat{\mathbf{n}}d\Omega.$$

We can now make use of the expressions for the flux traversing the interface upward and downward, Eqs. (8.1)–(8.2), accounting for the fact that we have two different media:

$$J^+(\mathbf{r}) = \int_{(2\pi)^+} [1 - R_{1\to 2}(\hat{\mathbf{s}} \cdot \hat{\mathbf{n}})]\, I_1(\mathbf{r}, \hat{\mathbf{s}})\hat{\mathbf{s}} \cdot \hat{\mathbf{n}}d\Omega, \tag{8.4}$$

$$J^-(\mathbf{r}) = \int_{(2\pi)^-} [1 - R_{2\to 1}(-\hat{\mathbf{s}} \cdot \hat{\mathbf{n}})]\, I_2(\mathbf{r}, \hat{\mathbf{s}})\hat{\mathbf{s}} \cdot (-\hat{\mathbf{n}})d\Omega, \tag{8.5}$$

We will now define the average intensity in the upper and lower media as U_1 and U_2, respectively, with the total flux traversing the interface given by $J_n(\mathbf{r})$ (see Eq. (8.3)). In order to obtain the boundary condition within the diffusion approximation we will follow the steps presented in [Aronson (1995)] and will introduce into the equations above the expression for the diffuse component of the specific intensity (see Chap. 4):

$$I_1(\mathbf{r}, \hat{\mathbf{s}}) \simeq \frac{1}{4\pi} U_1(\mathbf{r}) + \frac{3}{4\pi} J_n(\mathbf{r}) \hat{\mathbf{n}} \cdot \hat{\mathbf{s}}, \tag{8.6}$$

$$I_2(\mathbf{r}, \hat{\mathbf{s}}) \simeq \frac{1}{4\pi} U_2(\mathbf{r}) + \frac{3}{4\pi} J_n(\mathbf{r}) \hat{\mathbf{n}} \cdot \hat{\mathbf{s}}. \tag{8.7}$$

in which case we obtain:

$$J^+(\mathbf{r}) = \frac{U_1(\mathbf{r})}{4\pi} \int_{(2\pi)^+} [1 - R_{1 \to 2}(\hat{\mathbf{s}} \cdot \hat{\mathbf{n}})] \hat{\mathbf{s}} \cdot \hat{\mathbf{n}} d\Omega +$$
$$\frac{J_n(\mathbf{r})}{4\pi} \int_{(2\pi)^+} 3 [1 - R_{1 \to 2}(\hat{\mathbf{s}} \cdot \hat{\mathbf{n}})] (\hat{\mathbf{s}} \cdot \hat{\mathbf{n}})^2 d\Omega, \tag{8.8}$$

for the flux traversing upwards and:

$$J^-(\mathbf{r}) = \frac{U_2(\mathbf{r})}{4\pi} \int_{(2\pi)^-} [1 - R_{2 \to 1}(-\hat{\mathbf{s}} \cdot \hat{\mathbf{n}})] \hat{\mathbf{s}} \cdot (-\hat{\mathbf{n}}) d\Omega -$$
$$\frac{J_n(\mathbf{r})}{4\pi} \int_{(2\pi)^-} 3 [1 - R_{2 \to 1}(-\hat{\mathbf{s}} \cdot \hat{\mathbf{n}})] (\hat{\mathbf{s}} \cdot \hat{\mathbf{n}})^2 d\Omega, \tag{8.9}$$

for the flux traversing downwards. In both equations we can recognize the following integrals, accounting for the fact that the solid angle can be written as $d\Omega = d(\cos\theta)d\phi$ and that there is a symmetry in ϕ (i.e. $d\Omega = 2\pi d(\cos\theta)$):

$$T_U^{j \to k} = \int_0^1 [1 - R_{j \to k}(\mu)] \mu d\mu, \tag{8.10}$$

$$T_J^{j \to k} = 3 \int_0^1 [1 - R_{j \to k}(\mu)] \mu^2 d\mu, \tag{8.11}$$

where we have introduced the change of variable $\mu = \cos\theta$. Note that in the index matched case the above reduce to $T_U^{j \to k} = 1/2$ and $T_J^{j \to k} = 1$. Using this relationship, the total flux, $J_n(\mathbf{r}) = J^+(\mathbf{r}) - J^-(\mathbf{r})$ becomes:

$$J_n(\mathbf{r}) = \frac{U_2(\mathbf{r})}{2} T_U^{2 \to 1} + \frac{J_n(\mathbf{r})}{2} T_J^{2 \to 1} - \frac{U_1(\mathbf{r})}{2} T_U^{1 \to 2} + \frac{J_n(\mathbf{r})}{2} T_J^{1 \to 2},$$

which results in $U_1 = U_2$ in the $n_1 = n_2$ case.

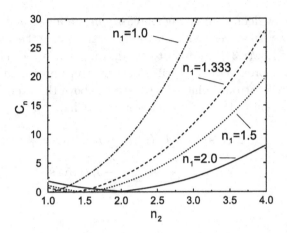

Fig. 8.2   Some values of C_n found numerically from Eq. (8.13).

We can now make use of the relationships for the reflectivity shown in Sec. 3.5 of Chap. 3, from which we find the following for $\mathcal{T}_U^{j\to k}$:

$$\mathcal{T}_U^{1\to 2} = \frac{n_2^2}{n_1^2}\mathcal{T}_U^{2\to 1}.$$

Introducing this into our expression for the total flux $J_n(\mathbf{r})$ we finally obtain:

$$U_2(\mathbf{r})\Big|_S - \left(\frac{n_2}{n_1}\right)^2 U_1(\mathbf{r})\Big|_S = C_n J_n(\mathbf{r}), \qquad (8.12)$$

where:

$$C_n = \frac{2 - \mathcal{T}_J^{2\to 1} - \mathcal{T}_J^{1\to 2}}{\mathcal{T}_U^{2\to 1}}, \qquad (8.13)$$

represents the index mismatch boundary coefficient for D-D interfaces. Accounting for Fick's law and flux conservation we additionally obtain:

$$J_n(\mathbf{r}) = -D_1\hat{\mathbf{n}}\cdot\nabla U_1(\mathbf{r})\Big|_S = -D_2\hat{\mathbf{n}}\cdot\nabla U_2(\mathbf{r})\Big|_S. \qquad (8.14)$$

Eqs. (8.12)–(8.14) represent the complete set of boundary conditions needed to solve any interaction at an arbitrary D-D interface.

   Building Block: **Boundary Conditions for D-D Interfaces**

These boundary conditions enable us to find the complete solution with arbitrary interfaces between two diffusive media with index of refraction mismatch.

An analytical expression for the boundary condition coefficient, C_n, shown in Eq. (8.13) can be found in [Aronson (1995)] as a polynomial expansion. Otherwise, a straightforward alternative is to calculate the $\mathcal{T}_U^{2\to1}$ and $\mathcal{T}_U^{2\to1}$ numerically (see Fig. 8.2), since their solution is very stable.

8.1.1.1 *Approximate Boundary Conditions*

Under certain assumptions, the boundary conditions across interfaces of index mismatched media given by Eqs. (8.12)–(8.14) can be simplified to the following:

$$U_2(\mathbf{r})\Big|_S \simeq \left(\frac{n_2}{n_1}\right)^2 U_1(\mathbf{r})\Big|_S,$$

$$-D_1\hat{\mathbf{n}} \cdot \nabla U_1(\mathbf{r})\Big|_S = -D_2\hat{\mathbf{n}} \cdot \nabla U_2(\mathbf{r})\Big|_S.$$

This approximation is valid as long as we can consider $(n_2/n_1)^2 U_1 \gg C_n J_n(\mathbf{r})$. The validity of this approximation is studied in [Ripoll and Nieto-Vesperinas (1999a)], and gives good results specifically when dealing with realistic values in biological tissue. As shown in [Bolin *et al.* (1989)], $1.3 < n < 1.5$ is the typical range of refractive indexes in biological media, and therefore the maximum expected value for $n_2 > n_1$ is $C_n \sim 5$. Note, however, that when working with scattering media such as phantom resins the value of the refractive index may be $n > 1.5$ in which case the above approximation might not hold.

8.1.1.2 *Index Matched Conditions*

In the particular case when we can consider two diffusive media with different optical properties but the same index of refraction we see that $C_n = 0$ (see Fig. 8.2) and we reach the simplest of boundary conditions:

$$U_2(\mathbf{r})\Big|_S = U_1(\mathbf{r})\Big|_S,$$

$$-D_1\hat{\mathbf{n}} \cdot \nabla U_1(\mathbf{r})\Big|_S = -D_2\hat{\mathbf{n}} \cdot \nabla U_2(\mathbf{r})\Big|_S.$$

8.1.2 *D-D Reflection and Transmission Coefficients*

In order to obtain the reflection and transmission coefficients for a D-D interface, we will again make use of the angular spectrum representation presented in Chap. 6:

$$U(\mathbf{R}, z) = \int_{-\infty}^{\infty} \mathcal{A}_{z_0}(\mathbf{K}) \exp(i\mathbf{K} \cdot \mathbf{R}) \exp(iq(\mathbf{K})(|z - z_0|)\mathrm{d}^2K,$$

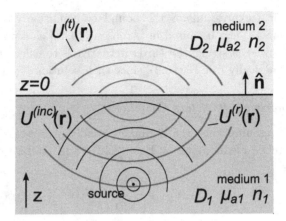

Fig. 8.3   Two semi-infinite diffusive media separated by a plane interface at $z = 0$, where U^{inc}, U^r and U^t represent the incident, reflected and transmitted average intensities, respectively.

where $\mathcal{A}_{z_0}$ is the angular spectrum at generic plane z_0. In order to use a notation similar to that use in classical optics, we will write the total average intensity in terms of an incident, a reflected, and a transmitted contribution.

Let us consider two semi-infinite media separated by a plane interface at $z = 0$ (see Fig. 8.3). For generality we will consider a point source with modulated intensity at frequency ω as the source distribution, always bearing in mind that, as we saw in the previous chapters (in particular see Chap. 7), we can account for any source distribution through a convolution with the solution for a point source. In other words, finding the solution for a point source gives us the Green function of the complete problem, i.e. the problem with an interface. Both media will have constant homogeneous but distinct optical properties given by μ_{a1}, D_1 and n_1 yielding the complex wavenumber κ_1 in the lower medium (see Fig. 8.3), and μ_{a2}, D_2 and n_2, which characterize the complex wavenumber κ_2 of the upper medium.

The point source is located at a distance z_s from the $z = 0$ plane, thus generating an incident diffuse wave which we will represent by U^{inc}. At any plane within medium 1, i.e. $z \leq 0$, we can express this incident wave by its angular spectrum representation of plane waves as:

$$U^{inc}(\mathbf{R}, z) = \int_{-\infty}^{\infty} \widetilde{U}^{(inc)}(\mathbf{K}, z) \exp(i\mathbf{K} \cdot \mathbf{R}) d^2 K, \forall z \leq 0$$

with $\widetilde{U}^{(inc)}$ given by:

$$\widetilde{U}^{(inc)}(\mathbf{K}, z) = \mathcal{A}_{z_s}(\mathbf{K}) \exp(iq(\mathbf{K})(|z - z_s|)), \forall z \leq 0 \qquad (8.15)$$

In the above equation $\mathcal{A}_{z_s}$ represents the angular spectrum of the incident wave (a point source in our case), and we must remember that $\exp(iq|z-z_s|)$ is the spatial transfer function. Similarly, we can define the reflected average intensity within medium 1 in K-space as:

$$\widetilde{U}^{(r)}(\mathbf{K}, z) = \mathcal{A}^{(r)}(\mathbf{K}) \exp(-iq(\mathbf{K})z), \forall z \leq 0 \qquad (8.16)$$

where $\mathcal{A}^{(r)}$ is the angular spectrum of the reflected average intensity defined at the interface, $z = 0$. The expression for the transmitted average intensity can be defined as:

$$\widetilde{U}^{(t)}(\mathbf{K}, z) = \mathcal{A}^{(t)}(\mathbf{K}) \exp(iq(\mathbf{K})z), \forall z \geq 0 \qquad (8.17)$$

where since z is defined for $z \geq 0$ we can remove the absolute value, $|z|$. It is important to understand that in order to reach this expression we have had to assume that the wave-vector projection on the z plane is the same for incident, reflected and transmitted intensities:

$$\mathbf{K}_i \cdot \mathbf{R} = \mathbf{K}_r \cdot \mathbf{R} = \mathbf{K}_t \cdot \mathbf{R}. \qquad (8.18)$$

Using the angular spectrum notation the total average intensity in the upper and lower media is defined respectively as:

$$\widetilde{U}_1(\mathbf{K}, z) = \widetilde{U}^{(inc)}(\mathbf{K}, z) + \widetilde{U}^{(r)}(\mathbf{K}, z), \qquad z \leq 0,$$
$$\widetilde{U}_2(\mathbf{K}, z) = \widetilde{U}^{(t)}(\mathbf{K}, z), \qquad z \geq 0.$$

If we now apply the boundary conditions we have recently derived, Eq. (8.12) and Eq. (8.14), at $z = 0$ we have:

$$\frac{n_2^2}{n_1^2}\left(\widetilde{U}^{(inc)}(\mathbf{K}, z)\Big|_{z=0} + \widetilde{U}^{(r)}(\mathbf{K}, z)\Big|_{z=0}\right) -$$
$$C_n D_1\left(\frac{\partial \widetilde{U}^{(inc)}(\mathbf{K}, z)}{\partial z}\Big|_{z=0} + \frac{\partial \widetilde{U}^{(r)}(\mathbf{K}, z)}{\partial z}\Big|_{z=0}\right) =$$
$$\widetilde{U}^{(t)}(\mathbf{K}, z)\Big|_{z=0}, \qquad (8.19)$$

for the boundary condition which accounts for the relationship between the average intensities in media 1 and 2, and:

$$D_1\left(\frac{\partial \widetilde{U}^{(inc)}(\mathbf{K}, z)}{\partial z}\Big|_{z=0} + \frac{\partial \widetilde{U}^{(r)}(\mathbf{K}, z)}{\partial z}\Big|_{z=0}\right) = D_2\frac{\partial \widetilde{U}^{(t)}(\mathbf{K}, z)}{\partial z}\Big|_{z=0}, \qquad (8.20)$$

which accounts for the flux continuity. We can now introduce the angular spectrum representation and apply the derivatives[1] in which case the

---

[1] Care must be taken with the sign of the derivative, always remembering that $\hat{n}$ points in the $+z$ direction. We must also account for the fact that the reflected wave propagates in the $-z$ direction, while both incident and reflected waves propagate in the $+z$ direction. This sign is of course reversed if $\hat{n}$ is changed.

boundary conditions translate into:

$$\frac{n_2^2}{n_1^2}\left(\mathcal{A}^{(inc)}(\mathbf{K}) + \mathcal{A}^{(r)}(\mathbf{K})\right) -$$

$$C_n D_1 \left(iq_1(\mathbf{K})\mathcal{A}^{(inc)}(\mathbf{K}) - iq_1(\mathbf{K})\mathcal{A}^{(r)}(\mathbf{K})\right) =$$

$$\mathcal{A}^{(t)}(\mathbf{K}), \quad (8.21)$$

and:

$$D_1\left(q_1(\mathbf{K})\mathcal{A}^{(inc)}(\mathbf{K}) - q_1(\mathbf{K})\mathcal{A}^{(r)}(\mathbf{K})\right) = D_2 q_2 \mathcal{A}^{(t)}(\mathbf{K}), \quad (8.22)$$

where $\mathcal{A}^{(inc)}(\mathbf{K})$ represents the angular spectrum of the incident average intensity at the interface:

$$\mathcal{A}^{(inc)}(\mathbf{K}) = \widetilde{U}^{(inc)}(\mathbf{K}, z = 0) = A_{z_s}(\mathbf{K}) \exp(iq(\mathbf{K})|z_s|).$$

In Eqs. (8.21)–(8.22) we have introduced the notation:

$$q_1 = \sqrt{\kappa_1^2 - \mathbf{K}^2},$$

$$q_2 = \sqrt{\kappa_2^2 - \mathbf{K}^2},$$

which represents the spatial frequency wave-vector in the z-direction (see Chap. 6) in medium 1 and 2 respectively pointing in the $+z$ direction (note that for propagation in the $-z$ direction the negative expression for q must be chosen — see Chap. 6. Eq. (6.12)). After some basic algebra, Eqs. (8.21)–(8.22) reduce to:

$$\mathcal{A}^{(r)}(\mathbf{K}) = \frac{D_1 q_1(\mathbf{K}) n_1^2 \left(1 + iC_n D_2 q_2(\mathbf{K})\right) - n_2^2 D_2 q_2(\mathbf{K})}{D_1 q_1(\mathbf{K}) n_1^2 \left(1 + iC_n D_2 q_2(\mathbf{K})\right) + n_2^2 D_2 q_2(\mathbf{K})} \mathcal{A}^{(inc)}(\mathbf{K}),$$

$$\mathcal{A}^{(t)}(\mathbf{K}) = \frac{2 n_1^2 D_1 q_1(\mathbf{K})}{D_1 q_1(\mathbf{K}) n_1^2 \left(1 + iCn D_2 q_2(\mathbf{K})\right) + n_2^2 D_2 q_2(\mathbf{K})} \mathcal{A}^{(inc)}(\mathbf{K}),$$

equations which relate the angular spectrum of the reflected and transmitted average intensities to that of the incident average intensity. Making use of these expressions, we can therefore define the reflection and transmission coefficients for D-D interfaces as:

$$\mathcal{R}_{dd}(\mathbf{K}) = \frac{D_1 q_1(\mathbf{K}) n_1^2 \left(1 + iC_n D_2 q_2(\mathbf{K})\right) - n_2^2 D_2 q_2(\mathbf{K})}{D_1 q_1(\mathbf{K}) n_1^2 \left(1 + iC_n D_2 q_2(\mathbf{K})\right) + n_2^2 D_2 q_2(\mathbf{K})}, \quad (8.23)$$

$$\mathcal{T}_{dd}(\mathbf{K}) = \frac{2 n_1^2 D_1 q_1(\mathbf{K})}{D_1 q_1(\mathbf{K}) n_1^2 \left(1 + iC_n D_2 q_2(\mathbf{K})\right) + n_2^2 D_2 q_2(\mathbf{K})}. \quad (8.24)$$

Building Block: **D-D Reflection and Transmission**

The reflection and transmission coefficients enable us not only to find the solution for any collection of planar interfaces, but give us intrinsic information on how much light is transmitted or reflected in a diffusive medium when encountering an interface such as an object.

Eqs. (8.23)–(8.24) have been defined so that $\mathcal{A}^{(t)}$ and $\mathcal{A}^{(r)}$ hold the following relationship with $\mathcal{A}^{(inc)}$:

$$\mathcal{A}^{(r)}(\mathbf{K}) = \mathcal{R}_{dd}(\mathbf{K})\mathcal{A}^{(inc)}(\mathbf{K}), \qquad (8.25)$$

$$\mathcal{A}^{(t)}(\mathbf{K}) = \mathcal{T}_{dd}(\mathbf{K})\mathcal{A}^{(inc)}(\mathbf{K}). \qquad (8.26)$$

Note that in the case where we have the surface normal $\hat{n}$ pointing downward the expressions for $\mathcal{R}_{dd}$ and $\mathcal{T}_{dd}$ are identical but replacing $(1 + iC_n D_2 q_2(\mathbf{K}))$ with $(1 - iC_n D_2 q_2(\mathbf{K}))$, in which case we reach the expression derived in [Ripoll and Nieto-Vesperinas (1999b)].

As can be seen, Eqs. (8.23)–(8.24) obey the conservation of total flux:

$$\mathcal{R}_{dd}(\mathbf{K}) + \frac{D_2 q_2(\mathbf{K})}{D_1 q_1(\mathbf{K})}\mathcal{T}_{dd}(\mathbf{K}) = 1.$$

Note that in the index matched case, $n_1 = n_2$, $\mathcal{R}_{dd}$ and $\mathcal{T}_{dd}$, also hold the following relationship:

$$\mathcal{T}_{dd}(\mathbf{K}) = \mathcal{R}_{dd}(\mathbf{K}) + 1.$$

In terms of these reflection and transmission coefficients the total average intensity reflected into medium 1 can be expressed as:

$$\widetilde{U}^{(r)}(\mathbf{K}, z) = \mathcal{R}_{dd}(\mathbf{K})\mathcal{A}^{(inc)}(\mathbf{K}) \exp(-iq_1(\mathbf{K})z), \, \forall z \leq 0 \qquad (8.27)$$

with the total average intensity inside medium 1 being given by:

$$\widetilde{U}_1(\mathbf{K}, z) = \widetilde{U}^{(inc)}(\mathbf{K}, z) + \mathcal{R}_{dd}(\mathbf{K})\mathcal{A}^{(inc)}(\mathbf{K}) \exp(-iq_1(\mathbf{K})z). \, \forall z \leq 0$$

Similarly, the total average intensity transmitted into medium 2 expressed in terms of the transmission coefficients becomes:

$$\widetilde{U}^{(t)}(\mathbf{K}, z) = \mathcal{T}_{dd}(\mathbf{K})\mathcal{A}^{(inc)}(\mathbf{K}) \exp(iq_2(\mathbf{K})z). \, \forall z \geq 0 \qquad (8.28)$$

The dependence of the reflection and transmission coefficients for D-D interfaces is shown in Fig. 8.4, where two clear cases can be seen: in the case where medium 2 has a stronger absorption, the reflection coefficient is *negative* indicating that the total average intensity in medium 1 is lower in the presence of medium 2 than in its absence (i.e. if medium 1 were infinite).

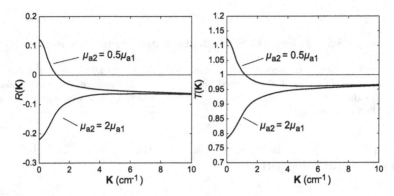

Fig. 8.4 Reflection and Transmission coefficients for D-D interfaces $\mu'_{s1} = \mu'_{s2} = 10cm^{-1}$, $\mu_a 1 = 0.025cm^{-1}$, $n1 = 1.333$, $n2 = 1.4$ for the cw case. Two plots are represented, one for $\mu_{a2} = 2\mu_{a1}$ and the other for $\mu_{a2} = \mu_{a1}/2$.

In the opposite case, if the second medium has a lower absorption than the first medium then the reflected intensity would be greater, indicating that we have more light in medium 1 now that the overall absorption is lower due to the presence of medium 2. Note also that for the $\mu_{a2} = 2\mu_{a1}$ case shown in Fig. 8.4 the reflection coefficient is zero for a certain $\mathbf{K}$ value. This zero reflection frequency is obtained by setting $\mathcal{R}_{dd}(\mathbf{K}) = 0$, and can be approximated to:

$$\mathbf{K}_{\text{zero}} \simeq \sqrt{\frac{n_1^4 D_1 \mu_{a1} - n_0^4 D_0 \mu_{a0}}{n_0^4 D_0^2 - n_1^4 D_1^2}}. \tag{8.29}$$

Note that the condition for zero reflectivity is qualitatively analogous to the Brewster modes (see [Born and Wolf (1999)]) we find in electromagnetism.

8.1.2.1 *Approximate reflection and transmission coefficients*

We can follow the same derivation as above but using the approximated boundary conditions we presented in Sec. 8.1.1.1. In this case we obtain:

$$\mathcal{R}_{dd}(\mathbf{K}) \simeq \frac{n_1^2 D_1 \sqrt{\kappa_1^2 - \mathbf{K}^2} - n_2^2 D_2 \sqrt{\kappa_2^2 - \mathbf{K}^2}}{n_1^2 D_1 \sqrt{\kappa_1^2 - \mathbf{K}^2} + n_2^2 D_2 \sqrt{\kappa_2^2 - \mathbf{K}^2}}$$

$$\mathcal{T}_{dd}(\mathbf{K}) \simeq \frac{2n_1^2 D_1 \sqrt{\kappa_1^2 - \mathbf{K}^2}}{n_1^2 D_1 \sqrt{\kappa_1^2 - \mathbf{K}^2} + n_2^2 D_2 \sqrt{\kappa_2^2 - \mathbf{K}^2}},$$

where we have expressed q_1 and q_2 explicitly. It is from these expressions that we obtained the expression for the zero reflection frequency $\mathbf{K}_{\text{zero}}$ presented in Eq. (8.29).

8.1.2.2 *Snell's law for Diffuse Waves*

As we saw on Eq. (8.18), in order to reach the reflection and transmission coefficients we had to assume that the transversal component of the wave-vector in both media is equivalent. This is necessary for conservation of momentum, analogous to the case in classical optics. It is from this equivalence that the law of reflection and Snell's law are obtained since we can write the projection of the wave-vector $\mathbf{k}$ onto the z plane as $\mathbf{K} = \kappa_0 \sin\theta$ with θ representing a complex angle, as we saw in Chap. 6. Represented this way, the law of reflection for diffuse waves becomes:

$$\mathbf{K}_i \cdot \mathbf{R} = \mathbf{K}_r \cdot \mathbf{R},$$
$$\kappa_1 \sin\theta_i = -\kappa_1 \sin\theta_r,$$

while Snell's law becomes:

$$\mathbf{K}_i \cdot \mathbf{R} = \mathbf{K}_t \cdot \mathbf{R},$$
$$\kappa_1 \sin\theta_i = \kappa_2 \sin\theta_t,$$

If in Snell's law for diffuse waves one assumes that both media have the same refractive index n and the same absorption coefficient μ_a, we obtain the expression presented in the classical paper on diffuse photon density waves, [O'Leary *et al.* (1992)]:

$$\frac{\sin\theta_t}{\sin\theta_i} = \sqrt{\frac{D_2}{D_1}}.$$

In the more general case, Snell's law can be represented as:

$$\frac{\sin\theta_t}{\sin\theta_i} = \sqrt{\frac{D_2}{D_1}} \sqrt{\frac{\mu_{a1}c - i\omega n_1}{\mu_{a2}c - i\omega n_2}},$$

which for most experimental values of the modulation frequency can be approximated to:

$$\frac{\sin\theta_t}{\sin\theta_i} \simeq \sqrt{\frac{D_2 \mu_{a1}}{D_1 \mu_{a2}}}.$$

8.1.3 *Frequency independent coefficients*

A very interesting case appears in the *cw* regime when we have index-matched boundary conditions. If the following relation between the optical properties of both media exists:

$$\frac{D_1}{D_2} = \frac{\mu_{a1}}{\mu_{a2}} = \gamma,$$

we then obtain that the wavenumber in each medium is *identical* even though both media are optically *different*. In this case, the expressions for the reflection and transmission coefficients for a D-D become frequency independent:

$$\mathcal{R}_{dd}(\mathbf{K}) = \mathcal{R}_{dd} = \frac{\gamma - 1}{\gamma + 1},$$

$$\mathcal{T}_{dd}(\mathbf{K}) = \mathcal{T}_{dd} = \frac{2\gamma}{\gamma + 1}.$$

This means that all frequency components incident on the interface are equally reflected and transmitted, independently of their frequency value. In practical terms, when having frequency independent coefficients we can expect the interface to act as diffusive mirror that is not a perfect reflector but that maintains the angular spectrum, just as a mirror does for electromagnetic waves. Also, if γ is replaced by $1/\gamma$, we obtain a 180° phase change in the reflected wave, the amplitude remaining unchanged. The transmitted wave would then travel inside the lower medium as if there were no interface, but with a lower amplitude reduced by the factor $\mathcal{T}_{dd}$, since it would then have the same phase as the incident wave. This frequency independent coefficient effect has been studied for electromagnetic waves (see for example [M. Lester (1996)]).

8.2   Diffusive/Non-diffusive (D-N) Interfaces

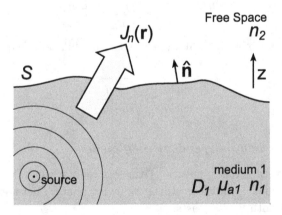

Fig. 8.5   A semi-infinite diffusive medium separated by an arbitrary surface S defined by its normal $\hat{\mathbf{n}}$, with only upward flux contributing to the total flux $J_n(\mathbf{r})$.

In order to find the boundary conditions and the reflection and transmission coefficients for Diffusive/Non-Diffusive (D-N) interfaces we can follow exactly the same steps as in the previous section, with the only difference that we will consider all light sources inside the diffusive medium (see Fig. 8.5). This effectively means that all the flux traversing the interface is due the flux traversing it upward, $J^+(\mathbf{r})$:

$$J_n(\mathbf{r}) = J^+(\mathbf{r}), \qquad \left(Watts/cm^2 \right) \tag{8.30}$$

with $J^+(\mathbf{r})$ given by Eq. (8.1).

8.2.1   *D-N Boundary Conditions*

Using the expressions we derived for the D-D interfaces, the flux traversing the interface upward can be written as (see Eq. (8.8)):

$$J^+(\mathbf{r}) = \frac{U_1(\mathbf{r})}{2}\mathcal{T}_U^{1\to2} + \frac{J_n(\mathbf{r})}{2}\mathcal{T}_J^{1\to2}, \tag{8.31}$$

and thus the expression for the total flux, Eq. (8.30), becomes:

$$J_n(\mathbf{r}) = \frac{U_1(\mathbf{r})}{2}\mathcal{T}_U^{1\to2} + \frac{J_n(\mathbf{r})}{2}\mathcal{T}_J^{1\to2}.$$

Grouping terms, we obtain:

$$U_1(\mathbf{r}) = \alpha J_n(\mathbf{r}), \tag{8.32}$$

where α is defined as:

$$\alpha = \frac{2 - \mathcal{T}_J^{1\to2}}{\mathcal{T}_U^{1\to2}}, \tag{8.33}$$

representing the boundary coefficient for D-N interfaces. Note that in the index-matched case, $n_1 = n_2$ we obtain $\alpha = 2$. Eq. (8.32) is termed the partial flux boundary condition and through Fick's law we can finally obtain an expression for the boundary condition of a D-N interface:

$$U_1(\mathbf{r})\Big|_S = -\alpha D_1 \nabla U_1(\mathbf{r})\Big|_S, \tag{8.34}$$

which in mathematics is termed a Robin boundary condition or third type boundary condition.

 Building Block: **Boundary Condition for D-N Interfaces**

This boundary condition enables us to find the complete solution with arbitrary interfaces between a diffusive medium and a non-diffusion medium (i.e. free space).

8.2.1.1 *The Extrapolated Boundary Condition*

Using the fact that the diffusion coefficient can be written as $D_1 = l_{\text{tr}}/3$, being l_{tr} the transport mean free path, we can express this boundary condition as:

$$U_1(\mathbf{r})\Big|_S = -\frac{\alpha}{3}l_{\text{tr}}\frac{\partial U_1(\mathbf{r})}{\partial \hat{\mathbf{n}}}\Big|_S. \tag{8.35}$$

The quantity $\alpha l_{tr}/3$ is usually called the extrapolated distance:

$$z_e = \alpha D_1 = \frac{\alpha}{3}l_{\text{tr}},$$

which in the index matched case, $n_1 = n_2$, reduces to:

$$z_e = \frac{2}{3}l_{\text{tr}}.$$

It must be stated that we have reached these values without any approximation, other than the diffusion approximation itself. However, the solution for the full RTE can be obtained for an index matched semi-infinite medium (termed the Milne problem) in which case we reach an expression equivalent to the extrapolated distance as $z_e \simeq 0.71041 l_{\text{tr}}$. This expression is slightly different to the $z_e = 0.6666 l_{\text{tr}}$ we obtain within the diffusion approximation and has been the reason for many studies regarding the validity of the diffusion approximation at boundaries. We will not dwell on this further, but always bear in mind when finding the limits of validity of the diffusion approximation by comparing it to the RTE that the RTE has limitations of its own.

The reason why the quantity z_e is usually termed the extrapolated distance is that in the case of a plane interface (at $z = 0$ for example) Eq. (8.35) becomes:

$$U_1(\mathbf{R}, z)\Big|_{z=0} = -\frac{\alpha}{3}l_{\text{tr}}\frac{\mathrm{d}U_1(\mathbf{R}, z)}{\mathrm{d}z}\Big|_{z=0},$$

which has a solution of the form:

$$U_1(\mathbf{R}, z) \sim \exp(-z/z_e).$$

We can approximate the exponential to $\exp(z/z_e) \simeq 1 - z/z_e + \mathcal{O}(z/z_e)^2$, valid as long as $z < z_e$:

$$U_1(\mathbf{R}, z) \sim U_1(\mathbf{R}, z = 0)\Big(1 - \frac{z}{z_e}\Big), \forall z < z_e$$

Clearly, this solution expresses that the average intensity U_1 is zero at the extrapolated distance z_e. Hence, an approximation to the Robin boundary condition presented in Eq. (8.34) can be expressed as:

$$U_1(\mathbf{R}, z)\Big|_{z=z_e} = 0. \qquad (8.36)$$

which is termed the extrapolated boundary condition.

 Building Block: **Extrapolated Boundary Condition**

The extrapolated boundary condition is very useful since it is very simple to implement and multiple interfaces can be introduced by using the method of images.

The extrapolated boundary condition is particularly useful since we can use the method of images originally derived for finding charge distribution in electrostatics (see [Jackson (1962)] for example), by distributing sources and light sinks so that the total average intensity is zero at the extrapolated boundary. This is the approach used in one of the pioneering papers on characterization of diffuse media by M. Patterson, B. Chance and B. Wilson [Patterson *et al.* (1989)]. For those cases when we can measure far away from the boundary (using a detector fiber, for example), it is possible to approximate the extrapolated boundary condition even further by assuming that the average intensity is zero at the boundary, i.e. $z_e \simeq 0$, further simplifying the derivation for multiple interfaces.

 Note: **About The Extrapolated Distance**

As we have seen, the extrapolated boundary condition consists on assuming that the intensity is zero at the extrapolated distance. It is very important to understand that this is an approximation to the complete Robin boundary condition, yielding equivalent values for the average intensity inside the diffusive medium. It does not mean in any way that the average intensity is in reality zero anywhere outside the diffusive medium. On the contrary, by means of the Robin boundary condition we have an expression for the flux traversing the diffusive/non-diffusive interface which we can use to obtain an expression for what the average intensity will be anywhere in free space. The only case where it would be zero is if there is a perfect absorber at the physical boundary.

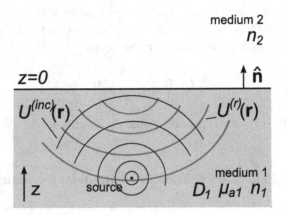

Fig. 8.6   A semi-infinite diffusive medium separated by a plane interface at $z = 0$, where U^{inc}, and U^r represent the incident and reflected average intensity, respectively.

8.2.2   *D-N Reflection and Transmission Coefficients*

Once we have established the boundary condition of D-N interfaces, we can proceed in exactly the same way we did for the D-D interfaces in order to obtain the expression for the reflection and transmission coefficients. Defining the average intensity in the angular spectrum representation in terms of the incident and reflected contributions the expressions for a planar interface at $z = 0$ become:

$$\tilde{U}_1(\mathbf{K}, z) = \tilde{U}^{(inc)}(\mathbf{K}, z) + \tilde{U}^{(r)}(\mathbf{K}, z), \forall z \leq 0.$$

If we now apply the Robin boundary conditions we have just derived, Eq. (8.34) at $z = 0$ we have:

$$\tilde{U}^{(inc)}(\mathbf{K}, z = 0) + \tilde{U}^{(r)}(\mathbf{K}, z = 0) = \alpha \tilde{J}_n(\mathbf{K}, z = 0),$$

with the total flux traversing the interface given by Fick's law:

$$\tilde{J}_n(\mathbf{K}, z = 0) = -D_1 \frac{\mathrm{d}}{\mathrm{d}z} \left(\tilde{U}^{(inc)}(\mathbf{K}, z) + \tilde{U}^{(r)}(\mathbf{K}, z) \right) \Bigg|_{z=0}.$$

We can now introduce the angular spectrum representation and apply the derivatives, again taking care of the sign of the derivative since the incident and reflected average intensities propagate in opposite directions:

$$\mathcal{A}^{(inc)}(\mathbf{K}) + \mathcal{A}^{(r)}(\mathbf{K}) =$$
$$- D_1 \left(i q_1(\mathbf{K}) \mathcal{A}^{(inc)}(\mathbf{K}) - i q_1(\mathbf{K}) \mathcal{A}^{(r)}(\mathbf{K}) \right), \quad (8.37)$$

where as a reminder $\mathcal{A}^{(inc)}(\mathbf{K})$ represents the angular spectrum of the incident average intensity at the interface:

$$\mathcal{A}^{(inc)}(\mathbf{K}) = \widetilde{U}^{(inc)}(\mathbf{K}, z = 0) = \mathcal{A}_{z_s}(\mathbf{K}) \exp(iq(\mathbf{K})|z_s|).$$

Grouping terms we finally obtain the following expressions for the reflection and transmission coefficients for D-N interfaces:

$$\mathcal{R}_{dn}(\mathbf{K}) = \frac{i\alpha D_1 q_1(\mathbf{K}) + 1}{i\alpha D_1 q_1(\mathbf{K}) - 1}, \tag{8.38}$$

$$\mathcal{T}_{dn}(\mathbf{K}) = \frac{2i\alpha D_1 q_1(\mathbf{K})}{i\alpha D_1 q_1(\mathbf{K}) - 1}, \tag{8.39}$$

$\mathcal{R}_{dn}$ and $\mathcal{T}_{dn}$ representing the frequency dependent reflection and transmission coefficients for D-N, respectively.

 Building Block: **D-N Reflection and Transmission**

The reflection and transmission coefficients for diffusive/non-diffusive interfaces will enable us to establish both the total flux which leaves the diffuse medium and the effect of the boundary on the average intensity distribution within the diffusive medium.

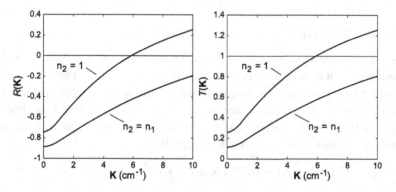

Fig. 8.7 Reflection and Transmission coefficients for an D-N interface between a diffusive medium of optical properties given by $\mu'_{s1} = 10cm^{-1}$, $\mu_{a1} = 0.025cm^{-1}$, and $n1 = 1.333$ for the *cw* case. Two plots are represented, one for for the index-matched case, $n_2 = n_1$, and the other for the outer medium having a refractive index of $n_2 = 1$.

As in the case for D-D interfaces both $\mathcal{R}_{nd}$ and $\mathcal{T}_{nd}$ are complex, and the sum of their moduli is not unity, holding the following relationship:

$$\mathcal{T}_{dn}(\mathbf{K}) = \mathcal{R}_{dn}(\mathbf{K}) + 1.$$

In terms of the reflection and transmission coefficients the total average intensity reflected into the diffusive medium can be expressed as:

$$\widetilde{U}^{(r)}(\mathbf{K}, z) = \mathcal{R}_{dn}(\mathbf{K})\mathcal{A}^{(inc)}(\mathbf{K}) \exp(-iq_1(\mathbf{K})z). \, \forall z \leq 0 \qquad (8.40)$$

With regards to the transmitted intensity, the important quantity we can relate with measurements is the total flux $\widetilde{J}_n$ rather than the total transmitted average intensity. Using the boundary condition, Eq. (8.32), the flux traversing the interface becomes:

$$\widetilde{J}_n(\mathbf{K}, z = 0) = \frac{1}{\alpha}\Big(\mathcal{A}^{(inc)}(\mathbf{K}) + \mathcal{R}_{dn}(\mathbf{K})\mathcal{A}^{(inc)}(\mathbf{K})\Big), \forall z \leq 0$$

or equivalently:

$$\widetilde{J}_n(\mathbf{K}, z = 0) = \frac{1}{\alpha}\mathcal{T}_{dn}(\mathbf{K})\mathcal{A}^{(inc)}(\mathbf{K}). \qquad (8.41)$$

As in the case for D-D interfaces, we can also find a value for which the reflection coefficient is zero, i.e. a zero-reflection frequency. This only takes place in the *cw* regime for which we find the the following expression for $\mathbf{K}_{\text{zero}}$:

$$\mathbf{K}_{\text{zero}} = \sqrt{\left(\frac{1}{\alpha D_1}\right)^2 - \frac{\mu_{a1}}{D_1}}. \qquad (8.42)$$

8.2.2.1 *Black interface*

As a particular case of a D-N, we could consider one which is a perfect absorber, such a black slab or a piece of black material. In this case we can consider that the non-scattering medium absorbs all light which reaches the interface and the total intensity at the interface can be considered to be zero. In this case we obtain that the relationship between the incident and reflected average intensities must be such that:

$$\widetilde{U}_1(\mathbf{K}, z = 0) = \widetilde{U}^{(inc)}(\mathbf{K}, z = 0) + \widetilde{U}^{(r)}(\mathbf{K}, z = 0) = 0,$$

which results in:

$$\widetilde{U}^{(r)}(\mathbf{K}, z = 0) = -\widetilde{U}^{(inc)}(\mathbf{K}, z = 0).$$

This would yield a constant reflection coefficient given by:

$$\mathcal{R}_{\text{black}}(\mathbf{K}) = -1,$$

equivalent to introducing $\alpha = 0$ in Eqs. (8.38)–(8.39).

8.3 Layered Diffusive Media

Once the reflection and transmission coefficients have been established for diffusive/diffusive (D-D) and diffusive/non-diffusive interfaces (D-N) in Sec. 8.1.2 and Sec. 8.2.2 respectively, we can use these coefficients to solve multiple layered media without any approximation (rather than the diffusion approximation itself), and in the limiting case solve for smoothly varying optical parameters. Before moving to the more general case of M slabs, we will first obtain the expression for a slab, equivalent to a three-layered medium.

8.3.1 *Expression for a Slab in a Diffusive medium*

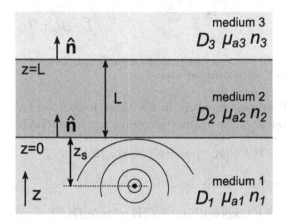

Fig. 8.8   Configuration for a slab of width L located at $0 \leq z \leq L$, where three different homogeneous media are distinguished.

So as to find the solution for a slab, we will consider the configuration depicted in Fig. 8.8, which consists of a slab of width L, located at $0 < z < L$, with certain optical properties D_2, μ_{a2} and n_2. This slab is flanked by two semi-infinite diffusive media at $z < 0$ and $z > L$ with optical properties given by D_1, μ_{a1}, n_1, and D_3, μ_{a3}, n_3, respectively. We will always consider the normal $\hat{n}$ at all the interfaces pointing in the z direction, i.e. $\hat{n} = (0, 0, 1)$. For generality we will consider a point source at $z = -z_s$, i.e. at distance of z_s from the first interface at $z = 0$.

In order to solve for this geometry, we will use the same approach used in physical optics to find the solution to a layered medium adding the multiple reflections from the boundaries in a series. Another option is to solve the system of equations, approach which we will use in Sec. 8.4 to solve for M slabs, but that provides less physical insight.

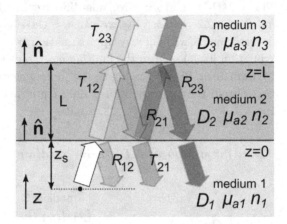

Fig. 8.9 Multiple reflections present at the diffusive slab, and how they contribute to the total reflected and transmitted wave.

The total average intensity in the three regions we have defined (see Fig. 8.8) in terms of the angular spectrum representation is given by:

$$\widetilde{U}_1(\mathbf{K}, z) = \widetilde{U}^{(inc)}(\mathbf{K}, z) + \mathcal{A}(\mathbf{K}) \exp(iq_1(\mathbf{K})|z + z_s|), \forall z < 0, \qquad (8.43)$$

for the first medium which contains the source,

$$\widetilde{U}_2(\mathbf{K}, z) = \mathcal{B}(\mathbf{K}) \exp(iq_2(\mathbf{K})z) + \mathcal{C}(\mathbf{K}) \exp(iq_2(\mathbf{K})(L - z)), \forall 0 \le z \le z, \qquad (8.44)$$

for the inside the slab and

$$\widetilde{U}_3(\mathbf{K}, z) = \mathcal{D}(\mathbf{K}) \exp(iq_3(\mathbf{K})(z - L)), \forall z \ge L \qquad (8.45)$$

for the medium where the intensity transmitted from the slab will propagate. Note that the z-component of the wave-vectors, q_i, have their usual meaning:

$$q_1(\mathbf{K}) = \sqrt{\kappa_1^2 - |\mathbf{K}|^2},$$

$$q_2(\mathbf{K}) = \sqrt{\kappa_2^2 - |\mathbf{K}|^2},$$

$$q_3(\mathbf{K}) = \sqrt{\kappa_3^2 - |\mathbf{K}|^2}.$$

Considering all multiple reflections from the interfaces (see Fig. 8.9), the total reflection and transmission coefficients for the slab are represented by:

$$\mathcal{R}^{Slab} = \mathcal{R}_{12} + \mathcal{T}_{12}\mathcal{R}_{23}\exp(2iq_2L)\mathcal{T}_{21} +$$
$$\mathcal{T}_{12}\mathcal{R}_{23}\exp(2iq_2L)\mathcal{R}_{21}\mathcal{R}_{23}\exp(2iq_2L)\mathcal{T}_{21} + ..., \quad (8.46)$$

for the total reflection of the slab, and:

$$\mathcal{T}^{Slab} = \mathcal{T}_{12}\exp(iq_2L)\mathcal{T}_{21} +$$
$$\mathcal{T}_{12}\mathcal{R}_{23}\exp(2iq_2L)\mathcal{R}_{21}\exp(iq_2L)\mathcal{T}_{21} + ..., \quad (8.47)$$

for its transmission, where $\mathcal{R}_{ij}$ and $\mathcal{T}_{ij}$ are the reflection and transmission coefficients while traversing the interface from medium i onto medium j defined by Eqs. (8.23)–(8.24). As can be seen, the above equations contain a geometric progression from which we obtain:

$$\mathcal{R}^{\text{Slab}}(\mathbf{K}) = \mathcal{R}_{12} + \frac{\mathcal{T}_{12}\exp(2iq_2L)\mathcal{R}_{23}\mathcal{T}_{21}}{1 - \mathcal{R}_{21}\mathcal{R}_{23}\exp(2iq_1L)}, \quad (8.48)$$

$$\mathcal{T}^{\text{Slab}}(\mathbf{K}) = \frac{\mathcal{T}_{12}\exp(iq_2L)\mathcal{T}_{23}}{1 - \mathcal{R}_{21}\mathcal{R}_{23}\exp(2iq_2L)}. \quad (8.49)$$

Using this approach $\mathcal{A}$, $\mathcal{B}$, $\mathcal{C}$ and $\mathcal{D}$ can be found as:

$$\mathcal{A}(\mathbf{K}) = \mathcal{R}^{\text{Slab}}(\mathbf{K})\mathcal{A}^{(inc)}(\mathbf{K}),$$

$$\mathcal{B}(\mathbf{K}) = \frac{\mathcal{T}_{12}}{1 - \mathcal{R}_{21}\mathcal{R}_{23}\exp(2iq_2L)}\mathcal{A}^{(inc)}(\mathbf{K}),$$

$$\mathcal{C}(\mathbf{K}) = \frac{\mathcal{T}_{12}\exp(iq_2L)\mathcal{R}_{23}}{1 - \mathcal{R}_{21}\mathcal{R}_{23}\exp(2iq_2L)}\mathcal{A}^{(inc)}(\mathbf{K}),$$

$$\mathcal{D}(\mathbf{K}) = \mathcal{T}^{\text{Slab}}(\mathbf{K})\mathcal{A}^{(inc)}(\mathbf{K}).$$

Eqs. (8.48)–(8.49) are equivalent to those obtained for a Fabry-Perot resonator with the condition $\mathcal{R}_{21}\mathcal{R}_{23}\exp(2iq_2L) = 1$ representing the oscillation condition. Unfortunately, diffuse waves are so damped that in general $\mathcal{R}_{21}\mathcal{R}_{23}\exp(2iq_2L) \ll 1$, as can be seen in Fig. 8.10 where we see that these expressions are smooth functions. Due to this fact the geometric progressions Eq. (8.46) and Eq. (8.47) can be truncated to first order without significant loss in accuracy:

$$\mathcal{R}^{\text{Slab}} \simeq \mathcal{R}_{12} + \mathcal{T}_{12}\exp(2iq_2L)\mathcal{R}_{23}\mathcal{T}_{21} \quad (8.50)$$

$$\mathcal{T}^{\text{Slab}} \simeq \mathcal{T}_{12}\exp(iq_2L)\mathcal{T}_{23}. \quad (8.51)$$

These approximations are very accurate, even for small slab width values, L. Additionally, as shown in Fig. 8.10, for values of $L > 3cm$ the reflection from the slab is simply $\mathcal{R}^{\text{Slab}} \simeq \mathcal{R}_{12}$. Therefore, $\mathcal{R}_{21}\mathcal{R}_{23}\exp(2iq_2L)$

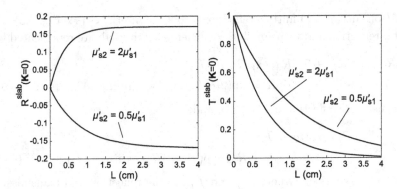

Fig. 8.10 Amplitude of $\mathcal{R}^{\text{Slab}}$ and $\mathcal{T}^{\text{Slab}}$ at $\mathbf{K} = 0$ for different slab widths L in the *cw* regime for $\mu_a = 0.025$ constant throughout the medium, constant index of refraction $n = 1.333$, and $\mu'_{s1} = \mu'_{s3} = 10cm^{-1}$. Two cases are shown, for $\mu'_{s2} = 2\mu'_{s1}$ and for $\mu'_{s2} = \mu'_{s1}/2$.

represents a way of determining the distance at which multiple reflections between the walls introduce a significant contribution to the measured average intensity. This is directly related to the maximum distance between the slab walls that can exist so that there is still a measurable contribution from the third medium when measuring in the reflection geometry. Clearly, as the slab width increases our measurement will tend to the solution of a single interface separating two semi-infinite media. So the question we can ask ourselves is: what is the maximum distance between the first and the second interface so that the third medium can be characterized? Since all the information from medium 3 is contained in the term $\mathcal{R}_{23}$, this value will depend on the relative weight of $\mathcal{T}_{12}\exp(2iq_2L)\mathcal{T}_{21}$ versus $\mathcal{R}_{12}$.

8.3.2 *Expression for a Slab in a Non-Diffusive medium*

Once we have understood how to account for multiple reflections in the case of a D-D interface, applying the same approach to a D-N is straightforward. So as to find the solution for a point source and thus the Green function of the system, let us consider the situation shown in Fig. 8.11 where we have a point source at z_s inside the diffusive medium, medium 2. In this case, following exactly the approach for Eq. (8.46) shown in Fig. 8.9 we obtain

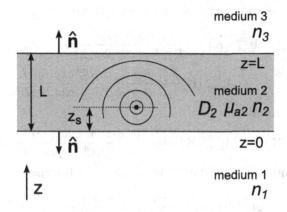

Fig. 8.11 Configuration for a slab of width L located at $0 \leq z \leq L$, surrounded by non-diffusive media of refractive indexes given by n_1 and n_2.

that the average intensity inside the diffusive slab of width L is given by:

$$
\widetilde{U}_2(\mathbf{K}) = \widetilde{U}^{(inc)}(\mathbf{K}, z) +
$$
$$
\mathcal{A}_{z=0}^{(inc)}(\mathbf{K})\Big(\mathcal{R}_{21}^{nd} \exp(iq_2 z) + \mathcal{R}_{21}^{nd} \exp(iq_2 L)\mathcal{R}_{23}^{nd} \exp(iq_2(L-z)) +
$$
$$
\mathcal{R}_{21}^{nd} \exp(iq_2 L)\mathcal{R}_{23}^{nd} \exp(iq_2 L)\mathcal{R}_{21}^{nd} \exp(iq_2 z) +
$$
$$
\mathcal{R}_{21}^{nd} \exp(iq_2 L)\mathcal{R}_{23}^{nd} \exp(iq_2 L)\mathcal{R}_{21}^{nd} \exp(iq_2 L)\mathcal{R}_{23}^{nd} \exp(iq_2(L-z)) ... \Big) +
$$
$$
\mathcal{A}_{z=L}^{(inc)}(\mathbf{K})\Big(\mathcal{R}_{23}^{nd} \exp(iq_2(L-z)) + \mathcal{R}_{23}^{nd} \exp(iq_2 L)\mathcal{R}_{21}^{nd} \exp(iq_2 z) +
$$
$$
\mathcal{R}_{23}^{nd} \exp(iq_2 L)\mathcal{R}_{21}^{nd} \exp(iq_2 L)\mathcal{R}_{23}^{nd} \exp(iq_2(L-z)) + ... \Big), \quad (8.52)
$$

where we can see three main contributions: the direct incident intensity and the multiple reflections starting at planes $z = 0$ and $z = L$ respectively. The contribution which begins at plane $z = 0$ can be written as:

$$
\widetilde{U}_{z=0}(\mathbf{K}, z) = \mathcal{A}_{z=0}^{(inc)}(\mathbf{K})\mathcal{R}_{21}^{nd} \exp(iq_2 z) \times
$$
$$
\Big(1 + \mathcal{R}_{21}^{nd}\mathcal{R}_{23}^{nd} \exp(2iq_2 L) + ... \Big) +
$$
$$
\mathcal{A}_{z=0}^{(inc)}(\mathbf{K})\mathcal{R}_{21}^{nd} \exp(iq_2 L)\mathcal{R}_{23}^{nd} \exp(iq_2(L-z)) \times
$$
$$
\Big(1 + \mathcal{R}_{21}^{nd}\mathcal{R}_{23}^{nd} \exp(2iq_2 L) + ... \Big). \quad (8.53)
$$

Grouping terms and identifying the geometric series we obtain:

$$\widetilde{U}_{z=0}(\mathbf{K}, z) = \mathcal{A}_{z=0}^{(inc)}(\mathbf{K}) \frac{\mathcal{R}_{21}^{nd} \exp(iq_2 z)}{1 - \mathcal{R}_{21}^{nd} \mathcal{R}_{23}^{nd} \exp(2iq_2 L)} +$$

$$\mathcal{A}_{z=0}^{(inc)}(\mathbf{K}) \frac{\mathcal{R}_{21}^{nd} \exp(iq_2 L) \mathcal{R}_{23}^{nd} \exp(iq_2(L - z))}{1 - \mathcal{R}_{21}^{nd} \mathcal{R}_{23}^{nd} \exp(2iq_2 L)}, \quad (8.54)$$

which finally results in:

$$\widetilde{U}_{z=0}(\mathbf{K}, z) = \mathcal{A}_{z=0}^{(inc)}(\mathbf{K}) \frac{\mathcal{R}_{21}^{nd} \exp(iq_2 z)(1 + \mathcal{R}_{23}^{nd} \exp(2iq_2 L))}{1 - \mathcal{R}_{21}^{nd} \mathcal{R}_{23}^{nd} \exp(2iq_2 L)}.$$

Equivalently, the contribution from the $z = L$ becomes:

$$\widetilde{U}_{z=L}(\mathbf{K}, z) = \mathcal{A}_{z=L}^{(inc)}(\mathbf{K}) \frac{\mathcal{R}_{23}^{nd} \exp(iq_2(L - z))}{1 - \mathcal{R}_{23}^{nd} \mathcal{R}_{21}^{nd} \exp(2iq_2 L)} +$$

$$\mathcal{A}_{z=L}^{(inc)}(\mathbf{K}) \frac{\mathcal{R}_{23}^{nd} \mathcal{R}_{21}^{nd} \exp(iq_2(L + z))}{1 - \mathcal{R}_{23}^{nd} \mathcal{R}_{21}^{nd} \exp(2iq_2 L)},$$

which grouping terms results in:

$$\widetilde{U}_{z=L}(\mathbf{K}, z) = \mathcal{A}_{z=L}^{(inc)}(\mathbf{K}) \frac{\mathcal{R}_{23}^{nd} \exp(iq_2(L - z))(1 + \mathcal{R}_{21}^{nd} \exp(2iq_2 z))}{1 - \mathcal{R}_{23}^{nd} \mathcal{R}_{21}^{nd} \exp(2iq_2 L)}.$$

We are now in the position to write the total average intensity inside medium 2 as:

$$\widetilde{U}_2(\mathbf{K}, z) = \widetilde{U}^{(inc)}(\mathbf{K}, z) +$$

$$\mathcal{A}_{z=L}^{(inc)}(\mathbf{K}) \frac{\mathcal{R}_{23}^{nd} \exp(iq_2(L - z))(1 + \mathcal{R}_{21}^{nd} \exp(2iq_2 z))}{1 - \mathcal{R}_{23}^{nd} \mathcal{R}_{21}^{nd} \exp(2iq_2 L)} +$$

$$\mathcal{A}_{z=0}^{(inc)}(\mathbf{K}) \frac{\mathcal{R}_{21}^{nd} \exp(iq_2 z)(1 + \mathcal{R}_{23}^{nd} \exp(2iq_2 L))}{1 - \mathcal{R}_{21}^{nd} \mathcal{R}_{23}^{nd} \exp(2iq_2 L)}. \quad (8.55)$$

In order to obtain the flux that traverses the interface we only need to set z in the right plane and apply the boundary condition, Eq. (8.32). For example, the total flux traversing the interface at $z = L$ between medium 2 and 3 will be given by:

$$\widetilde{J}_n(\mathbf{K}, z = L) = \frac{1}{\alpha} \widetilde{U}_2(\mathbf{K}, z = L), \quad (8.56)$$

which results in:

$$\widetilde{J}_n(\mathbf{K}, z = L) = \frac{1}{\alpha} \mathcal{A}_{z=L}^{(inc)}(\mathbf{K}) \frac{\mathcal{T}_{23}^{nd}}{1 - \mathcal{R}_{23}^{nd} \mathcal{R}_{21}^{nd} \exp(2iq_2 L)} +$$

$$\frac{1}{\alpha} \mathcal{A}_{z=0}^{(inc)}(\mathbf{K}) \frac{\mathcal{R}_{21}^{nd} \exp(iq_2 L)(1 + \mathcal{R}_{23}^{nd} \exp(2iq_2 L))}{1 - \mathcal{R}_{21}^{nd} \mathcal{R}_{23}^{nd} \exp(2iq_2 L)},$$

where we have made use of the expression $\mathcal{T}_{23}^{nd} = \mathcal{R}_{23}^{nd} + 1$ we found in Sec. 8.2.2. Note that even though we have filled two pages with formulas in order to arrive to this expression, the derivation is quite straightforward and simple (just take care not lose any coefficient on the way!). The expression for the flux traversing outward of the interface at $z = 0$ between medium 2 and medium 1 becomes:

$$\widetilde{J}_n(\mathbf{K}, z = 0) = \frac{1}{\alpha} \mathcal{A}_{z=0}^{(inc)}(\mathbf{K}) \frac{\mathcal{T}_{21}^{nd}}{1 - \mathcal{R}_{23}^{nd}\mathcal{R}_{21}^{nd} \exp(2iq_2 L)} +$$
$$\frac{1}{\alpha} \mathcal{A}_{z=L}^{(inc)}(\mathbf{K}) \frac{\mathcal{R}_{23}^{nd} \exp(iq_2 L)\mathcal{T}_{21}^{nd}}{1 - \mathcal{R}_{21}^{nd}\mathcal{R}_{23}^{nd} \exp(2iq_2 L)}.$$

If the solution for a point source is what we are looking for, we would only need to include its expression in K-space at $z = 0$ and $z = L$ (see Sec. 6.2.1 of Chap. 6):

$$\mathcal{A}_z^{(inc)}(\mathbf{K}) = \frac{S_0}{4\pi D} \frac{i}{2\pi q(\mathbf{K})} \exp(iq(\mathbf{K})|z_s - z|).$$

8.4 Multiple layered media

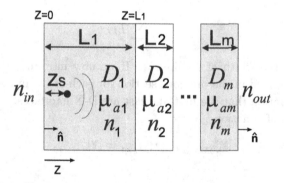

Fig. 8.12 Multiple layered configuration of M slabs, where n_{in} and n_{out} are the refractive indexes of the input and output non-scattering media, respectively. In all cases, the normal to each interface is considered to point into the $+z$ direction, i.e. along the propagation direction of the incident average intensity.

In the previous sections we addressed how the solution for the total average intensity could be found in the presence of D-D and D-N interfaces, and we also considered the case of a slab. In order to obtain the expression for slabs, however, we presented the total contribution as an infinite series containing all the reflection and transmission events that the original incident average intensity would receive on propagating through this slab. This approach is intuitive and simple (even though, as we have seen, quite lengthy) and lends itself to simple approximations by truncating the number of reflections to any order wished. In the case of multiple slabs, however, if we do not stay within very low orders the solution based on an infinite series becomes extremely difficult to handle. It is in those cases where solving numerically for multiple layered media is the best option.

In order to present an example of how a multiple layered medium could be solved, consider the M-layered system, shown in Fig. 8.12. In this figure we have M-slabs of diffusive media flanked by a non-scattering (free-space) medium with refractive index n_{in} which contains the source and a non-scattering medium with index of refraction n_{out} where detection is performed. An equivalent solution for multiple layered dielectric media can be found in [Born and Wolf (1999)]. In any of the jth inner diffusive media, the total average intensity in the angular spectrum representation would be given by:

$$U_j = \mathcal{A}_j \exp(iq_j(z - z_j)) + \mathcal{B}_j \exp(iq_j(z_{j+1} - z)), \quad z_j \leq z \leq z_{j+1},$$

where z_j represents the position of j th interface:

$$z_j = \sum_{k=1}^{k=j} L_k,$$

where L_k is the width of medium k. If we now introduce the boundary conditions derived in the previous section for each of the interfaces, taking into consideration that the first and last are D-N interfaces, we obtain the following set of equations:

$$[\mathsf{M}]_{2M \times 2M} \cdot [\mathsf{X}]_{2M \times 1} = [\mathsf{Y}]_{2M \times 1}, \tag{8.57}$$

where:

$$[\mathsf{M}] \equiv \begin{pmatrix} f_{in} & -g_{in}E_1 & 0 & 0 & \cdots & 0 & 0 \\ a_{12}E_1 & b_{12} & -1 & -E_2 & \cdots & 0 & 0 \\ q_1D_1E_1 & -q_1D_1 & -q_2D_2 & q_2D_2E_2 & \cdots & 0 & 0 \\ 0 & 0 & 0 & 0 & \ddots & \vdots & \vdots \\ \vdots & \vdots & \vdots & \vdots & \cdots & -q_MD_M & q_MD_ME_M \\ 0 & 0 & 0 & 0 & \cdots & -g_{out}E_M & f_{out} \end{pmatrix}$$

$$
[\mathcal{X}] \equiv \begin{pmatrix} \mathcal{A}_1 \\ \mathcal{B}_1 \\ \vdots \\ \vdots \\ \mathcal{A}_M \\ \mathcal{B}_M \end{pmatrix} \; ; \quad [\mathcal{Y}] \equiv \begin{pmatrix} g_{in}\widetilde{U}^{(inc)}(\mathbf{K}, z = 0) \\ -a_{12}\widetilde{U}^{(inc)}(\mathbf{K}, z = L_1) \\ -q_1 D_1 \widetilde{U}^{(inc)}(\mathbf{K}, z = L_1) \\ 0 \\ \vdots \\ 0 \end{pmatrix} ,
$$

where we have introduced:

$$
E_j = \exp[iq_j L_j] ,
$$

the coefficients a_{jk} and b_{jk} being given by (see Eq. (8.19)):

$$
a_{jk} = \left(\frac{n_k}{n_j}\right)^2 + i\mathcal{C}_{jk} D_j q_j , \tag{8.58}
$$

$$
b_{jk} = \left(\frac{n_k}{n_j}\right)^2 - i\mathcal{C}_{jk} D_j q_j . \tag{8.59}
$$

Eqs. (8.58)–(8.59) are the coefficients for the boundary conditions between the jth and $k = j + 1$ diffusive media, which in the case $n_j = n_k$ have the values $a_{jk} = b_{jk} = 1$ (see Sec. 8.1.1.2). For the input and output interfaces with the non-scattering media we have defined:

$$
f_{in} = i\alpha_{in} q_1 D_1 - 1 ; \quad g_{in} = i\alpha_{in} q_1 D_1 + 1 ,
$$

$$
f_{out} = i\alpha_{out} q_M D_M - 1 ; \quad g_{out} = i\alpha_{out} q_M D_M + 1 .
$$

Note that when the input interface is black ($\alpha_{in} = 0$) these yield $f_{in} = -1$, and $g_{in} = 1$. Also, notice that $g_x/f_x = \mathcal{R}_{nd}$ represents the D-N reflection coefficient given by Eq. (8.38).

Solving the linear system of equations given by Eq. (8.57) for each frequency component $\mathbf{K}$ will give us the expression for the coefficients $\mathcal{A}_j$ and $\mathcal{B}_j$ with which we can find the average intensity anywhere in the medium. Note, however, that the way $[\mathcal{Y}]$ in Eq. (8.57) is defined for the incident average intensity $\widetilde{U}^{(inc)}$ refers to the case where the source is located *inside* the first slab. The non-zero elements of $[\mathcal{Y}]$ represent the average intensity at the first boundary (one single boundary condition imposed) and on the second boundary (where two boundary conditions are needed, one for the average intensity and one for the conservation of flux). If we place the source elsewhere, we would have to change this matrix accordingly. Note that in the limiting case of very small slabs we can use the matrix approach we just derived to model smoothly varying optical parameters.

8.5   The Detected Power in Diffuse Media

Once we have established how the average intensity can be defined at any point of the diffusive medium's volume, the next important question we need to ask ourselves is: how does this relate to our measurements? The answer to this question is directly linked to what we presented in Sec. 3.3 of Chap. 3, which found an expression for measurements at a detector in terms of the energy flow through the detector's surface. Let us consider the same situation we depicted in Fig. 3.5, where we have a detector at a certain point $\mathbf{r}$ (could be a fiber, for example), defined by its surface A and its surface normal $\hat{\mathbf{n}}_d$. As was explained there, what a detector measures is the total flux that traverses its interface, i.e. $J_n(\mathbf{r})$:

$$J_n(\mathbf{r}) = J^+(\mathbf{r}) - J^-(\mathbf{r}),$$

where the positive direction is defined by the surface normal $\hat{\mathbf{n}}_d$. If we can neglect the effect that the detector has on light propagation, inside a diffusive medium the flux that will traverse surface A is defined directly by Fick's law:

$$J_n(\mathbf{r}) = -D\hat{\mathbf{n}}_d \cdot \nabla U(\mathbf{r}), \qquad \left(Watts/cm^2\right)$$

and the total detected power will therefore be:

$$P_{det}(\mathbf{r}) = -D \int_A \hat{\mathbf{n}}_d \cdot \nabla U(\mathbf{r})dS. \qquad \left(Watts\right)$$

If the area of our detector is small compared to the transport mean free path (the characteristic distance of our diffusive medium), we can express the total detected power as:

$$P_{det}(\mathbf{r}) \simeq -D\hat{\mathbf{n}}_d \cdot \nabla U(\mathbf{r})A.$$

It is convenient to express the flux in terms of the complete Green function of the medium since we can write (see Eq. (5.41) in Chap. 5) :

$$\nabla U(\mathbf{r}) = \nabla \int_V S_0(\mathbf{r}')G(\mathbf{r} - \mathbf{r}')d^3r',$$

where S_0 represents the source density which we had found in Chap. 5 to be given in terms of the reduced average intensity by $S_0(\mathbf{r}) = \mu_s' U_{ri}(\mathbf{r})$. Since the gradient is applied to $\mathbf{r}$ we can directly include it in the volume integral in which case the expression for the flux at $\mathbf{r}$ is given by:

$$J_n(\mathbf{r}) = -D \int_V S_0(\mathbf{r}')\hat{\mathbf{n}}_d \cdot \nabla_{\mathbf{r}} G(\mathbf{r} - \mathbf{r}')d^3r'.$$

In an infinite homogeneous medium we can use directly the expression obtained for ∇G in Chap. 5 (see Eq. (5.70)):

$$J_n(\mathbf{r}) = -D \int_V S_0(\mathbf{r}') \frac{\exp(i\kappa_0 |\mathbf{r} - \mathbf{r}'|)}{4\pi D |\mathbf{r} - \mathbf{r}'|} \left(i\kappa_0 - \frac{1}{|\mathbf{r} - \mathbf{r}'|} \right) \hat{\mathbf{u}}_{\mathbf{r}-\mathbf{r}'} \cdot \hat{\mathbf{n}}_d d^3 r',$$

which for a point source at $\mathbf{r}_s \in V$, $S_0(\mathbf{r}') = S_0 \delta(\mathbf{r} - \mathbf{r}_s)$, would give:

$$J_n(\mathbf{r}) = S_0 \frac{1}{4\pi |\mathbf{r} - \mathbf{r}_s|} \left(\frac{1}{|\mathbf{r} - \mathbf{r}_s|} - i\kappa_0 \right) \hat{\mathbf{u}}_{\mathbf{r}-\mathbf{r}_s} \cdot \hat{\mathbf{n}}_d.$$

As an important and useful comparison, consider the above equation for the flux in a diffusive medium which has no absorption, i.e. $\kappa_0 = \sqrt{-\mu_a/D} = 0$. In this case the solution for a point source results in:

$$J_n(\mathbf{r}) = \frac{S_0 \cos \theta}{4\pi |\mathbf{r} - \mathbf{r}_s|^2},$$

with $\cos \theta = \hat{\mathbf{n}} \cdot \hat{\mathbf{u}}_{\mathbf{r}-\mathbf{r}_s}$. That is, we recover the $1/r^2$ dependence expected for the power with no dependence whatsoever on the scattering component, equivalent to the one we defined for a non-scattering medium in Eq. (3.32) of Chap. 3. The total measured power at a detector of small area A becomes:

$$P_{det}(\mathbf{r}) = \frac{S_0 A \cos \theta}{4\pi |\mathbf{r} - \mathbf{r}_s|^2}.$$

So far, we have derived an expression for our measurements in real space, which is very handy when dealing either with infinite media or with arbitrary boundaries for which the solution in terms of diffuse plane waves and the reflection and transmission coefficients is not too efficient. If, on the other hand, we are dealing with plane interfaces we can make use of the angular spectrum representation, following the notation we have been using in this chapter. In this case we can express the flux in K-space by assuming the detector is located at z_d and its surface normal pointing in the $+z$ direction, and obtain an expression for the flux at the detector as:

$$\widetilde{J}_n(\mathbf{K}, z_d) = -D \frac{\partial \widetilde{U}(\mathbf{K}, z)}{\partial z} \bigg|_{z_d}.$$

From this expression it is straightforward to obtain a closed form solution by introducing the value of $\widetilde{U}(\mathbf{k})$ in terms of the reflection and transmission coefficients we have derived in this chapter. As an example, for a point source in an otherwise infinite medium we would obtain:

$$\widetilde{J}_n(\mathbf{K}, z_d) = -D i q(\mathbf{K}) \mathcal{A}_\delta(\mathbf{K}) \exp(i q(\mathbf{K}) |z_s - z_d|),$$

with $\mathcal{A}_\delta$ being the angular spectrum of a point source.

In the case where instead of having the detector immersed in the diffusive medium we have a detector *in contact* with a D-N interface given by $\mathbf{r} \in S$, we can use directly the boundary condition Eq. (8.32). This means that the power measured is directly proportional to the average intensity at the location of the detector:

$$J_n(\mathbf{r}) = \frac{1}{\alpha}U(\mathbf{r}) = \frac{1}{\alpha}\int_V S_0(\mathbf{r}')G(\mathbf{r}-\mathbf{r}')\mathrm{d}^3r', \forall \mathbf{r} \in S$$

and we can thus express all the measurements in terms of the average intensity directly, considerably simplifying our expressions.

8.5.1 *Accounting for the Detector Profile*

In this chapter we have established the main relation between the intensity at any plane of measurement z and a general source profile at z_s. We can now easily account for the detector profile in a manner completely analogous to what we did for a source profile in Sec. 7.4 of Chap. 7. For that, we shall consider the case of a contact detector at a D-N interface for which we can use Eq. (8.32). Note that the following derivation can also be done for a detector embedded in a diffusive medium by using the expressions presented in the previous subsection. Instead of assuming a Gaussian sensitivity function, which would represent collection from a fiber and thus be very similar to the derivation for the sources, we will now consider a square detector of area $\Delta_x \times \Delta_y$. Assuming our detector spans the range $x = [x_d - \Delta_x/2, x_d + \Delta_x/2]$ and $y = [y_d - \Delta_y/2, y_d - y/2]$, and is located at z_d the total *power* measured at the detector would be given by:

$$P_{det}(\mathbf{r}_d) = \frac{1}{\alpha}\int_{x_d-\Delta_x/2}^{x_d+\Delta_x/2}\int_{y_d-\Delta_y/2}^{y_d+\Delta_y/2} U(\mathbf{R}, z_d)\mathrm{d}y\mathrm{d}x. \quad \left(Watts\right)$$

Expressing U in terms of the angular spectrum representation we obtain:

$$P_{det}(\mathbf{R}_d, z_d) = \frac{1}{\alpha}\int_{-\infty}^{\infty} \widetilde{U}(\mathbf{K}, z_d)\times$$
$$\left[\int_{x_d-\Delta_x/2}^{x_d+\Delta_x/2}\int_{y_d-\Delta_y/2}^{y_d+\Delta_y/2} \exp(i\mathbf{K}\cdot\mathbf{R})\mathrm{d}y\mathrm{d}x\right]\mathrm{d}^2K.$$

The quantity within brackets can be solved directly through:

$$\int_{x_d-\Delta_x/2}^{x_d+\Delta_x/2} \int_{y_d-\Delta_y/2}^{y_d+\Delta_y/2} \exp(i\mathbf{K}\cdot\mathbf{R})\mathrm{d}y\mathrm{d}x =$$

$$\exp(i\mathbf{K}\cdot\mathbf{R}_d) \int_{-\Delta_x/2}^{\Delta_x/2} \int_{-\Delta_y/2}^{\Delta_y/2} \exp(i\mathbf{K}\cdot\mathbf{R})\mathrm{d}y\mathrm{d}x =$$

$$\exp(i\mathbf{K}\cdot\mathbf{R}_d)\left[\frac{1}{iK_x}\left(\exp(iK_x\Delta_x/2)-\exp(-iK_x\Delta_x/2)\right)\times\right.$$

$$\left.\frac{1}{iK_y}\left(\exp(iK_y\Delta_y/2)-\exp(-iK_y\Delta_y/2)\right)\right],$$

where we can identify the sinc function:

$$\mathrm{sinc}(Ka) = \frac{\sin(Ka)}{Ka} = \frac{1}{Ka}\frac{\exp(iKa)-\exp(-iKa)}{2i}.$$

Making use of the expression for the sinc function the total power measured at the detector becomes:

$$P_{det}(\mathbf{R}_d, z_d) = \frac{\Delta_x\Delta_y}{\alpha}\int_{-\infty}^{\infty}\widetilde{U}(\mathbf{K}, z_d)\exp(i\mathbf{K}\cdot\mathbf{R})\times$$

$$\left[\mathrm{sinc}\left(\frac{K_x\Delta_x}{2}\right)\mathrm{sinc}\left(\frac{K_y\Delta_y}{2}\right)\right]\mathrm{d}^2K,$$

which, as in the case we derived for the sources in Chap. 7, represents the convolution of the average intensity at z_d with a square function centered at the detector with widths Δ_x and Δ_y. For convenience, we can represent the detector function in K-space as:

$$\widetilde{f}_{det}(\mathbf{K}) = \mathrm{sinc}\left(\frac{K_x\Delta_x}{2}\right)\mathrm{sinc}\left(\frac{K_y\Delta_y}{2}\right), \tag{8.60}$$

in which case the power measured at z_d can be expressed in the spatial frequency domain as:

$$\widetilde{P}_{det}(\mathbf{K}, z_d) = \frac{1}{\alpha}\Delta_x\Delta_y\widetilde{U}(\mathbf{K}, z_d)\widetilde{f}_{det}(\mathbf{K}). \tag{8.61}$$

Obviously, combining a generic source profile with a generic detector profile simply results in:

$$P_{det}(\mathbf{R}, z_d) = A\int_{-\infty}^{\infty}\widetilde{U}(\mathbf{K}, z_d)\times$$

$$\widetilde{f}_{det}(\mathbf{K})\widetilde{f}_{src}(\mathbf{K})\exp(i\mathbf{K}\cdot\mathbf{R})\mathrm{d}^2R, \qquad \left(Watts\right) \tag{8.62}$$

where A is the area of the detector. The above set of equations highlights the advantage of solving the equations in the Fourier domain. As can be seen from Eq. (8.62) working in K-space enables us to recover a very simple expression which accounts for both source and detector profiles, consisting only of a product of functions in Fourier space.

8.6 Non-contact Measurements

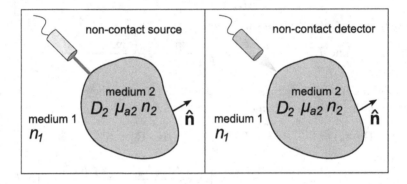

Fig. 8.13   Non-contact geometries for collimated sources and detectors.

All the examples we presented in the previous sections dealt with the case where we can consider the source and detector to be either directly inside the diffusive medium or in contact with its surface. In many practical applications, however, it is simpler to have either sources, detectors or both in a non-contact configuration. In this way we can make use of collimated sources such as lasers (a non-contact source) which we can guide through mirrors, and CCD cameras, for example, as a non-contact detector (see Fig. 8.13). Considering a laser a diffraction free source, including the free-space propagation of the laser is not necessary and we can directly model the incoming flux at the interface for a specific intensity profile. Doing this will be the first part of this section. A different matter is when we have non-contact detectors: in this case, since the light that leaves the diffusive interface is represented by a large number of angular components, if we wish to account for free-space propagation we would need to model this properly. This will be the focus of the second part of this section.

8.6.1 *Free-space source*

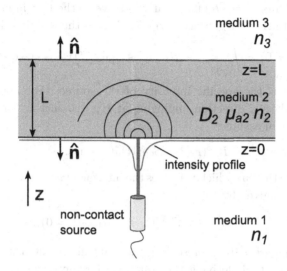

Fig. 8.14 Non-contact geometries for collimated sources and detectors.

We will now consider the case of a non-contact source illuminating a diffusive medium. In order to compare with the derivation we just completed in the previous sections we will consider the case of Sec. 8.3.2 for a diffusive slab surrounded by non-scattering media (free-space) when the source hits on the surface from one of the non-diffusive regions (lets assume medium 1). An example of this could be a laser impinging on the interface as shown in Fig. 8.14. In order to solve for this we have three options: a) to apply the boundary condition $U_2 = \alpha J_n(\mathbf{r})$ in order to find the average intensity at the interface and calculate the multiple reflections within the slab similar to what we did in the previous section; b) to use the point source approximation presented in Chap. 7 for which the solution is exactly the expressions we just derived; or c) to solve rigorously using the expression for an exponentially decaying (collimated) source. We will here consider the results for the first case, since it is the most intuitive. Note, however, that the complete solution (case c) can be found by applying the proper angular spectrum at each z position in the equations presented in the previous section and taking great care with the integral which appears on the exponential distribution of the source we derived in Sec. 7.2 of Chap. 7 (see Eq. (7.21)).

As a starting point, we need to define the flux that traverses the interface, now that we have a source incident from free space. Accounting for this external flux, the *total* flux that will traverse the first interface will be given by (note that the surface normal points in the $-z$ direction):

$$J^{src}(\mathbf{R}, z = 0) = \int_{(2\pi)^-} [1 - R_{1\rightarrow 2}(-\hat{\mathbf{s}} \cdot \hat{\mathbf{n}})] I^{src}(\mathbf{r}, \hat{\mathbf{s}}) \hat{\mathbf{s}} \cdot (-\hat{\mathbf{n}}) d\Omega, \quad (8.63)$$

where $I(\mathbf{r}, \hat{\mathbf{s}})$ is the specific intensity of the source. For a pencil beam pointing in the $+z$ direction impinging at $\mathbf{R}_s = 0$ such as that shown in Fig. 8.14 we can assume:

$$I^{src}(\mathbf{r}, \hat{\mathbf{s}}) = S_0 \delta(\mathbf{R}) \delta(\hat{\mathbf{s}} - \mathbf{u}_z).$$

and therefore the flux which traverses the interface *inward* due to the source in free-space is given by:

$$J^{src}(\mathbf{r})\Big|_{z=0} = S_0 \delta(\mathbf{R})(1 - R_{1\rightarrow 2}(\theta = 0)),$$

with $(1 - R_{1\rightarrow 2}(\theta = 0))$ representing the transmitted intensity for normal incidence (note that deriving the equations for any angle of incidence is straightforward).

In this case the total flux which traverses the interface will be given by:

$$J_n(\mathbf{r})\Big|_{z=0} = J^+(\mathbf{r})\Big|_{z=0} - J^{src}(\mathbf{r})\Big|_{z=0}, \quad (8.64)$$

which introducing our expression for $J^+(\mathbf{r})$ derived in Eq. (8.31) becomes:

$$J_n(\mathbf{r})\Big|_{z=0} = \mathcal{T}_U^{1\rightarrow 2} \frac{U_2(\mathbf{r})}{2}\Big|_{z=0} + \mathcal{T}_J^{1\rightarrow 2} \frac{J_n(\mathbf{r})}{2}\Big|_{z=0} - J^{src}(\mathbf{r})\Big|_{z=0}.$$

Grouping terms we obtain:

$$U_2(\mathbf{r})\Big|_{z=0} = \alpha J_n(\mathbf{r})\Big|_{z=0} + \frac{2}{\mathcal{T}_U^{1\rightarrow 2}} J^{src}(\mathbf{r})\Big|_{z=0}, \quad (8.65)$$

which applying Fick's law becomes (note the change of sign on Fick's law due to the direction of the normal $\hat{\mathbf{n}}$):

$$U_2(\mathbf{r})\Big|_{z=0} = D_2 \alpha \frac{\partial U_2(\mathbf{r})}{\partial z}\Big|_{z=0} + \frac{2}{\mathcal{T}_U^{1\rightarrow 2}} J^{src}(\mathbf{r})\Big|_{z=0}.$$

We can now express this in the angular spectrum representation, following the same procedure we presented in Sec. 8.2.2:

$$\widetilde{U}_2(\mathbf{K}, z = 0) = \alpha D_2 i q_2(\mathbf{K}) \widetilde{U}_2(\mathbf{K}, z = 0) + \frac{2}{\mathcal{T}_U^{1\rightarrow 2}} \widetilde{J}^{src}(\mathbf{K}, z = 0),$$

giving the angular spectrum of the average intensity at $z = 0$ as:

$$\mathcal{A}_{z=0}^{(inc)} = \tilde{U}_2(\mathbf{K}, z = 0) = \frac{2}{\mathcal{T}_U^{1\to2}} \frac{\tilde{J}^{src}(\mathbf{K}, z = 0)}{1 - i\alpha D_2 q_2(\mathbf{K})},$$

where, since we have not yet introduced the contribution from the second interface, we have termed the expression for the average intensity at $z = 0$ as $\mathcal{A}_{z=0}^{(inc)}$, equivalent to the incident average intensity we defined in Eq. (8.37). We can now account for the multiple reflections at the interface and measure the total average intensity at any point within the slab as:

$$\tilde{U}_2(\mathbf{K}, z) = \mathcal{A}_{z=0}^{(inc)}(\mathbf{K}) \Big(\exp(iq_2 z) +$$

$$\exp(iq_2 L)\mathcal{R}_{23}^{nd} \exp(iq_2(L-z)) +$$

$$\exp(iq_2 L)\mathcal{R}_{23}^{nd} \exp(iq_2 L)\mathcal{R}_{21}^{nd} \exp(iq_2 z) +$$

$$\exp(iq_2 L)\mathcal{R}_{23}^{nd} \exp(iq_2 L)\mathcal{R}_{21}^{nd} \exp(iq_2 L)\mathcal{R}_{23}^{nd} \exp(iq_2(L-z)) + ... \Big).$$

We can once again identify the geometric series and regroup to obtain:

$$\tilde{U}_2(\mathbf{K}, z) = \mathcal{A}_{z=0}^{(inc)}(\mathbf{K}) \frac{\exp(iq_2 z) + \mathcal{R}_{23}^{nd} \exp(iq_2(2L-z))}{1 - \mathcal{R}_{21}^{nd}\mathcal{R}_{23}^{nd} \exp(2iq_2 L)}.$$

This expression accounts for all the multiple reflections from both interfaces of the slab, yielding the complete expression for the average intensity *inside* the slab, i.e. for $0 \leq z \leq L$. The flux that would exit the slab from $z = L$ can now be found making use of the boundary condition $\tilde{U}_2(\mathbf{K}, z = L) = \alpha \tilde{J}_n(\mathbf{R}, z = L)$:

$$\tilde{J}_n(\mathbf{K}, z = L) = \frac{1}{\alpha} \mathcal{A}_{z=0}^{(inc)}(\mathbf{K}) \exp(iq_2 L) \frac{\mathcal{T}_{23}^{nd}}{1 - \mathcal{R}_{21}^{nd}\mathcal{R}_{23}^{nd} \exp(2iq_2 L)}.$$

Introducing the expression for $\mathcal{A}^{(inc)}$ in terms of the incident flux, J^{src} we obtain:

$$\tilde{J}_n(\mathbf{K}, z = L) = \frac{2}{\alpha \mathcal{T}_U^{1\to2}} \frac{\tilde{J}^{src}(\mathbf{K}, z = 0)\exp(iq_2 L)}{(1 - i\alpha D_2 q_2(\mathbf{K}))} \frac{\mathcal{T}_{23}^{nd}}{1 - \mathcal{R}_{21}^{nd}\mathcal{R}_{23}^{nd} \exp(2iq_2 L)}.$$

This expression represents the total flux *transmitted* by the diffusive slab of width L. We can reach a similar expression for the *reflected* flux from the slab taking into account that the total flux is given by Eq. (8.31):

$$\tilde{J}_n(\mathbf{K}, z = 0) = \frac{2}{\alpha \mathcal{T}_U^{1\to2}} \frac{\tilde{J}^{src}(\mathbf{K}, z = 0)}{(1 - i\alpha D_2 q_2(\mathbf{K}))} \times$$

$$\frac{1 + \mathcal{R}_{23}^{nd} \exp(2iq_2 L)}{1 - \mathcal{R}_{21}^{nd}\mathcal{R}_{23}^{nd} \exp(2iq_2 L)} - \frac{2}{\alpha \mathcal{T}_U^{1\to2}} \tilde{J}^{src}(\mathbf{K}, z = 0).$$

By using the expressions we have derived for the source profile (see Sec. 7.4 of Chap. 7) and for the detector profile we just derived in Sec. 8.5.1 we can now account for any source and detector profile directly in the above expression for the transmitted flux.

8.6.2　Free-space detector

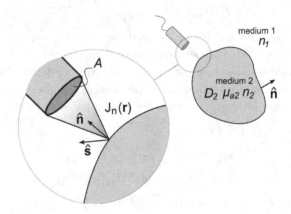

Fig. 8.15　Non-contact geometries for collimated sources and detectors.

In the previous derivation we assumed that the incident source was collimated and we thus did not need to account free space propagation. In the detector case, however, the light that exits the interface is diffuse and therefore radiated in many (if not all) possible angles (see Fig. 8.15) . In order to account for this radiation and include it in our model we first need to find an expression for the specific intensity $I(\mathbf{r}, \hat{\mathbf{s}})$ at the interface. The expression for the specific intensity inside the diffusive medium we can assume known, since it was the basis for deriving the diffusion approximation (see Chap. 4):

$$I_2(\mathbf{r}, \hat{\mathbf{s}}) \simeq \frac{U_2(\mathbf{r})}{4\pi} + \frac{3}{4\pi} J_n(\mathbf{r}) \hat{\mathbf{s}} \cdot \hat{\mathbf{n}}(\mathbf{r}), \forall \mathbf{r} \in S$$

where $\hat{\mathbf{n}}$ represents the surface normal at the interface S between the diffusive medium 2 and free-space[2]. We can now apply the boundary conditions for D-N interface we recently derived in Sec. 8.2, $U_2(\mathbf{r}) = \alpha J_n(\mathbf{r})$, and obtain:

$$I_2(\mathbf{r}, \hat{\mathbf{s}}) \simeq \frac{\alpha + 3\hat{\mathbf{s}} \cdot \hat{\mathbf{n}}(\mathbf{r})}{4\pi} J_n(\mathbf{r}), \forall \mathbf{r} \in S$$

which, accounting for the reflected specific intensity at the boundary gives the following approximation for the specific intensity radiated from the

---

[2]Note that in the above equation we could also include the contribution of the reduced intensity $I_{ri}(\mathbf{r}, \hat{\mathbf{s}})$, which we will not do in order to maintain the derivation as simple as possible. Including it, however, is straightforward as long as the reflection and transmission coefficients for $I_{ri}(\mathbf{r}, \hat{\mathbf{s}})$ are taken into account at the interface.

surface at a point $\mathbf{r}$ into a solid angle $\hat{\mathbf{s}}$:

$$I_1(\mathbf{r}, \hat{\mathbf{s}}) \simeq (1 - R_{2\to1}(\hat{\mathbf{s}} \cdot \hat{\mathbf{n}})) \frac{\alpha + 3\hat{\mathbf{s}} \cdot \hat{\mathbf{n}}(\mathbf{r})}{4\pi} J_n(\mathbf{r}), \forall \mathbf{r} \in S \qquad (8.66)$$

Of course, if we integrate $I_1(\mathbf{r}, \hat{\mathbf{s}})$ over all transmission angles we obtain the outward flux $J^+(\mathbf{r})$ (see Eq. (8.8)):

$$\int_{(2\pi)^+} I_1(\mathbf{r}, \hat{\mathbf{s}})\hat{\mathbf{s}} \cdot \hat{\mathbf{n}}d\Omega = J_n(\mathbf{r}) = J^+(\mathbf{r}).$$

Once we have an approximate expression for the specific intensity in the non-scattering medium, we can find the power received at a detector at a given position $\mathbf{r_d}$ defined by its area A and its surface normal $\hat{\mathbf{n}}_d$. For that we will make use of the expressions derived in Chap. 3, Sec. 3.3. The total power that reaches the detector can be expressed as (see Eq. (3.22)):

$$P_{det}(\mathbf{r_d}) = \int_A dA' \int_{(2\pi)^-} f_{det}(\hat{\mathbf{s}} \cdot \hat{\mathbf{n}}_d)I_1(\mathbf{r}', \hat{\mathbf{s}})\hat{\mathbf{s}} \cdot \hat{\mathbf{n}}_d d\Omega, \mathbf{r}' \in A \qquad \left(Watts\right)$$
$$(8.67)$$

where A is the surface of the detector and f_{det} its angular response. We can now write the solid angle in terms of the surface area it projects at the interface:

$$d\Omega = \mathbf{u}_{\mathbf{r}'-\mathbf{r}} \cdot \hat{\mathbf{n}} \frac{dS}{|\mathbf{r}' - \mathbf{r}|^2},$$

in which case the expression for the power becomes:

$$P_{det}(\mathbf{r_d}) = \int_A dA' \int_S f_{det}(\hat{\mathbf{s}} \cdot \hat{\mathbf{n}}_d)I_1(\mathbf{r}', \hat{\mathbf{s}}) \times$$
$$(\hat{\mathbf{s}} \cdot \hat{\mathbf{n}}_d)(\mathbf{u}_{\mathbf{r}'-\mathbf{r}} \cdot \hat{\mathbf{n}}) \frac{dS}{|\mathbf{r}' - \mathbf{r}|^2}, \mathbf{r}' \in A, \mathbf{r} \in S \qquad (8.68)$$

We can now make use of the invariance of the specific intensity which relates point $\mathbf{r}'$ to a point in the surface $\mathbf{r}$ by:

$$I_1(\mathbf{r}', \hat{\mathbf{s}}) = I_1(\mathbf{r}, \hat{\mathbf{s}})\delta(\hat{\mathbf{s}} - \mathbf{u}_{\mathbf{r}'-\mathbf{r}}),$$

obtaining the general expression for the total power at the detector as:

$$P_{det}(\mathbf{r_d}) = \int_A dA' \int_S f_{det}(\mathbf{u}_{\mathbf{r}'-\mathbf{r}} \cdot \hat{\mathbf{n}}_d)I_1(\mathbf{r}, \mathbf{u}_{\mathbf{r}'-\mathbf{r}}) \times$$
$$(\mathbf{u}_{\mathbf{r}'-\mathbf{r}} \cdot \hat{\mathbf{n}}_d)(\mathbf{u}_{\mathbf{r}'-\mathbf{r}} \cdot \hat{\mathbf{n}}) \frac{dS}{|\mathbf{r}' - \mathbf{r}|^2}, \mathbf{r}' \in A, \mathbf{r} \in S \qquad (8.69)$$

In those cases where the size of the detector is small compared to the distance from the surface, $A \ll |\mathbf{r_d} - \mathbf{r}|$, there is no need to perform the integral over the detector area in which case the total power becomes:

$$P_{det}(\mathbf{r_d}) = A \int_S f_{det}(\mathbf{u}_{\mathbf{r'}-\mathbf{r}} \cdot \hat{\mathbf{n}}_d) I_1(\mathbf{r}, \mathbf{u}_{\mathbf{r}-\mathbf{r}_d}) \times$$

$$(\mathbf{u}_{\mathbf{r'}-\mathbf{r}} \cdot \hat{\mathbf{n}}_d)(\mathbf{u}_{\mathbf{r'}-\mathbf{r}} \cdot \hat{\mathbf{n}}(\mathbf{r})) \frac{\mathrm{d}S}{|\mathbf{r_d} - \mathbf{r}|^2}, \mathbf{r} \in S \quad (8.70)$$

In any of the above equations we can now introduce our approximated expression for I_1, obtaining for the $A \ll |\mathbf{r_d} - \mathbf{r}|$ case:

$$P_{det}(\mathbf{r_d}) = \frac{A}{4\pi} \int_S J_n(\mathbf{r}) f_{det}(\mathbf{u}_{\mathbf{r}-\mathbf{r}_d} \cdot \hat{\mathbf{n}}_d) \left(\left(1 - R_{2 \to 1}(\mathbf{u}_{\mathbf{r}-\mathbf{r}_d} \cdot \hat{\mathbf{n}})\right) \times \right.$$

$$\left. \left(\alpha + 3\mathbf{u}_{\mathbf{r}-\mathbf{r}_d} \cdot \hat{\mathbf{n}}\right)\right) (\mathbf{u}_{\mathbf{r}-\mathbf{r}_d} \cdot \hat{\mathbf{n}}_d)(\mathbf{u}_{\mathbf{r}-\mathbf{r}_d} \cdot \hat{\mathbf{n}}) \frac{\mathrm{d}S}{|\mathbf{r_d} - \mathbf{r}|^2}, \mathbf{r} \in S \quad (8.71)$$

A particularly useful and simple case is the one in which when the detector is directly above point $\mathbf{r}$ with the surface normal $\hat{\mathbf{n}}_d$ pointing in the $-z$ direction, $\hat{\mathbf{n}}_d = -\mathbf{u}_z$ and we assume very small numerical aperture in the detector, i.e. $f_{det}(\mathbf{u}_{\mathbf{r}-\mathbf{r}_d} \cdot \hat{\mathbf{n}}_d) \simeq \delta(\mathbf{u}_{\mathbf{r}-\mathbf{r}_d} + \hat{\mathbf{n}}_d)$. In this case the power at the detector is expressed as:

$$P_{det}(\mathbf{r_d}) = \frac{J_n(\mathbf{r}) A}{4\pi |z_d - z|^2} \left(1 - R_{2 \to 1}(\cos\theta)\right) \left(\alpha + 3\cos\theta\right) \cos\theta,$$

with $\cos\theta = \mathbf{u}_z \cdot \hat{\mathbf{n}}$, and $\mathbf{r}$ and $\mathbf{r_d}$ being represented as $\mathbf{r} = (\mathbf{R}, z)$ and $\mathbf{r_d} = (\mathbf{R}_d, z_d)$, respectively.

Finally, we can further simplify the equations by assuming that the specific intensity at the interface radiates *isotropically*, i.e. as a Lambertian source. This approximation was considered in Chap. 3, Sec. 3.4 (see Eq. (3.33)), and is expressed by replacing $I_1(\mathbf{r}, \hat{\mathbf{s}})$ by:

$$I_1(\mathbf{r}, \hat{\mathbf{s}}) = \frac{J_n(\mathbf{r})}{\pi}.$$

Introducing this expression in Eq. (8.69) considerably simplifies the equations, but it is important to remember that it does not account for the angular dependence accurately. A more accurate expression can be easily found by introducing the expression for I_1 presented in Eq. (8.66).

8.6.2.1 *Free-space detection through a system of lenses*

An important situation arises when we have a collection of lenses which projects a virtual detector at the interface (or, in other words, collect intensity from a localized region of the interface). This is the case that represents how a camera with an objective works and is the most useful for small animal imaging where typically CCD cameras and photography lenses are used to collect the diffuse light by focusing at the interface. When in focus, we can assume the total power at each pixel will be given by:

$$P_{det}(\mathbf{r_d}) = 2\pi \int_{A_v} dA' \int_0^{\cos\theta_A} f_{det}(\theta) I_1(\mathbf{r'}, \theta) \cos\theta \, d(\cos\theta),$$

where θ_A represents the maximum acceptance angle of the imaging system (see Chap. 3, Sec. 3.3.1), and A_v is the area of the virtual detector (i.e. the image of the pixel at the interface). If we can consider the surface radiating like a Lambertian source this expression simplifies further, more so if we consider the virtual size of the detector to be in the order of the scattering mean free path (note that a more accurate solution can easily be reached by introducing the more complete expression for I_1 given by Eq. (8.66)):

$$P_{det}(\mathbf{r_d}) = 2J_n(\mathbf{r})A_v \int_0^{\cos\theta_A} f_{det}(\theta) \cos\theta \, d(\cos\theta).$$

This expression is extremely handy since it relates one single pixel in the camera to a single differential area of the interface, assigning a *constant* weight due to the acceptance angle of the objective. This means that through normalization we can effectively remove this weight and our non-contact measurements have a one to one relationship with virtual contact detectors (more on this in the next subsection). Obviously, if the numerical aperture is close to one, $\theta_A = \pi$, and the angular acceptance is equal to one for all angles, $f_{det} = 1$, the above equation results in:

$$P_{det}(\mathbf{r_d}) = 2J_n(\mathbf{r})A_v \int_0^1 \cos\theta \, d(\cos\theta) = J_n(\mathbf{r})A_v,$$

as expected.

8.6.2.2 *Normalized Fluorescence in Free-space*

As we have just mentioned in the context of the use of lenses, it is clear that the general equation for the power at a non-contact detector Eq. (8.71) can be written in terms of the total flux traversing the interface times a weight factor which has to do with the surface and detector geometries and positions:

$$P_{det}(\mathbf{r_d}) = \int_S J_n(\mathbf{r}) \mathcal{G}_{nc}(\mathbf{r}, \mathbf{r}_d) dS,$$

with $\mathcal{G}$ defined as:

$$\mathcal{G}_{nc}(\mathbf{r}, \mathbf{r}_d) = \int_A \frac{dA'}{4\pi} f_{det}(\mathbf{u}_{\mathbf{r'-r}} \cdot \hat{\mathbf{n}}_d) \Bigg(\Big(1 - R_{2\to 1}(\mathbf{u}_{\mathbf{r'-r}} \cdot \hat{\mathbf{n}}) \Big) \times$$

$$\Big(\alpha + 3\mathbf{u}_{\mathbf{r'-r}} \cdot \hat{\mathbf{n}} \Big) \Bigg) (\mathbf{u}_{\mathbf{r'-r}} \cdot \hat{\mathbf{n}}_d)(\mathbf{u}_{\mathbf{r'-r}} \cdot \hat{\mathbf{n}}) \frac{1}{|\mathbf{r'} - \mathbf{r}|^2}.$$

If we now consider that we have a fluorescence and an excitation measurement, assuming no change in the optical properties due to difference in wavelength the total fluorescence power detected would be:

$$P_{det}^{fl}(\mathbf{r_d}) = \int_S J_n^{fl}(\mathbf{r}) \mathcal{G}_{nc}(\mathbf{r}, \mathbf{r}_d) dS,$$

with $\mathcal{G}$ being defined exactly as above. As we shall see in Sec. 9.4 of Chap. 9, normalizing measurements is very beneficial specially for removing experimental-dependent quantities. In the non-contact case normalized measurements could be represented as:

$$P_{det}^{norm}(\mathbf{r_d}) = \frac{P_{det}^{fl}(\mathbf{r_d})}{P_{det}^{exc}(\mathbf{r_d})} = \frac{\int_S J_n^{fl}(\mathbf{r}) \mathcal{G}_{nc}(\mathbf{r}, \mathbf{r}_d) dS}{\int_S J_n^{exc}(\mathbf{r}) \mathcal{G}_{nc}(\mathbf{r}, \mathbf{r}_d) dS},$$

where P_{det}^{exc} represents the measured power due to the excitation source. In principle these integrals have to be solved separately in order to normalize the measurements. However, if one considers the case of *focused* measurements and the pixel size can be considered significantly small so that $f_{det}(\mathbf{u}_{\mathbf{r'-r}} \cdot \hat{\mathbf{n}}_d)$ is close to a delta function, representing the image of the detectors projected onto the N-D interface as $\mathbf{r}_d^v$ the above equation becomes:

$$P_{det}^{norm}(\mathbf{r_d}) = \frac{P_{det}^{fl}(\mathbf{r_d})}{P_{det}^{exc}(\mathbf{r_d})} \simeq \frac{J_n^{fl}(\mathbf{r}_d^v)}{J_n^{exc}(\mathbf{r}_d^v)},$$

further stressing the importance of normalizing data. Unless dealing with large detectors or out-of-focus data, the above normalized expression yields very good results and gives a one-to-one relationship between a measurement in our camera ($\mathbf{r}_d$) and a position on the object's surface ($\mathbf{r}_d^v$).

Key points

From this chapter the following important points should be remembered:

- **Boundary Conditions**: We derived the boundary conditions at Diffusive/Diffusive (D-D) interfaces and at Diffusive/Non-Diffusive (D-N) interfaces based on the relationship between the specific intensity and the flux within the diffusion approximation.
- **Reflection and Transmission Coefficients**: Once we found the boundary conditions, we derived the expressions for the reflection and transmission coefficients at D-D and D-N interfaces based on the angular spectrum representation. These coefficients can be used to solve any number of plane-parallel interfaces in a manner similar to what is done in classical optics resulting in an infinite geometric series. We also saw that a system of M slabs can be tackled without any approximation by solving a linear system of equations for each spatial frequency $\mathbf{K}$. Additionally, since all the expressions have been derived for a point source in the angular spectrum representation, following what we learned in Chap. 7 we can easily incorporate either the exact (within the diffusion approximation) expression for a collimated source or any generic source profile.
- **Detector Profile**: Similarly to the expressions found for the sources, we can account for the geometry of the detectors in a simple manner by working in the Fourier domain with the angular spectrum representation.
- **Non-contact sources**: We saw how modeling collimated sources consists on correctly accounting for the total flux which will traverse the interface in both directions (inward and outward). These results in a change in the boundary conditions.
- **Non-contact detectors**: The case of non-contact detectors was a bit more cumbersome since it involved taking into account the angular dependence of the specific intensity at the interface. We saw, however, that in cases where we make use of focused detection the normalized values of the fluorescence with the excitation measurements are equivalent to having measurements taken directly at the interface with a virtual detector.

Further Reading

This chapter has covered several very important topics for which there was not enough space to consider in great detail. Amongst these we studied the interaction of diffuse light at interfaces which in the case of diffusive/non-diffusive interfaces has been a heavily studied subject due to the expected limitations of the diffusion approximation at boundaries. There are many works dealing with boundary conditions that I would recommend for further reading but above all I recommend the paper by Aronson [Aronson (1995)]. I would suggest following the references you will find there in order to get a complete picture of the issues involved in the interface between diffusive and non-scattering media. With regards to the use of the method of images, I would recommend the original paper by Patterson, Chance and Wilson [Patterson *et al.* (1989)], and the famous paper by Haskell and colleagues [Haskell *et al.* (1994)] where an alternative to the use of the angular spectrum representation for solving the Robin boundary condition can be found. The case of the boundary conditions between diffusive media with index mismatch was studied thoroughly in [Ripoll and Nieto-Vesperinas (1999a)], for further insight in the subject I recommend this paper and the references therein, in particular [Walker *et al.* (1998)].

Additionally to studying the boundary conditions in this chapter we presented the expressions for the reflection and transmission coefficients. More insight on these can be found in [Ripoll and Nieto-Vesperinas (1999b)] for a single plane and [Ripoll *et al.* (2001b)] for multiple layered geometries.

Since this book is dealing with the principles of diffuse light propagation and is not too focused on the applications, we have only covered the basic interaction of diffuse light at planar interfaces. There are, of course, a great deal of works focused on solving more general geometries. In particular I suggest one of the first papers published on diffuse light propagation by Arridge, Cope and Delpy, [Arridge *et al.* (1992)], which presents general solutions for slabs, cylinders and spheres. The solution for spheres was also considered in terms of the Mie coefficients for diffuse light and measured experimentally in [Boas *et al.* (1994)]. With regards to more general geometries, there is an option to use an approximation based on the reflection and transmission coefficients based on the Kirchhoff approximation [Ripoll *et al.* (2001a)], which can be improved iteratively for a homogeneous medium with arbitrary geometries [Ripoll and Ntziachristos (2003)].

Another approach which has not been considered in this chapter is the possibility to remove the contribution of the boundary in order to make use of the solution for infinite homogeneous media. Details on this can be found in [Ripoll and Ntziachristos (2006)].

Finally, non-contact measurements have been only briefly introduced in this chapter. For further insight I recommend [Ripoll *et al.* (2003)] and [Ripoll and Ntziachristos (2004)], where more general cases of apertures and out-of-focus measurements are presented.

Chapter 9

Fluorescence and Bioluminescence in Diffuse Media: An ill-posed problem

Summary. In this chapter we shall analyze the reason why imaging in diffusive media involves solving what is termed an 'ill-posed' problem. We shall attempt to uncover the physics behind this ill-posedness and explain not only its broader meaning but the implications it has on imaging in diffusive media. In order to showcase this effect, we will concentrate on fluorescence and bioluminescence, two light emitting processes with direct applications in biology and medicine which, as we shall see, have profound differences in terms of the level of ill-posedness each one represents. In order to present these differences we will rely on the angular spectrum representation and the transfer function introduced in Chap. 6.

9.1 Fluorescence in Diffuse Media

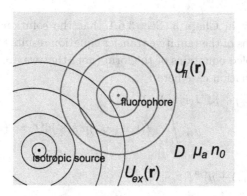

Fig. 9.1 Excitation of a fluorophore by an isotropic source in a diffusive medium of optical properties given by μ_a, D and n_0.

In this section we will derive the equations that predict the measurement of the fluorescence emission from a collection of fluorophores or fluorescent proteins within a diffusive medium. We shall consider only the *cw* case, since it is the simplest and the one that more clearly conveys the message behind the ill-posed nature of the problem. Additionally, it is the regime where fluorescence and bioluminescence are directly comparable. The basics of fluorescence were covered in Chap. 1 where we saw that the emitted power had the following dependence (see Sec. 1.4.2, Eq. (1.7)) in the *cw* regime:

$$P_{fl} = U_0 \sigma_a [C]_{cw} \Phi V, \qquad \left(Watts \right) \qquad (9.1)$$

where, as a reminder, U_0 represented the incident average intensity at the location of the fluorophore, σ_a its absorption cross-section, $[C]_{cw}$ the concentration of this fluorophore, Φ was the quantum yield and V the volume that these fluorophores occupy (i.e. $[C]_{cw}V$ represents the total number of fluorophores). Let us now consider that this concentration of fluorophores has a certain spatial distribution, $[C(\mathbf{r})]$ (we shall drop the subscript from now on), in an otherwise infinite diffusive medium. The spatial distribution of the concentration of fluorophores is what ultimately we would like to recover in our imaging experiment. This diffusive medium will be characterized by a constant diffusion coefficient, D, a constant absorption coefficient μ_{a0} and an index of refraction n_0. In this case, assuming a distribution of excitation intensity $U_{ex}(\mathbf{r})$ (see Fig. 9.1), the energy density emitted by this collection of sources can be represented as (assuming all fluorophores are identical and thus have the same absorption cross-section σ_a):

$$S_{fl}(\mathbf{r}) = \frac{P_{fl}(\mathbf{r})}{V} = U_{ex}(\mathbf{r}) \sigma_a \Phi [C(\mathbf{r})]. \qquad \left(Watts/cm^3 \right) \qquad (9.2)$$

We also saw in Chap. 3, Sec. 3.6.1 that the solution to fluorescence emission in terms of the radiative transfer equation results from the solution to a set of coupled equations of the form (note that we are considering the constant illumination case here):

$$\hat{\mathbf{s}} \cdot \nabla I(\mathbf{r}, \hat{\mathbf{s}}; \lambda_{ex}) + \mu_t^{ex} I(\mathbf{r}, \hat{\mathbf{s}}; \lambda_{ex})$$

$$- \mu_t^{ex} \int_{(4\pi)} I(\mathbf{r}, \hat{\mathbf{s}}'; \lambda_{ex}) p(\hat{\mathbf{s}}, \hat{\mathbf{s}}') d\Omega' = \epsilon(\mathbf{r}, \hat{\mathbf{s}}; \lambda_{ex}); \qquad (9.3)$$

and:

$$\hat{\mathbf{s}} \cdot \nabla I(\mathbf{r}, \hat{\mathbf{s}}; \lambda_{em}) + \mu_t^{em} I(\mathbf{r}, \hat{\mathbf{s}}; \lambda_{em})$$

$$- \mu_t^{em} \int_{(4\pi)} I(\mathbf{r}, \hat{\mathbf{s}}'; \lambda_{em}) p(\hat{\mathbf{s}}, \hat{\mathbf{s}}') d\Omega' = \sigma_a [C(\mathbf{r})] \Phi(\lambda_{em}) I(\mathbf{r}, \hat{\mathbf{s}}; \lambda_{ex}). \qquad (9.4)$$

This set of equations represents the complete problem which we will approximate by solving two coupled diffusion equations. In order to do so, however, we must not forget that $\sigma_a[C(\mathbf{r})]$ represents the local absorption coefficient at $\mathbf{r}$, i.e. $\mu_a^{fl}(\mathbf{r})$. That is, the presence of the fluorophore or fluorescent protein changes the spatial distribution of absorption: even though we have an infinite homogeneous diffusive medium, by introducing these fluorophores we now have a spatially dependent distribution of absorption. Since the fluorophores do not introduce, in principle, any changes in scattering we can still assume that the diffusion coefficient is constant. This will considerably simplify the expression for the diffusion equation in inhomogeneous media shown in Sec. 5.6 of Chap. 5, Eq. (5.57):

$$-D_{ex}\nabla^2 U_{ex}(\mathbf{r}) + \mu_{a0}^{ex}U_{ex}(\mathbf{r}) = \mu'_{s0}U_{ri}(\mathbf{r}) + \mathcal{Q}_l(\mathbf{r}) + \mathcal{Q}_{nl}(\mathbf{r}; U),$$

where now the linear source contribution, $\mathcal{Q}_l(\mathbf{r})$, is given by (compare with Eq. (5.58)):

$$\mathcal{Q}_l(\mathbf{r}) = -(\sigma_a[C(\mathbf{r})] - \mu_{a0})U_{ri}(\mathbf{r}),$$

while the non-linear contribution (compare with Eq. (5.59)) becomes:

$$\mathcal{Q}_{nl}(\mathbf{r}; U) = -(\sigma_a[C(\mathbf{r})] - \mu_{a0})U_{ex}(\mathbf{r}).$$

To simplify the derivation, and since it holds no relationship with the ill-posedness of the problem, in the above equations we have assumed that the excitation source is isotropic $(I_{ri}(\mathbf{r}, \hat{\mathbf{s}}) = I_0(\mathbf{r})$, see Fig. 9.1), in which case the contribution of the reduced intensity flux, $\mathbf{J}_{ri}(\mathbf{r})$, is zero:

$$\mathbf{J}_{ri}(\mathbf{r})(\mathbf{r}) = \int_{(4\pi)} \hat{\mathbf{s}}I_{ri}(\mathbf{r}, \hat{\mathbf{s}})d\Omega = I_0(\mathbf{r})\int_{(4\pi)} \hat{\mathbf{s}}d\Omega = 0.$$

When considering the emission wavelength we can assume that the fluorophore does not introduce a relevant change in absorption. Bear in mind that in order to be effective the fluorophore or fluorescent protein requires a very high absorption cross-section and a high quantum yield for the excitation energies but not in the emission wavelengths, therefore its absorption perturbation usually needs only to be accounted for at the excitation wavelength. We can now write the set of coupled diffusion equations for excitation and emission as:

$$-D_{ex}\nabla^2 U_{ex}(\mathbf{r}) + \mu_{a0}^{ex}U_{ex}(\mathbf{r}) =$$
$$\mu'_s U_{ri}(\mathbf{r}) - \Big(\sigma_a[C(\mathbf{r})] - \mu_{a0}\Big)U_{ri}(\mathbf{r}) - \Big(\sigma_a[C(\mathbf{r})] - \mu_{a0}\Big)U_{ex}(\mathbf{r}), \quad (9.5)$$

and:

$$-D_{em}\nabla^2 U^{fl}(\mathbf{r}) + \mu_{a0}^{em}U^{fl}(\mathbf{r}) = S_{fl}(\mathbf{r}). \quad (9.6)$$

In the above equation S_{fl} represents the energy density due to the fluorescent source distribution which is given by:

$$S_{fl}(\mathbf{r}) = \int_{(4\pi)} \sigma_a[C(\mathbf{r})]\Phi(\lambda_{em})I(\mathbf{r},\hat{\mathbf{s}};\lambda_{ex})\mathrm{d}\Omega, \qquad \left(Watts/cm^3\right)$$

or equivalently,

$$S_{fl}(\mathbf{r}) = \sigma_a[C(\mathbf{r})]\Phi(\lambda_{em})\int_{(4\pi)} I(\mathbf{r},\hat{\mathbf{s}};\lambda_{ex})\mathrm{d}\Omega =$$

$$\sigma_a[C(\mathbf{r})]\Phi(\lambda_{em})U_{ex}(\mathbf{r}), \quad (9.7)$$

thus recovering the same expression we recently derived in Eq. (9.2).

Using our definition for the Green function (see Chap. 5), the average intensity emitted at a specific emission wavelength λ_{em} at any point in space $\mathbf{r}$ will be given by:

$$U^{fl}(\mathbf{r}) = \sigma_a\Phi(\lambda_{em})\int_{-\infty}^{\infty} G_{cw}^{em}(\mathbf{r}-\mathbf{r}')[C(\mathbf{r}')]U_{ex}(\mathbf{r}')\mathrm{d}^3r, \qquad (9.8)$$

with $U_{ex}(\mathbf{r})$ being the solution to:

$$U_{ex}(\mathbf{r}) = \int_{-\infty}^{\infty} G_{cw}^{ex}(\mathbf{r}-\mathbf{r}')\Big(\mu_s' + \mu_{a0} - \sigma_a[C(\mathbf{r})]\Big)U_{ri}(\mathbf{r}')\mathrm{d}^3r$$

$$- \int_{-\infty}^{\infty} G_{cw}^{ex}(\mathbf{r}-\mathbf{r}')\Big(\sigma_a[C(\mathbf{r})] - \mu_{a0}\Big)U_{ex}(\mathbf{r}')\mathrm{d}^3r, \quad (9.9)$$

where we have explicitly written the dependence of the Green function on the absorption coefficient at the excitation and emission wavelengths. At this point we have all the necessary tools to find our complete solution but, as can be seen in the above equations we have to solve a set of coupled *nonlinear* equations, quite a difficult task. There is, however, an approximation we can make at this point to linearize the problem without compromising the result in most practical experimental situations: we can assume that the change in overall absorption due to the presence of the fluorophores can be neglected, for example assuming (correctly, I must say) that *in-vivo* the average background absorption is quite high. In doing so we are also implying that the presence of the fluorophore does not affect the distribution of excitation intensity:

$$U_{ex}(\mathbf{r}) \simeq \int_{-\infty}^{\infty} G_{cw}^{ex}(\mathbf{r}-\mathbf{r}')\mu_s'U_{ri}(\mathbf{r}')\mathrm{d}^3r, \qquad (9.10)$$

in which case our equation for the emission wavelength, Eq. (9.8), remains unchanged:

$$U^{fl}(\mathbf{r}) = \int_{-\infty}^{\infty} G_{cw}^{em}(\mathbf{r}-\mathbf{r}')S_{fl}(\mathbf{r}')\mathrm{d}^3r, \qquad (9.11)$$

where we made use of Eq. (9.7). We will base our demonstration on the ill-posed nature of diffuse light propagation on this simple equation for the fluorescence, Eq. (9.11). As we shall soon see, our main concern for recovering the spatial distribution of fluorophores, $[C(\mathbf{r})]$, will be solving this equation.

9.2 Bioluminescence in Diffuse Media

We can proceed in a manner analogous to the previous derivation for obtaining the total power emitted by a collection of bioluminescent reporters, with one significant difference: the total number of molecules which reach the excited state does not depend on the average intensity as in the case of fluorescence, but on the total number of efficient chemoluminescent reactions. As was mentioned in Sec. 1.4.3 of Chap. 1, one common bioluminescent reaction is:

luciferin + luciferase + ATP + O_2 + [...] $\longrightarrow$ [...] + oxyluciferin + light.

In this case we might expect the total power emitted to depend on the total number of substrate (luciferin) which has reacted with the enzyme (the luciferase in this case) times the efficiency of this conversion. This is analogous to the quantum yield defined for fluorescence, where in the case of bio or chemoluminescence it is defined as the probability of light emission via the reaction of a single substrate molecule and is determined experimentally by dividing the absolute emitted light (note that measuring this quantitatively is extremely complicated) by the number of consumed substrate molecules. We shall denote the bioluminescence quantum yield as Φ_{bio} and the concentration of the substrates which performed the reaction, $[C]_{bio}$. It is important to remember that during the experiment one usually injects the substrate and looks at *where* this substrate reacts with the enzyme. In principle this enzyme is only produced at the site we are interested in studying, such as cancer cells which have been transfected to express luciferase, for example. In this context, the total bioluminescent emitted power will be:

$$P_{bio} = E_{bio} k_r^{bio} [C]_{bio} \Phi_{bio} V, \qquad \left(Watts \right) \qquad (9.12)$$

where E_{bio} is the energy of the emission transition (equivalent to E_{em} used for fluorescence) which has units of Joules, k_r^{bio} is the rate of emission from this excited state and V is the volume occupied by the reaction. Note that, by defining in this manner the bioluminescence quantum yield, Φ_{bio} it is independent of the emission wavelength, being this dependence represented by $E_{bio}k_r^{bio}$. We can write the emission rate in terms of the emission lifetime, $\tau_{bio} = 1/k_r^{bio}$, in which case the spatially-dependent energy density emitted by the collection of substrates which undergo the reaction is:

$$S_{bio}(\mathbf{r}) = \frac{P_{bio}(\mathbf{r})}{V} = \frac{E_{bio}}{\tau_{bio}}\Phi_{bio}[C(\mathbf{r})]_{bio}. \qquad \left(Watts/cm^3\right) \qquad (9.13)$$

Once we have found the distribution of energy density emission due to bioluminescence, the total average intensity at the emission wavelength is given by:

$$U_{bio}(\mathbf{r}) = \int_{-\infty}^{\infty} G_{cw}^{em}(\mathbf{r} - \mathbf{r}')S_{bio}(\mathbf{r}')d^3r, \qquad (9.14)$$

which is formally identical to Eq. (9.11) for the fluorescence case. As we shall see, however, it is the fact that S_{fl} depends on a quantity which we can modify externally (i.e. the amount of intensity that reaches the fluorophore, $U_{ex}(\mathbf{r})$) what reduces the ill-posedness of the imaging problem in the case of fluorescence. But first let us look at the reasons behind the ill-posed nature of diffuse light imaging.

9.3 Why is imaging in diffuse media an ill-posed problem?

Following the definition of J. Hadamard, a well-posed problem is such that a solution exists, and this solution is unique and stable. Similarly, we can define an ill-posed problem as one for which more than one solution is possible. In this case, why is imaging in diffuse media an ill-posed problem? The answer to this question was actually addressed, even though in a concealed manner, when analyzing the spatial transfer function in Chap. 6. In fact, the ill-posed nature of imaging in diffuse media is fundamentally represented by the way information is lost on propagation and the function which is responsible for this is, as we saw in Chap. 6, the transfer function:

$$\widetilde{H}(\mathbf{K}, z - z') = \exp(iq(\mathbf{K})|z - z'|). \qquad (9.15)$$

We shall now concentrate in analyzing under which conditions we might reach a unique solution for our equation:

$$U^{fl}(\mathbf{r}) = \int_{-\infty}^{\infty} G_{cw}^{em}(\mathbf{r} - \mathbf{r}')S_{fl}(\mathbf{r}')d^3r,$$

which is equivalent to that of bioluminescence. By using what we derived in Chap. 6 (see Eq. (6.29)) we can write the Green function as:

$$G_{cw}(\mathbf{r} - \mathbf{r}') = \int_{-\infty}^{\infty} \frac{1}{4\pi D} \frac{i}{2\pi q(\mathbf{K})} \times$$

$$\exp(i\mathbf{K}(\mathbf{R} - \mathbf{R}')) \exp(iq(\mathbf{K})|z - z'|)\mathrm{d}^2 K, \quad (9.16)$$

with $q(\mathbf{K})$ given by:

$$q(\mathbf{K}) = \sqrt{\kappa_0^2 - \mathbf{K}^2}, \quad (9.17)$$

where the wavelength dependence of κ_0 (remember that we are dealing with the emission wavelength at this point) has been omitted. If we introduce Eq. (9.16) in our expression for $U^{fl}(\mathbf{r})$ (note that the same identical expression will appear for $U_{bio}(\mathbf{r})$, no need to duplicate this derivation) the complete solution for the average emitted intensity $U^{fl}(\mathbf{r})$, becomes:

$$U^{fl}(\mathbf{r}) = \frac{1}{4\pi D} \int_{-\infty}^{\infty} \left[\int_V S_{fl}(\mathbf{R}', z') \frac{i}{2\pi q(\mathbf{K})} \times \right.$$

$$\left. \exp(i\mathbf{K}(\mathbf{R} - \mathbf{R}')) \exp(iq(\mathbf{K})|z - z'|)\mathrm{d}^3 r' \right] \mathrm{d}^2 K, \quad (9.18)$$

where we have simply written $S_{fl}(\mathbf{r})$ in cylindrical coordinates, $S_{fl}(\mathbf{R}, z)$. Since the volume of the diffusive medium, V, extends to infinity, we can identify terms within the volume integral as the Fourier transform of the sources S_{fl}:

$$\widetilde{S}_{fl}(\mathbf{K}, z) = \frac{1}{4\pi^2} \int_{-\infty}^{\infty} S_{fl}(\mathbf{R}', z) \exp(-i\mathbf{K} \cdot \mathbf{R}')\mathrm{d}^2 K, \quad (9.19)$$

in which case we can express Eq. (9.18) as:

$$U^{fl}(\mathbf{r}) = \frac{1}{4\pi D} \int_{-\infty}^{\infty} \frac{i}{2\pi q(\mathbf{K})} \left[\int_{-\infty}^{\infty} 4\pi^2 \widetilde{S}_{fl}(\mathbf{K}, z') \times \right.$$

$$\left. \exp(iq(\mathbf{K})|z - z'|)\mathrm{d}z' \right] \exp(i\mathbf{K} \cdot \mathbf{R})\mathrm{d}^2 K. \quad (9.20)$$

In order to simplify things, let us now consider that the distribution of emitting sources (could be either fluorescent or bioluminescent) is located at a single plane $z = z_0$. In this case the integral over z' in the equation above becomes:

$$U^{fl}(\mathbf{r}) = \frac{1}{4\pi D} \int_{-\infty}^{\infty} \frac{i}{2\pi q(\mathbf{K})} 4\pi^2 \widetilde{S}_{fl}(\mathbf{K}) \exp(iq(\mathbf{K})|z - z_0|) \exp(i\mathbf{K} \cdot \mathbf{R})\mathrm{d}^2 K,$$

$$(9.21)$$

where we can now identify the angular spectrum of the fluorescent (or bioluminescent) sources (see Sec. 6.2 of Chap. 6):

$$\mathcal{A}_{fl}(\mathbf{K}) = \frac{1}{4\pi D}\frac{i}{2\pi q(\mathbf{K})}4\pi^2 \widetilde{S}_{fl}(\mathbf{K}). \tag{9.22}$$

In terms of its angular spectrum, we saw that the total fluorescence intensity at any plane of measurement z would be given by (compare with Eq. (6.16) in Chap. 6):

$$U^{fl}(\mathbf{r}) = \int_{-\infty}^{\infty} \mathcal{A}_{fl}(\mathbf{K}) \exp(iq(\mathbf{K})|z - z_0|) \exp(i\mathbf{K} \cdot \mathbf{R}) d^2 K, \tag{9.23}$$

where we see our spatial transfer function once again. As discussed in Chap. 6 what this equation represents is the fact that on propagation spatial frequencies originally present in our collection of sources are damped in different proportions according to the spatial transfer function.

9.3.1 *Recovering size and position in diffuse media*

We are now in the position to analyze how feasible it is to recover information that has left our collection of sources, in particular the distribution of fluorophore concentration (or luciferin substrate which has undergone a chemoluminescent reaction). Let us consider we are performing a measurement of the total average intensity emitted by our collection of sources located at $z = z_0$ at a measurement plane $z_d > z_0$. In this case the expression for the emitted average intensity would be given in Fourier space by:

$$\widetilde{U}_0^{fl}(\mathbf{K}, z) = \mathcal{A}_{fl}^{z_0}(\mathbf{K}) \exp(iq(\mathbf{K})(z_d - z_0)), \tag{9.24}$$

where now $\mathcal{A}_{fl}^{z_0}$ stands for the angular spectrum of the distribution of sources at $z_d = z_0$. Similarly, we can have a different collection of sources S_{fl}' at $z_1 < z_d$ with an angular spectrum $\mathcal{A}_{fl}^{z_1}$ which would yield in principle a different distribution of light intensity at z:

$$\widetilde{U}_1^{fl}(\mathbf{K}, z) = \mathcal{A}_{fl}^{z_1}(\mathbf{K}) \exp(iq(\mathbf{K})(z_d - z_1)). \tag{9.25}$$

In order to showcase the ill-posed nature of diffuse light propagation let us now consider the following: I have not yet imposed any specific form to our source distribution, neither at z_0 nor at z_1. And, in principle, in any imaging experiment our objective is to recover the location of our emitters (in case we are interested in fluorescence or bioluminescence, that is, the same holds for imaging absorption), from which we do not know anything *a priori.* Our objective here is to see if there is any (any whatsoever) possible combination of source distributions which would give me the *same* measurement at z_d for two different distributions at z_0 and z_1. If this combination exists, this reflects the fact that with this single measurement at z it would be impossible to distinguish between them, reflecting the nature of the ill-posed problem.

The conditions under which Eq. (9.24) and Eq. (9.25) give the same result can be obtained directly as (assuming $z_1 > z_0$):

$$\mathcal{A}_{fl}^{z_1}(\mathbf{K}) = \mathcal{A}_{fl}^{z_0}(\mathbf{K}) \exp(iq(\mathbf{K})(z_1 - z_0)), \qquad (9.26)$$

where it can be easily seen that substitution of this expression into Eq. (9.25) yields Eq. (9.24). This expression, however, is still not proof enough; we need to see if this solution exists for $z_1 \neq z_0$. If not, then only the $z_1 = z_0$ solution exists, which means that the problem is well-posed (we have a single source distribution represented in each individual measurement). The solution to this question lies in the transfer function, $\exp(iq(\mathbf{K})(z_1 - z_0))$.

9.3.1.1 *Ill-posed nature of Propagating Scalar Waves*

So as to better understand the effect of the transfer function, let us first consider the case for propagating scalar waves we already encountered in Chap. 6 when discussing the issue of resolution. In this case κ_0 was real, $\kappa_0 = 2\pi/\lambda$, with λ being the wavelength of the propagating wave. Considering that we are measuring far enough from the sources to neglect the near-field contribution (i.e. at $z > \lambda$), the transfer function is simply a band-pass filter for all frequencies $|\mathbf{K}| < 2\pi/\lambda$. In fact, in this case we may write the wave-vector pointing into z as $q(\mathbf{K}) = K_z = \kappa_0 \cos\theta$. Additionally, since all other frequencies $|\mathbf{K}| > 2\pi/\lambda$ are not measured at z our equation Eq. (9.26) still holds. The transfer function for these propagating frequencies (we shall represent their angular spectrum as $\mathcal{A}_{prop}$) is therefore simply a phase factor:

$$\mathcal{A}_{prop}^{z_1}(\mathbf{K}) = \mathcal{A}_{prop}^{z_0}(\mathbf{K}) \exp\left(i\frac{2\pi(z_1 - z_0)\cos\theta}{\lambda}\right), \theta \in [-\pi/2, \pi/2], \quad (9.27)$$

with $\mathbf{K}$ given by $|\mathbf{K}| = \kappa_0 \sin\theta$, which as we saw in Chap. 6 is what gives the angular spectrum its name. Eq. (9.27) represents that in the case of propagating scalar waves, if we are capable of measuring the amplitude *and* the phase at z there is an infinite set of source distributions which would give the same measurement having all the same spatial frequencies and a phase relation between them. Even in this simple case of propagating waves, we see that there is an ambiguity with depth and we have an infinite set of solutions. However, and this is a very important point, there is *no change* in the amplitude of the angular spectrum, only on its phase — once we have neglected the evanescent components, of course. We might not be able to recover the depth from a single measurement, but Eq. (9.27) tells us that the solution is our measurement propagated into z with only a change of phase. This, in fact, is the basis for x-ray tomography, where we assume that the angular spectrum does not change and we simply backpropagate it through z (note that in this case we only measure intensity, and have to find a solution without any phase information). In other words, if the source distribution is a disc of radius R, the average intensity throughout z will also have a half-width of R. We of course need different projections to find the 3D distribution or, equivalently, enough measurements to cover the complete $\mathbf{K}$ space, but at least our spatial frequency distribution is constrained. Note also that in the case we just presented, we have lost all frequencies $|\mathbf{K}| > 2\pi/\lambda$, having access only to those within the range $|\mathbf{K}| < 2\pi/\lambda$, within what is termed the Ewald sphere [Nieto-Vesperinas (2006); Born and Wolf (1999)]. In this case, the imaging approach is ill-posed but not severely. Note that there is still a great deal of work on what is termed the phase retrieval problem since, as we just mentioned, what can be obtained in a more or less straightforward manner is the amplitude of the angular spectrum, but not its phase.

9.3.1.2   *Ill-posed nature of Diffuse Waves*

As we just saw for propagating scalar waves the distribution of the spatial frequencies of the angular spectrum does not change once we neglect the evanescent components. This is in strong contrast to what happens to waves of diffuse light, where there is always a complex component to κ_0 in which case not only all frequencies get attenuated but they do so in different amounts. We can now perform a simple back-of-the-envelope calculation to show how it is not possible to recover not only the depth but also the diameter (angular spectrum really) of our object after light has propagated through a diffusive medium. Let us consider the simple case where we have a Gaussian distribution of the sources with halfwidth w_g at $z = 0$:

$$S_{fl}(\mathbf{R}, z) = \frac{S_0}{\pi w_g^2} \exp\left(-\frac{\mathbf{R}^2}{w_g^2} \right) \delta(z).$$

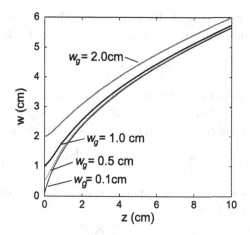

Fig. 9.2   Width, $w(z)$, of equivalent angular spectrum which yields the same intensity distribution as a Gaussian source distribution of half-width w_g at z=0.

In this case, its expression in Fourier space would be given by (see Appendix A):

$$\widetilde{S}_{fl}(\mathbf{K}, z) = \frac{S_0}{4\pi^2} \exp\left(-\frac{|\mathbf{K}|^2}{\widetilde{w}_g^2}\right)\delta(z),$$

with $\widetilde{w}_g = 2/w_g$.

Using Eq. (9.22), the angular spectrum of this distribution of sources is given by:

$$\mathcal{A}_{fl}(\mathbf{K}) = \frac{S_0}{4\pi D}\frac{i}{2\pi q(\mathbf{K})}\exp\left(-\frac{|\mathbf{K}|^2}{\widetilde{w}_g^2}\right),$$

and the fluorescence measurement at a plane z_d would be given by:

$$\widetilde{U}^{fl}(\mathbf{K}, z_d) = \frac{S_0}{4\pi D}\frac{i}{2\pi q(\mathbf{K})}\exp\left(-\frac{|\mathbf{K}|^2}{\widetilde{w}_g^2}\right)\exp(iq(\mathbf{K})z_d).$$

We can now define a set of source distributions at $z < z_d$ which would give the same measurement at z_d (see Eq. (9.24)) as:

$$\mathcal{A}_{fl}^z(\mathbf{K}) = \frac{S_0}{4\pi D}\frac{i}{2\pi q(\mathbf{K})}\exp\left(-\frac{|\mathbf{K}|^2}{\widetilde{w}_g^2}\right)\exp(iq(\mathbf{K})z).$$

So far no approximations have been made. However, we can approximate the transfer function to a Gaussian of full width at half maximum

$\Delta|\mathbf{K}|$ which decays exponentially as $\kappa_0 z$ (note that the $\mathbf{K} = 0$ case represents the maximum value of the transfer function, and $q(\mathbf{K} = 0) = \kappa_0$):

$$\widetilde{H}(\mathbf{K}, z) \simeq \exp\left(-\frac{|\mathbf{K}|^2}{\Delta|\mathbf{K}|^2/4}\right) \exp(i\kappa_0 z),$$

with the expression for $\Delta|\mathbf{K}|$ given by Eq. (6.52) in Chap. 6 (note that $\Delta|\mathbf{K}|$ represents the full width at half max, and w_g represents the half width). Using this approximation the expression for $\mathcal{A}_{fl}^z$ becomes:

$$\mathcal{A}_{fl}^z(\mathbf{K}) \simeq \frac{S_0}{4\pi D} \frac{i}{2\pi q(\mathbf{K})} \exp\left(-\frac{|\mathbf{K}|^2}{\widetilde{w}(z)^2}\right) \exp(i\kappa_0 z),$$

where now the halfwidth of the angular spectrum has been approximated to a product of two Gaussians:

$$\widetilde{w}(z) = \frac{\widetilde{w}_g \Delta|\mathbf{K}|/2}{\sqrt{\widetilde{w}_g^2 + \Delta|\mathbf{K}|^2/4}},$$

with $\Delta|\mathbf{K}|$ given by:

$$\Delta|\mathbf{K}| = 2\sqrt{\left(\kappa_{\Im m} + \frac{\ln 2}{z}\right)^2 + \kappa_{\Re e}^2 - \kappa_{\Im m}^2 - \frac{\kappa_{\Im m}^2 \kappa_{\Re e}^2}{(\kappa_{\Im m} + \ln 2/z)^2}}.$$

Note that as $z \to 0$, we have that $\Delta|\mathbf{K}| \to \infty$, in which case $\widetilde{w} = \widetilde{w}_g$ as expected.

Within this approximation the half-width in real space of the angular spectrum at z, $\mathcal{A}_{fl}^z$, will be given by:

$$w(z) = \frac{2}{\widetilde{w}(z)} = w_g \sqrt{1 + \left(\frac{4}{w_g \Delta|\mathbf{K}|}\right)^2},$$

where we have made use of $w_g = 2/\widetilde{w}_g$. Note that this is the half-width of the angular spectrum which gives the same measurement at z_d than our original or 'real' source distribution at $z = 0$, $\mathcal{A}_{fl}$. In the cw case we were initially considering in this chapter we had that $\kappa_0 = i\kappa_{\Im m}$ in which case $\Delta|\mathbf{K}|$ becomes:

$$\Delta|\mathbf{K}| = 2\sqrt{2\kappa_{\Im m}\frac{\ln 2}{z} + \left(\frac{\ln 2}{z}\right)^2}.$$

Introducing this expression into $w(z)$, rearranging terms and writing $\kappa_{\Im m}$ as $\kappa_{\Im m} = \sqrt{\mu_a/D}$ we can express $w(z)$ as:

$$w(z) = w_g \sqrt{1 + \left(\frac{2z}{w_g}\right)^2 \frac{1}{2\sqrt{\mu_a/D}\ln 2z + (\ln 2)^2}}, \qquad (9.28)$$

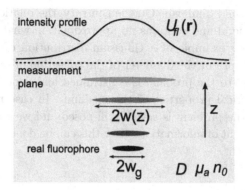

Fig. 9.3 Representation of how different source distributions with different intensities yield the same average intensity profile at a measurement plane. The evolution of the half-width at each z position is shown in Fig. 9.2

which behaves as expected for $z = 0$ (giving $w(z = 0) = w_g$) and for $z \to \infty$ (yielding $\lim_{z \to \infty} w(z) \sim \sqrt{z}$). The shape of $w(z)$ for several half-widths w_g is shown Fig. 9.2 (note that in the case of propagating waves $w(z) = w_g$ would be constant throughout z).

Let us now try to understand the deeper meaning of this simple derivation. In practical terms, if we have a distribution of sources of diameter $2w_g$ and intensity S_0 at $z = 0$, Eq. (9.28) tells us that any distribution of sources at $z > z_0$ with diameter $2w(z)$ and intensity $S_z = S_0 \exp(i\kappa_0 z)$ will give the same intensity distribution at our measurement plane $z_d > z$ (see Fig. 9.3). This is remarkable, given the fact that we are assuming we know the optical properties of the diffusive medium, and it tells us that from a single measurement it is impossible to determine both the depth and the size of a collection of sources *even* if we know the background optical properties.

Several points should be highlighted before moving on. First of all, this result is equivalent for sources, scatterers and absorbers. Since it has been formulated in terms of the angular spectrum, it applies identically to absorption, fluorescence and bioluminescence measurements. Secondly, this ambiguity exists even if we assume we know the average optical properties of the medium. In case these are not known (as in most practical instances) this introduces an even larger uncertainty. Another issue which should not be forgotten is that the expression for $w(z)$ shown in Eq. (9.28) is an approximation assuming we have a collection of sources with a Gaussian distribution. This does not mean that in other more complex source distributions this ambiguity could diminish. On the contrary, the relationship shown in Eq. (9.26) is general and contains no approximation whatsoever. Finally, going back to our example of a Gaussian distribution of fluorophores in case either the depth or the diameter of the collection of sources is known, then the solution to the problem is constrained and depends on how well we know the optical properties of the medium. In case neither of them is known, then the problem is severely ill-posed and we somehow need to increase the amount of information. How this can be done will be discussed in the next section.

9.4   Reducing Ill-posedness

As we saw in the previous section, there exists a fundamental difference between propagating scalar waves and diffuse waves based on their difference in the spatial transfer function. In the case of propagating waves we studied how their angular spectrum was only affected by a phase and that this angular spectrum could be recovered by increasing the number of measurements in the K-space, at least for the frequencies within the Ewald sphere. This can be done, for example, by rotating the sample as in X-ray computed tomography, where the amplitude of the angular spectrum is retrieved. Note that for propagating scalar waves the phase cannot usually be retrieved unless measurements of both phase and amplitude take place. In the case of diffuse waves, however, we saw that the angular spectrum changes as it propagates and recovering it is not as straightforward as in the case of propagating scalar waves. Exploring the different methods which increase the amount of information and thus reduce the ill-posed nature of the problem is the purpose of this section; we will be considering the following main approaches: introducing a spatial dependence on the emission (scanning the excitation source, for example), normalizing the data, multispectral imaging, having distinct background absorption features and introducing phase information (time resolved or frequency modulated).

9.4.1   *Introducing a spatial dependence on the emission*

One way of reducing the ill-posed nature of the problem is by changing, *in a controlled manner*, the intensity of the emission so that a spatial-dependence is introduced. In the case of fluorescence this can be done directly through the excitation source, $U_{ex}(\mathbf{r})$:

$$S_{fl}(\mathbf{r}) = \sigma_a[C(\mathbf{r})]\Phi(\lambda_{em})U_{ex}(\mathbf{r}).$$

Note that it is not sufficient to simply change the intensity of the excitation source, but the actual profile of the excitation intensity within the diffusive volume. This can be done by scanning an excitation point source at different positions, for example, or by generating different excitation patterns within the volume. In this case finding the fluorophore distribution results in solving the following set of equations, one for of each source position $\mathbf{r}_{si}$:

$$U^{fl}(\mathbf{r})(\mathbf{r_d}, \mathbf{r}_{si}) = \sigma_a\Phi(\lambda_{em})\int_V G_{cw}^{em}(\mathbf{r_d} - \mathbf{r}')[C(\mathbf{r}')]U_{ex}(\mathbf{r}', \mathbf{r}_{si})\mathrm{d}^3r', \quad (9.29)$$

with $U^{fl}(\mathbf{r})(\mathbf{r_d}, \mathbf{r}_{di})$ representing the emission measurement at $\mathbf{r_d}$ due to the source at $\mathbf{r}_{si}$ and $U_{ex}(\mathbf{r}', \mathbf{r}_{si})$ being the excitation average intensity at a point $\mathbf{r}'$ within the diffusive volume generated by a source (or source pattern) $\mathbf{r}_{si}$.

Solving the above set of equations for $[C(\mathbf{r})]$ will be the motivation of the next chapter and is termed solving the inverse-problem. For the time being, it is sufficient to know that by acquiring a large number of independent measurements we are reducing the ill-posed nature of the problem since all these measurements correspond to the same fluorophore distribution. A similar approach can be done in absorption imaging, since the absorbed light will depend on the position of the sources. However, and this is extremely important, such an option does not currently exist for bioluminescence: there is no way to externally introduce a spatial dependence on the emission of a bio or chemoluminescent reaction. It is difficult in itself to control that there is enough substrate present where the enzyme (the luciferase, for example) is expressed, let alone control externally at will this amount or the efficiency of the reaction. This is one of the major drawbacks of bioluminescence in the sense that there is no way (at least currently) to introduce a spatial dependence and thus greatly reduce the ill-posed nature of the problem no matter how many external measurements we take.

9.4.2 *Normalized measurements*

One approach which is extremely effective specifically when dealing with experimental data is our ability to normalize the measurements. In the case of fluorescence it is straightforward since we can use the measurement of the excitation light and combine it with the spatial dependence on the emission we just introduced:

$$U_n(\mathbf{r_d}, \mathbf{r}_{si}) = \frac{U^{fl}(\mathbf{r})(\mathbf{r_d}, \mathbf{r}_{si})}{U_{ex}(\mathbf{r_d}, \mathbf{r}_{si})} =$$

$$\sigma_a \Phi(\lambda_{em}) \int_V G_{cw}^{em}(\mathbf{r_d} - \mathbf{r}')[C(\mathbf{r}')] \frac{U_{ex}(\mathbf{r}', \mathbf{r}_{si})}{U_{ex}(\mathbf{r_d}, \mathbf{r}_{si})} d^3 r'. \quad (9.30)$$

This approach does not reduce significantly the ill-posedness of the problem; however, it greatly reduces uncertainty present in an experiment. In particular, it takes care of unknowns related to camera efficiency, emission filters and excitation source values. Additionally, and I would say more importantly, it reduces problems due to the fact that we do not know accurately the background optical properties. Normalizing the emission by the excitation greatly reduces the uncertainty in our measurements since we are in fact dividing two transfer functions: even if we do not know the optical properties accurately, the resulting division would reduce the contribution of the unknown values greatly. This can be seen directly if we consider a point source located at $z = 0$ which excites a point fluorophore of concentration $[C(\mathbf{r})] = [C]\delta(\mathbf{R})\delta(z - z_0)$ at $z = z_0$, placing our measurement plane at $z = z_d$. In this case the fluorescence average intensity in the spatial Fourier domain would be given by:

$$\widetilde{U}^{fl}(\mathbf{K}, z_d) = \widetilde{S}_{fl}(\mathbf{K}) \left(\frac{1}{4\pi D} \frac{i}{2\pi q(\mathbf{K})} \right) \exp(iq(\mathbf{K})(z_d - z_0)),$$

which introducing the expression for $\widetilde{S}_{fl}$ in and the expression for $U_{ex}(\mathbf{r})$ in the angular spectrum representation gives:

$$\widetilde{U}^{fl}(\mathbf{K}, z_d) = S_0 \sigma_a \Phi(\lambda_{em})[C] \left(\frac{1}{4\pi D} \frac{i}{2\pi q(\mathbf{K})} \right)^2 \exp(iq(\mathbf{K})z_d).$$

Clearly, if we normalize the above by the excitation average intensity we obtain:

$$\frac{\widetilde{U}^{fl}(\mathbf{K}, z_d)}{\widetilde{U}^{ex}(\mathbf{K}, z_d)} = \sigma_a \Phi(\lambda_{em})[C] \left(\frac{1}{4\pi D} \frac{i}{2\pi q(\mathbf{K})} \right),$$

which as can be seen has removed the transfer function and the source excitation intensity dependency. Note, of course, that several approximations have been introduced to illustrate the efficiency of normalizing the data: first of all, we have considered the absorption properties identical for excitation and emission, which is rarely the case. Secondly and most importantly, this is the solution to a point source, not the general solution. However, this simple expression clearly shows how normalizing the emission with the excitation accounts for errors in our modeling of the background optical properties. Note that a similar approach can also be used for absorption measurements, normalizing the total scattered intensity by the signal we would measure in the absence of the absorbing or scattering object (if this measurement is available).

9.4.3 *Multispectral imaging*

Another way to increase the information of the system is to make use of the few things we know for certain, being one of them the emission spectrum[1]. In principle, since we selected the fluorescent probe or protein we know its emission profile for a large number of emission wavelengths. We can use this to our advantage, assuming we somehow know the values of the optical properties of the medium, and introduce more information into our system of equations by adding relationships of the form:

$$U^{fl}(\mathbf{r_d}, \lambda_{em}^i) = \sigma_a \Phi(\lambda_{em}^i) \int_V G_{cw}^{em}(\mathbf{r_d} - \mathbf{r}'; \lambda_{em}^i)[C(\mathbf{r}')]U_{ex}(\mathbf{r}')\mathrm{d}^3r', \quad (9.31)$$

where λ_{em}^i represents the different spectral wavelengths measured. The same approach can be used in this case for bioluminescence:

$$U_{bio}(\mathbf{r_d}, \lambda_{em}^i) = \frac{E_{bio}(\lambda_{em}^i)}{\tau_{bio}(\lambda_{em}^i)} \Phi_{bio} \int_V G_{cw}^{em}(\mathbf{r_d} - \mathbf{r}'; \lambda_{em}^i)[C(\mathbf{r}')]_{bio}\mathrm{d}^3r'. \quad (9.32)$$

Note, however, that the amount of information gained is very limited since all these measurements are strongly correlated (they are practically related by a constant). To express this point more clearly, consider the situation in which the average background optical properties do not change with wavelength, i.e. μ_{a0} has no wavelength dependence. In this case $G_{cw}^{em}(\mathbf{r_d} - \mathbf{r}')$ is identical for all cases. Clearly we have then that for two emission wavelengths λ_{em}^i and λ_{em}^j:

$$\frac{U^{fl}(\mathbf{r_d}, \lambda_{em}^i)}{U^{fl}(\mathbf{r_d}, \lambda_{em}^j)} = \frac{\Phi(\lambda_{em}^i)}{\Phi(\lambda_{em}^j)}, \quad (9.33)$$

and:

$$\frac{U_{bio}(\mathbf{r_d}, \lambda_{em}^i)}{U_{bio}(\mathbf{r_d}, \lambda_{em}^j)} = \frac{E_{bio}(\lambda_{em}^i)\tau_{bio}(\lambda_{em}^j)}{E_{bio}(\lambda_{em}^j)\tau_{bio}(\lambda_{em}^i)}, \quad (9.34)$$

---

[1]This is not entirely accurate: you might remember that we mentioned in Chap. 1 that the emission spectrum of a fluorophore or a bioluminescent reaction depends very much on the local environment, and that shifts in the emission are actually used to quantify local changes in pH.

which clearly states that this collection of measurements does not offer additional information. This approach works very well when combined with the previous approach (scanning the excitation source or generating different emission intensity patterns) for imaging fluorescent probes and it is especially useful for getting rid of unwanted signal such as autofluorescence.

9.4.3.1 *Distinct background absorption features*

As we just saw, multi-spectral measurements on their own do not offer enough information to determine the problem, even though they might be used to reduce the contribution of the autofluorescence. This is due to the fact that, in itself, multi-spectral information does not introduce any spatial dependence on the emission profile. However, if there are distinct features on the *absorption* spectrum for those emission wavelengths, some depth information can be inferred, even if it is not too determinant. In this case, the functions $G_{cw}^{em}(\mathbf{r_d} - \mathbf{r}')$ are indeed different for each wavelength and the new measurements do introduce additional information. How different these functions are will depend on the distinct features of the background absorption: smooth changes will reflect in very subtle differences in the transfer function, whereas sharp changes will reflect higher information content through distinct transfer functions. This can be seen directly on our expression for the fluorescence (or bioluminescence) in the Fourier domain measured at z_d at wavelength λ_{em}^i:

$$\widetilde{U}^{fl}(\mathbf{K}, z_d, \lambda_{em}^i) = \mathcal{A}_{fl}^i(\mathbf{K}) \exp(iq^i(\mathbf{K})z_d),$$

in which case comparing two measurements taken at λ_{em}^i and λ_{em}^j yields:

$$\frac{\widetilde{U}^{fl}(\mathbf{K}, z_d, \lambda_{em}^i)}{\widetilde{U}^{fl}(\mathbf{K}, z_d, \lambda_{em}^j)} = \frac{\Phi(\lambda_{em}^i)}{\Phi(\lambda_{em}^j)} \exp\left(i(q^i(\mathbf{K}) - q^j(\mathbf{K}))z_d\right),$$

with:

$$q^{i,j}(\mathbf{K}) = \sqrt{\frac{-\mu_{a0}(\lambda_{em}^{i,j})}{D} - \mathbf{K}^2}.$$

Similarly, for bioluminescence we would obtain:

$$\frac{\widetilde{U}_{bio}(\mathbf{K}, z_d, \lambda_{em}^i)}{\widetilde{U}_{bio}(\mathbf{K}, z_d, \lambda_{em}^j)} = \frac{E_{bio}(\lambda_{em}^i)\tau_{bio}(\lambda_{em}^j)}{E_{bio}(\lambda_{em}^j)\tau_{bio}(\lambda_{em}^j)} \exp\left(i(q^i(\mathbf{K}) - q^j(\mathbf{K}))z_d\right),$$

Clearly, the more distinct q^i and q^j are, the more information on z we will be able to obtain by using spectral measurements. With regards to small animal imaging we can then use to our advantage the absorption spectrum of blood which for visible frequencies has very distinct features, diminishing as we move towards the near infrared. This approach is being exploited in several commercial imaging setups and is based on the *a priori* knowledge of the amount of blood present. Obviously, we cannot know with great accuracy these values but they are at least a good indicator. However, note that this by no means resolves the ill-posed nature of the problem and that results, both in terms of quantitation and localization/resolution, should be taken with a pinch of salt if this is the only additional information used. A similar approach has been used by the group of S. Andersson-Engels [Axelsson *et al.* (2007)] to generate spatially varying regularization maps based on the ratio between two wavelength measurements. Whichever approach is used, however, it is important to remember that the closer μ_a^i is to μ_a^j the closer we are to not providing any additional information at all.

9.4.4   *Phase Information*

Additional information can be provided if we use time-resolved or frequency modulated sources and are capable of measuring either the spatial distribution of the intensity in time or the amplitude and phase of the spatial distribution. In this case, since we know the phase of the source the dephase we measure will have implicit some depth information. In particular, if we modulate the intensity the diffusion equation in the case of fluorescence becomes:

$$U^{fl}(\mathbf{r}, \omega) = \int_{-\infty}^{\infty} G_\omega^{em}(\mathbf{r} - \mathbf{r}')S_{fl}(\mathbf{r}', \omega)\mathrm{d}^3r,$$

with $S_{fl}(\mathbf{r}, \omega)$ given by:

$$S_{fl}(\mathbf{r}, \omega) = \frac{1}{\tau} \int_0^\infty S_{fl}(\mathbf{r}, t)\exp(i\omega t)\mathrm{d}t,$$

which yields (see Eq. (1.2) in Chap. 1):

$$S_{fl}(\mathbf{r}, \omega) = \sigma_a \Phi[C(\mathbf{r})]U_{ex}(\mathbf{r}, \omega)\frac{1}{1 - i\omega\tau}, \qquad \left(Watts/cm^3\right) \qquad (9.35)$$

solution which we may reach as long as we assume $\omega \ll 1/\tau$, implying that the dynamics of probe excitation and emission occur at time scales much shorter than the changes in the average intensity $U_{ex}(\mathbf{r})$ that excites the fluorophores. In terms of the angular spectrum and the transfer function, there is now a phase relationship when measuring at z_d, and thus when we compare two possible solutions which would yield the same measurement we can express their relationship as (see Eq. (9.26)):

$$\mathcal{A}_{fl}^{z_1}(\mathbf{K}) = \left(\mathcal{A}_{fl}^{z_0}(\mathbf{K}) \exp(-q_{\Im m}(\mathbf{K})(z_1 - z_0)) \right) \exp(iq_{\Re e}(\mathbf{K})(z_1 - z_0))$$

and in this case the amount of additional information will depend on how distinct this dephase is for different depths. Note that this dephase does not reduce the ill-posedness implicit in the complex part of $q(\mathbf{K})$ but still it introduces additional (and very useful) information, begin equivalent to the phase relation case we found for propagating scalar waves. As an approximation consider that the $\mathbf{K} = 0$ frequency case is $q_{\Re e} = \kappa_{\Re e}$ with $\kappa_{\Re e}$ being defined as (see Chap. 6, Eq. (6.6)):

$$\kappa_{\Re e} = \frac{1}{\sqrt{2}} \sqrt{\left(\frac{\omega^2}{D^2 c_0^2} + \frac{\mu_a^2}{D^2} \right)^{\frac{1}{2}} - \frac{\mu_a}{D}}.$$

In this case, the dephase would be given by:

$$\delta = \frac{2\pi}{\lambda_0}(z_1 - z_0),$$

with $\lambda_0 = 2\pi/\kappa_{\Re e}$ representing the wavelength of the diffuse wave. As we saw in Chap. 6, we can approximate λ_0 for low modulation frequencies in which case the dephase becomes:

$$\delta \simeq \frac{\omega}{c_0} \frac{z_1 - z_0}{2\sqrt{D\mu_{a0}}}.$$

This dephase, found for $\mathbf{K} = 0$, shows that the lower the modulation frequency the less the additional information. It is clear that if our diffuse wavelength is in the order of the size of the system (as would typically be the case for a modulation frequency in the 200MHz range), then the dephase introduced is quite small and only large $z_1 - z_0$ distances offer significant and measurable changes. Typical values for 200MHz modulation and optical properties as those used throughout this book would yield dephase values in the range of 30-40 degrees per cm. Equivalently, we can say that changes of 1mm in depth result in a 3 to 4 degree dephase, which is quite probably near the detection limits. This does not mean that the phase measured cannot be used to improve our inverse problem. On the contrary, it simply points to the fact that reducing the ill-posed nature of the problem would require very high frequencies or equivalently temporal measurements with very high dynamic range produced by very short pulses.

In the case where we can use temporal measurements directly there are also commercially available setups which make use of the fact that different depths yield a temporal profile with shifts in the time where maximum intensity is detected. Using this shift of the intensity profile with time some depth information can be recovered, similar to what we just discussed for multi-spectral measurements. Depth information obtained in this manner, however, should be regarded with great caution.

At this point we can sum up all the contributions and clearly state that the greatest amount of information can be obtained when we combine all possible approaches: source scanning (or pattern illumination) to introduce some spatial information, normalized measurements, multi-spectral measurements with, if possible, some spectral features in the absorption spectrum, and making use of time-resolved or frequency modulated measurements. Once the measurements have been performed, the amount of information we have obtained can be further improved if we can make use of *a priori* knowledge.

Table 9.1 Approaches to reduce the ill-posed nature of diffuse light

Source of Contrast	Spatial	Normalized	Multispectral	Phase
Absorption/scattering	Yes	Yes	No[a]	Yes
Fluorescence	Yes	Yes	Yes	Yes
Bioluminescence	No	No	Yes	No

[a] Yes if the object has a significantly different spectral absorption signature when compared to the background tissue and measurements are taken at several wavelengths.

Table 9.1 shows different approaches that can be used to reduce the ill-posed nature of diffuse waves. Ideally, combining the most of them will undoubtedly increase the information in our measurements to the maximum and thus reduce the ill-posedness significantly.

9.4.5 Background Signal

Table 9.2 Contribution of background signal to the measurement

Source of Contrast	Background Signal
Absorption/scattering	Changes in background absorption and/or scattering
Fluorescence	Autofluorescence and fluorescent probe unspecificity
Bioluminescence	No background signal

So far we have been dealing with the physical reasons behind the ill-posed nature of diffuse waves, and we saw that they can all be studied based on the spatial transfer function. This consideration was done assuming that there is no additional (and unwanted) source of contrast present in our measurement. There are, however, additional factors which introduce uncertainty in our measurements and thus increase the ill-posed nature of the inverse problem. One of them is noise, which will always be present in our measurements and which we can experimentally control to be kept to a minimum. Sources of noise are typically the readout noise of our detectors (a CCD camera, for example), thermal noise (which we can reduce by cooling the detector system), and shot noise, amongst others. On the excitation or illumination side, we can have uncontrolled fluctuations of our laser or lamp which can be partially accounted for by always having a readout of the power of our excitation sources. In those cases where we scan the sources or generate an excitation pattern, uncertainty in the position of the scanning beam also increases the ill-posed nature of our problem.

However, there is another source of uncertainty in our measurement which is more difficult to get rid of, which is the background signal contribution. What is meant by background signal in this case is a signal which is barely (if at all) distinguishable from the signal originated by the object we want to recover. In the case where we intend to measure an object based on its absorption or scattering properties, the background signal would be changes in the background scattering or absorbing properties which we are not accounting for. The only way to reduce this contribution is to improve our knowledge of the background optical properties, or by using a contrast probe that will enhance the absorption only (or mainly) at the site of the object we want to image. This was the approach used by V. Ntziachristos from the groups of B. Chance and A. Yodh in one of the first studies in humans using diffuse light to recover in 3D the spatial distribution of absorption and relate it to breast cancer [Ntziachristos *et al.* (2000)]. In this case Indocyanine Green (ICG) was used to increase the absorption contrast (note that ICG is also a flurophore, but no fluorescence was measured in this case). Since then several other studies in humans have made use of this source of contrast, mainly due to the fact that it is one of the few that have FDA approval. Note that as in any case where our source of contrast is unspecific, ICG will be distributed all over, and present a longer clearing rate at the tumor site due to the complex vascular network. In fact, this clearing rate can also be used to increase the information of our system (see [Hillman and Moore (2007)] for an example where this approach is used to superficially distinguish different organs based on their clearance rates).

In the case of fluorescence, the main source of background signal is tissue autofluorescence. As was mentioned in Chap. 1, this contribution can be reduced but never completely removed by feeding the animals a chlorophyll free diet. On the down side, this means that the low signal we are interested in measuring will have to compete with the presence of autofluorescence. On the up side, the emission spectrum of intrinsic fluorophores is usually quite distinguishable from that of the fluorescent probe or fluorescent protein used. This means that using multi-spectral imaging approaches can help in reducing the contribution of autofluorescence. Additionally to the autofluorescence present, another source of unwanted signal is related to how specific your fluorescent protein or probe is. In other words, if we are using an activatable probe and this probe is activated (i.e it fluoresces) in sites other that the one we want to image, there is absolutely no way of distinguishing one from the other, since both are generated by the same probe. The only way around this is through the development of brighter and more specific probes.

Finally, the case of bioluminescence is the most favorable in terms of the background contribution. Since the enzyme which catalyzes the bioluminescent reaction is only present at those sites we are interested in (typically cells which express luciferase, for example), we will only get signal from those sites as long as the substrate is capable of reaching them. This effectively means that there is no background signal in the case of bioluminescence, being this the main reason it is used in small animal imaging.

9.4.6 *Prior Information*

Another approach to reduce the ill-posed nature of our problem is to introduce prior information when it is available. This information could be of many sorts as, for example:

- **The size of the collection of sources**: If we know, for example, that the source distribution is a single point source and we can assume some optical properties for the surrounding medium, then we can localize this point source within the volume.
- **The number of the collection of point sources**: If we can assume our sources act as point sources and we introduce the actual number N of these sources, our solution is very constrained. Note, however, that this number is generally not known.

- **The position of the collection of sources**: If we know where our collection of sources are, then what is left is to find its size and concentration. Depending on the application we may have this type of information available, but in principle our aim of obtaining the distribution of fluorescent or bioluminescent probes involves the recovery of the position which is usually unknown.
- **Anatomical priors**: As techniques based on diffuse light imaging are evolving, we are seeing how other methods which offer anatomical information such as X-ray computed tomography or magnetic resonance imaging are being combined with these diffuse optical tomography setups. Since we have information on the anatomy, we can increase the amount of information by assigning optical properties to each organ and thus represent a much better model of the true transfer function, introducing additional information similar to having distinct background absorption values in multi-spectral imaging. The use of anatomical priors is, at least at the time I'm typing these words, one of the areas of optical molecular imaging which is attracting the most attention due to the advantages it represents not only in terms of the quality of the reconstructed images, but also as the anatomical reference diffuse light imaging lacked in the past.
- **Spatially constraining the solution**: The next step when we have available anatomical priors is to constrain where the distribution of sources should be. A great example is shown in [Hyde *et al.* (2009)] from the group of V. Ntziachristos where, besides of the anatomical priors which are available through X-ray CT, the solution has a greater weight at the cortex of the brain where they expect the fluorescent probe to activate.

Key points

From this chapter the following important points should be remembered:

- **Ill-posed nature**: We saw that the ill-posed nature of diffuse waves is significantly more severe than that of propagating waves of an equivalent wavelength. The main culprit behind this is the transfer function, and ultimately the diffuse wavenumber which always has an imaginary part.

- **Reducing ill-posedness**: There are several approaches that can be used to reduce the ill-posed nature of the problem, the most important of all being the ability to introduce some spatial information on the system. This can be done by either scanning the sources or generating illumination patterns. Since this is lacking for bioluminescence, we saw that there is a significant difference between fluorescence and bioluminescence in terms of the severity of their ill-posed nature. Additionally, multi-spectral and time or frequency modulated measurements will increase the amount of information we can extract from the system.

- **Noise and Background Contribution**: Apart from the physics which dictates how ill-posed the problem is, there are other factors which increase the uncertainty of our measurements. In particular, we saw that background signal is different for different modalities, and that it is the lack of background signal which makes bioluminescence so attractive.

- **Use of priors**: If somehow (through an additional anatomical image, for example) we have access to *a priori* information, we can use this knowledge both to improve how we predict light propagation in our specimen and to constrain our solution to a specific organ (when possible). This will boost both the quantitation and the resolution of our imaging experiment.

Further Reading

This chapter barely skims over the very complex and extense issue of ill-posed problems. In order to gain more general insight in the context of partial differential equations, I recommend the original works by Hadamard, [Hadamard (2003)]. In the more specific case of diffuse light, I recommend reading [Arridge and Lionheart (1998)] and the reviews by Simon Arridge, in particular [Arridge (1999)] and [Arridge (2011)] and the one written together with John Schotland [Arridge and Schotland (2009)]. The issue of reducing the ill-posed nature of diffuse waves by scanning the laser will be covered in greater detail in the next chapter on the inverse-problem. However, for the more novel approach of structured illumination to induce a known (or predictable at least) spatial distribution of intensity there are currently several interesting works on the topic, since it is attracting great interest. I suggest for example [Ducros *et al.* (2010)] and [Lukic *et al.*

(2009)], as a start. With respect to using priors, it is quite a novel approach and there is still much to be done. However, a good place to start is [Hyde *et al.* (2009)].

Chapter 10

Imaging in Diffusive Media: The Inverse Problem

Summary. In this last chapter we present the main features of the forward and inverse problems and the most commonly used approximations and procedures to solve them. Finally, we present a direct inversion formula based on the angular spectrum representation.

10.1 The Forward and Inverse Problem

Throughout this book we have been studying how to model light propagation in diffuse media, including those cases where we have fluorophores or fluorescent proteins present. In all these cases our goal was to derive the appropriate expressions, and obtain approximations whenever possible, in order to predict what the average intensity or the flux would look like at a certain position. In order to do so, we derived the boundary conditions and reflection and transmission coefficients enabling us to find expressions which now account for the presence of boundaries. All the above can be considered solving the forward problem, i.e. finding the average intensity or flux given a certain distribution of sources, scatterers and absorbers. One such equation that would represent the complete solution was derived in Chap. 5, Sec. 5.6 and could be written in the cw case as:

$$-D_0\nabla^2 U(\mathbf{r}) + \mu_{a0}U(\mathbf{r}) = S_0(\mathbf{r}) - \overline{\mu}_a(\mathbf{r})U(\mathbf{r}), \qquad (10.1)$$

where we had defined the spatially varying absorption as $\mu_a(\mathbf{r}) = \mu_{a0} + \overline{\mu}_a(\mathbf{r})$ and where we have assumed that there is only change in absorption, not scattering, in the medium. To simplify the notation we have grouped in S_0 all the reduced intensity contribution as a single source. We can find a general solution to the above equation as:

$$U(\mathbf{r}) = U^{inc}(\mathbf{r}) - \int_V G(\mathbf{r}',\mathbf{r})\bar{\mu}_a(\mathbf{r}')U(\mathbf{r}')\mathrm{d}^3r', \qquad (10.2)$$

where:

$$U^{inc}(\mathbf{r}) = \int_V S_0(\mathbf{r}')G(\mathbf{r}',\mathbf{r})\mathrm{d}^3r', \qquad (10.3)$$

with $G(\mathbf{r},\mathbf{r}')$ representing the Green function of the complete problem including the boundaries (but excluding the presence of the absorption inhomogeneity). Solving the non-linear expression in Eq. (10.2) represents the solution to the *forward problem* whilst making use of Eq. (10.2) to obtain the spatial distribution of $\bar{\mu}_a(\mathbf{r})$ would be solving the *inverse problem*. As we saw in Chap. 9, Sec. 9.3, solving the inverse problem is severely ill-posed since more than one distribution of $\bar{\mu}_a(\mathbf{r})$ can yield the same measurement $U(\mathbf{r})$[1].

We will now derive expressions which are simpler to work with for the inverse problem by first of all linearizing Eq. (10.2). Two approaches are commonly used for this purpose: the Born approximation and the Rytov approximations.

10.2   The Born Approximation

The Born approximation, named after Max Born (physics Nobel prize in 1954 and co-author of the all-time classic in optics [Born and Wolf (1999)]) is a perturbation method which has been applied to practically all instances where a linearization of a non-linear problem is needed. It was developed originally in the context of quantum mechanics but quickly diffused to all scattering-related problems. The basics behind the Born approximation consists on finding a solution to Eq. (10.2) as a series by first introducing the homogeneous solution as the zeroth order approximation:

$$U^{(0)}(\mathbf{r}) \simeq U^{inc}(\mathbf{r}). \qquad (10.4)$$

Introducing this expression into Eq. (10.2) we obtain the first order approximation:

$$U^{(1)}(\mathbf{r}) \simeq U^{inc}(\mathbf{r}) - \int_V G(\mathbf{r}',\mathbf{r})\bar{\mu}_a(\mathbf{r}')U^{(0)}(\mathbf{r}')\mathrm{d}^3r', \qquad (10.5)$$

---

[1]Note that in Chap. 9 we considered the ill-posed nature of fluorescence, but the results were general for light propagation in diffuse media.

which is commonly termed the first order Born approximation or simply the Born approximation. We can of course proceed further to any order N:

$$U^{(N)}(\mathbf{r}) \simeq U^{inc}(\mathbf{r}) - \int_V G(\mathbf{r}', \mathbf{r}) \overline{\mu}_a(\mathbf{r}') U^{(N-1)}(\mathbf{r}') \mathrm{d}^3 r', \qquad (10.6)$$

but it is not common to reach orders higher than two due to the complexity of their solutions. The complete perturbation series is termed the *Dyson equation* (typically found for the electric and magnetic fields) and is usually represented by Feynman diagrams whereas the equivalent for the electromagnetic intensity (or the correlation function) is termed the Bethe-Salpeter equation. Using the above, the first order Born approximation for an absorption perturbation becomes:

$$U(\mathbf{r}) \simeq U^{inc}(\mathbf{r}) - \int_V G(\mathbf{r}', \mathbf{r}) \overline{\mu}_a(\mathbf{r}') U^{inc}(\mathbf{r}') \mathrm{d}^3 r', \qquad (10.7)$$

which is an expression easy to deal with since U^{inc} is assumed known inside the medium.

We can proceed in a similar manner to obtain an expression for the contribution of fluorescence by approximating the excitation intensity at the fluorophore as the average intensity in the homogeneous medium, $U^{ex} = U^{inc}$:

$$U^{fl}(\mathbf{r}) \simeq \int_V G(\mathbf{r}', \mathbf{r}; \lambda_{em}) F(\mathbf{r}', \lambda_{em}) U^{inc}(\mathbf{r}', \lambda_{ex}) \mathrm{d}^3 r', \qquad (10.8)$$

where F accounts for the fluorophore concentration (see Eq. (9.8) from Chap. 9):

$$F(\mathbf{r}, \lambda_{em}) = \sigma_a \Phi(\lambda_{em})[C(\mathbf{r})]. \qquad \left(cm^{-1}\right) \qquad (10.9)$$

10.3 The Rytov Approximation

Another approach for linearizing Eq. (10.2) is the Rytov approximation, developed in the mid 1900's by Sergei Mikhalovich Rytov originally for the study of diffraction of light by ultrasound and extensively used to model light propagation in random media, including atmospheres[2]. The Rytov approximation linearizes our problem by assuming that the perturbation

---

[2]We should not forget that most of the progress in wave propagation in turbid media has been taken place on the subject of statistical radiophysics or wave propagation in the atmosphere in the decades from the 1950's to the 1970's. Note that this period coincides with the Cold War and Space Race.

due to the presence of the inhomogeneity only changes the phase of the incident wave:

$$U(\mathbf{r}) = U^{inc}(\mathbf{r}) \exp(\Phi(\mathbf{r})), \qquad (10.10)$$

where Φ could be complex (in the frequency modulated case as in the original derivation) but it is real in the *cw* regime we are currently interested in. As a reminder U^{inc} is the solution to:

$$-D_0 \nabla^2 U^{inc} + \mu_{a0} U^{inc} = S_0(\mathbf{r}). \qquad (10.11)$$

Additionally, we will write the homogeneous average intensity also as a phase:

$$U^{inc}(\mathbf{r}) = \exp(\Phi_0(\mathbf{r})), \qquad (10.12)$$

or equivalently:

$$\Phi_0(\mathbf{r}) = \log(U^{inc}(\mathbf{r})), \qquad (10.13)$$

so that the total average intensity is now defined as:

$$U(\mathbf{r}) = \exp(\Phi_0(\mathbf{r}) + \Phi(\mathbf{r})). \qquad (10.14)$$

The solution to the homogeneous equation in terms of Φ_0 is found by substituting Eq. (10.12) into Eq. (10.11) and dividing by U^{inc} as:

$$-D_0 \nabla^2 \Phi_0(\mathbf{r}) - D_0(\nabla\Phi_0(\mathbf{r}))^2 + \mu_{a0} = S_0(\mathbf{r}) \exp(-\Phi_0(\mathbf{r})). \qquad (10.15)$$

The total average intensity in this representation will be a solution to the full equation:

$$- D_0 \nabla^2 \exp(\Phi_0(\mathbf{r}) + \Phi(\mathbf{r})) + \mu_{a0} \exp(\Phi_0(\mathbf{r}) + \Phi(\mathbf{r})) = \\ S_0(\mathbf{r}) - \overline{\mu}_a(\mathbf{r}) \exp(\Phi_0(\mathbf{r}) + \Phi(\mathbf{r})). \qquad (10.16)$$

If we apply the Laplace operator on the expression for $U(\mathbf{r})$ and regroup we obtain:

$$\exp(\Phi_0(\mathbf{r}) + \Phi(\mathbf{r})) \left[- D_0 \nabla^2 \Phi_0(\mathbf{r}) - D_0(\nabla\Phi_0(\mathbf{r}))^2 - \right.$$

$$\left. D_0 \nabla^2 \Phi(\mathbf{r}) - D_0(\nabla\Phi(\mathbf{r}))^2 - 2D_0 \nabla\Phi_0(\mathbf{r}) \cdot \nabla\Phi(\mathbf{r}) \right] +$$

$$\mu_{a0} \exp(\Phi_0(\mathbf{r}) + \Phi(\mathbf{r})) = \\ S_0(\mathbf{r}) - \overline{\mu}_a(\mathbf{r}) \exp(\Phi_0(\mathbf{r}) + \Phi(\mathbf{r})), \qquad (10.17)$$

where we can now introduce the relation Eq. (10.15) and divide by $U(\mathbf{r})$:

$$S_0(\mathbf{r})\exp(-\Phi_0(\mathbf{r})) - D_0\nabla^2\Phi(\mathbf{r}) - D_0(\nabla\Phi(\mathbf{r}))^2 - 2D_0\nabla\Phi_0(\mathbf{r})\cdot\nabla\Phi(\mathbf{r}) =$$
$$S_0(\mathbf{r})\exp(-\Phi_0(\mathbf{r}) - \Phi(\mathbf{r})) - \overline{\mu}_a(\mathbf{r}), \quad (10.18)$$

where the μ_{a0} terms have canceled out. Grouping terms we finally obtain:

$$-D_0\nabla^2\Phi(\mathbf{r}) - D_0(\nabla\Phi(\mathbf{r}))^2 - 2D_0\nabla\Phi_0(\mathbf{r})\cdot\nabla\Phi(\mathbf{r}) = -\overline{\mu}_a(\mathbf{r}),$$

where since Φ represents a perturbation we have approximated $\exp(-\Phi_0(\mathbf{r}))(\exp(-\Phi(\mathbf{r})) - 1) \simeq 0$, which is strongly reinforced by the fact that the contribution at the source is expected to be mainly due to U^{inc}. The main assumption in the Rytov approximation consists now on neglecting the $(\nabla\Phi(\mathbf{r}))^2$ contribution (the conditions under which this assumption is valid are quite extensive and can be found in [Brown. Jr. (1966)]), so that we obtain:

$$-D_0\nabla^2\Phi(\mathbf{r}) - 2D_0\nabla\Phi_0(\mathbf{r})\cdot\nabla\Phi(\mathbf{r}) = -\overline{\mu}_a(\mathbf{r}), \quad (10.19)$$

We now need to assume that $\overline{\mu}_a(\mathbf{r})$ represents a *small* perturbation of the background. We can ensure this by representing $\overline{\mu}_a(\mathbf{r})$ as:

$$\overline{\mu}_a(\mathbf{r}) = \delta\overline{\mu}_a(\mathbf{r}) \quad (10.20)$$

with δ begin a unit-less constant which ensures that the absorption coefficeint does not deviate too much from μ_{a0} (remember that $\overline{\mu}_a(\mathbf{r}) = \mu_a(\mathbf{r}) - \mu_{a0}$). In this representation, the Rytov expansion can be found as a perturbation series in terms of the small parameter δ of the form:

$$\Phi(\mathbf{r}) = \sum_{j=0}^{\infty} \delta^j \Phi_j(\mathbf{r}).$$

If we include this series expansion into Eq. (10.19), we obtain:

$$-D_0\sum_{j=0}^{\infty}\delta^j\nabla^2\Phi_j(\mathbf{r}) - 2D_0\sum_{j=0}^{\infty}\delta^j\nabla\Phi_j(\mathbf{r})\cdot\nabla\Phi_0(\mathbf{r}) + \delta\overline{\mu}_a(\mathbf{r}) = 0. \quad (10.21)$$

We now have for each power of δ a separate equation (otherwise the only solution would be $\delta = 0$). For the zeroth order case ($j = 0$) we obtain the homogeneous solution to Eq. (10.19):

$$-D_0\nabla^2\Phi_0(\mathbf{r}) - 2D_0\nabla\Phi_0(\mathbf{r})\cdot\nabla\Phi_0(\mathbf{r}) = 0. \quad (10.22)$$

The order of interest to us is the first order approximation which is found for $j = 1$ as:

$$-D_0\nabla^2\Phi_1(\mathbf{r}) - 2D_0\nabla\Phi_1(\mathbf{r})\cdot\nabla\Phi_0(\mathbf{r}) = -\overline{\mu}_a(\mathbf{r}). \quad (10.23)$$

The solution to this equation can be found by noticing that the Laplacian of $U^{inc}(\mathbf{r})\Phi_1(\mathbf{r})$ is given by:

$$-D_0\nabla^2(U^{inc}\Phi_1) = -D_0\Phi_1(\mathbf{r})\nabla^2 U^{inc}(\mathbf{r})-$$
$$2D_0 U^{inc}(\mathbf{r})\nabla\Phi_1(\mathbf{r})\cdot\nabla\Phi_0(\mathbf{r}) - D_0 U^{inc}(\mathbf{r})\nabla^2\Phi_1(\mathbf{r}),$$

where we have used the relation $\nabla U^{inc}(\mathbf{r}) = U^{inc}(\mathbf{r})\nabla\Phi_0(\mathbf{r})$. Making use of the fact that as long as $\mathbf{r}$ is not at the location where the source $S(\mathbf{r})$ is defined we have (see Eq. (10.11)):

$$-D_0\nabla^2 U^{inc}(\mathbf{r}) = -\mu_{a0}U^{inc}(\mathbf{r}),$$

we obtain that the Laplacian of $U^{inc}(\mathbf{r})\Phi_1(\mathbf{r})$ is:

$$-D_0\nabla^2(U^{inc}(\mathbf{r})\Phi_1(\mathbf{r})) = -\mu_{a0}U^{inc}(\mathbf{r})\Phi_1(\mathbf{r})-$$
$$2D_0 U^{inc}(\mathbf{r})\nabla\Phi_1(\mathbf{r})\cdot\nabla\Phi_0(\mathbf{r}) - D_0 U^{inc}(\mathbf{r})\nabla^2\Phi_1(\mathbf{r}),$$

which can be rearranged to give:

$$2D_0 U^{inc}(\mathbf{r})\nabla\Phi_1(\mathbf{r})\cdot\nabla\Phi_0(\mathbf{r}) + D_0 U^{inc}(\mathbf{r})\nabla^2\Phi_1(\mathbf{r}) =$$
$$D_0\nabla^2(U^{inc}(\mathbf{r})\Phi_1(\mathbf{r})) - \mu_{a0}U^{inc}(\mathbf{r})\Phi_1(\mathbf{r}).$$

Multiplying by $U^{inc}(\mathbf{r})$ our first-order Rytov approximation, Eq. (10.23), we can identify terms and obtain:

$$-D_0\nabla^2(U^{inc}\Phi_1(\mathbf{r})) + \mu_{a0}U^{inc}(\mathbf{r})\Phi_1(\mathbf{r}) = -\overline{\mu}_a(\mathbf{r})U^{inc}(\mathbf{r}). \qquad (10.24)$$

We can now find the solution to this equation in the usual way for $U^{inc}\Phi_1$ as:

$$U^{inc}(\mathbf{r})\Phi_1(\mathbf{r}) = -\int_V \overline{\mu}_a(\mathbf{r}')U^{inc}(\mathbf{r}')G(\mathbf{r}',\mathbf{r})\mathrm{d}^3 r',$$

which finally gives us an expression for Φ in the first-order Rytov approximation:

$$\Phi_1(\mathbf{r}) = -\frac{1}{U^{inc}(\mathbf{r})}\int_V \overline{\mu}_a(\mathbf{r}')U^{inc}(\mathbf{r}')G(\mathbf{r}',\mathbf{r})\mathrm{d}^3 r',$$

that we can relate directly to the total average intensities as:

$$\log\left(\frac{U(\mathbf{r})}{U^{inc}(\mathbf{r})}\right) \simeq -\frac{1}{U^{inc}(\mathbf{r})}\int_V \overline{\mu}_a(\mathbf{r}')U^{inc}(\mathbf{r}')G(\mathbf{r},\mathbf{r}')\mathrm{d}^3 r'. \qquad (10.25)$$

The above equation is a linear solution to our complete equation. Note how similar it is when compared to the first-order Born approximation we derived in the previous section, pointing to the fact that the Born and Rytov approximations are equivalent. This issue has been very well established (see [Brown. Jr. (1966)]) and was discussed in the context of Diffuse waves in the PhD dissertation of V. Ntziachristos [Ntziachristos (2000)]. The relation between the Rytov and Born approximation can be seen directly if we take into account that $U(\mathbf{r}) = U^{inc}(\mathbf{r}) \exp(\Phi_1(\mathbf{r}))$ can be written as:

$$U(\mathbf{r}) \simeq U^{inc}(\mathbf{r})\Big(1 + \Phi_1(\mathbf{r}) + \mathcal{O}(\Phi_1^2)\Big)$$

which when including the expression for $\Phi_1(\mathbf{r})$ gives:

$$U(\mathbf{r}) \simeq U^{inc}(\mathbf{r}) - \int_V \bar{\mu}_a(\mathbf{r}')U^{inc}(\mathbf{r}')G(\mathbf{r}',\mathbf{r})\mathrm{d}^3r',$$

which is the Born Approximation[3]. In the case of fluorescence we can consider the total average intensity $U(\mathbf{r})$ as the fluorescence average intensity, and the intensity that excites the fluorophore to be $U^{ex} = U^{inc}$:

$$\log\left(\frac{U^{fl}(\mathbf{r})(\mathbf{r}, \lambda_{em})}{U^{ex}(\mathbf{r}, \lambda_{ex})}\right) \simeq \frac{1}{U^{inc}(\mathbf{r}, \lambda_{ex})} \int_V F(\mathbf{r}', \lambda_{em})U^{inc}(\mathbf{r}', \lambda_{ex})G(\mathbf{r}', \mathbf{r}, \lambda_{em})\mathrm{d}^3r'.$$

10.4 The Normalized Born Approximation and the Sensitivity Matrix

A very important issue we discussed in Sec. 9.4.2 of Chap. 9 is how the use of normalized measurements significantly improves our data quality not only because they get rid of experimental parameters such as detector response curves and source intensities, but because they additionally reduce the ill-posedness. With regards to the inverse problem, one common approach is to use the *normalized Born approximation* which simply consists on normalizing Eq. (10.7) by the homogeneous contribution:

$$U^{nB}(\mathbf{r}) \simeq \frac{1}{U^{inc}(\mathbf{r})}\left(U^{inc}(\mathbf{r}) - \int_V G(\mathbf{r}', \mathbf{r})\bar{\mu}_a(\mathbf{r})U^{inc}(\mathbf{r}')\mathrm{d}^3r'\right), \quad (10.26)$$

and in the case of fluorescence it becomes (remember $U^{ex} = U^{inc}$ at the location of the fluorophore):

$$U^{nB}_{fl}(\mathbf{r}, \lambda_{em}) \simeq \frac{1}{U^{inc}(\mathbf{r}, \lambda_{ex})} \int_V G(\mathbf{r}, \mathbf{r}', \lambda_{em})F(\mathbf{r}', \lambda_{em})U^{inc}(\mathbf{r}, \lambda_{ex})\mathrm{d}^3r'. \quad (10.27)$$

---

[3]Note that in order to reach this expression we had to truncate the Taylor series of $\exp(\Phi_1(\mathbf{r}))$, which might suggest that Rytov is a slightly better approximation than Born. However, we must not forget that the Rytov series were derived under the assumption that $(\nabla\Phi)^2$ could be neglected. Under the conditions for which this assumption is valid, it so happens that both Born and Rytov produce the same results[Brown. Jr. (1966)].

Note: **The Normalized Born Approximation**

It is important to understand that by simply normalizing the measurements we are *not* applying the normalized Born approximation: we could go to higher orders of approximation using normalized measurements, consistently increasing the accuracy of the solution. The Born approximation comes about by assuming that the total average intensity inside the absorbing object can be approximated to U^{inc} (or U^{ex} in the fluorescence case), ignoring the presence of the absorbing object. How we solve the integral is another issue, unrelated to the Born or Rytov approximations.

Once we have the above equations linearized in terms of the homogeneous contributions, the next step is to relate them to actual measurements. As has been mentioned throughout this book, what is measured experimentally is the flux through the surface of the detector, and not the average intensity which Eqs. (10.26)–(10.27) represent. This was discussed in Sec. 8.5 of Chap. 8 where we saw that in the specific case when we are measuring *inside* an infinite diffusive medium, we can use directly Fick's law, $J_n(\mathbf{r}) = -D\hat{\mathbf{n}}_d \cdot \nabla U(\mathbf{r})$, being $\hat{\mathbf{n}}_d$ the surface normal of our detector. Making use of the expressions for the gradient of Green's function shown at the end of Chap. 5:

$$\hat{\mathbf{n}} \cdot \nabla g(\mathbf{r} - \mathbf{r}') = g(\mathbf{r} - \mathbf{r}') \left(i\kappa_0 - \frac{1}{|\mathbf{r} - \mathbf{r}'|} \right) \hat{\mathbf{n}}_d \cdot \hat{\mathbf{u}}_{\mathbf{r}-\mathbf{r}'}$$

the Born approximation for the normalized flux becomes:

$$J^{nB}(\mathbf{r}) \simeq \frac{1}{J^{inc}(\mathbf{r})} \left(J^{inc}(\mathbf{r}) - \int_V g(\mathbf{r}', \mathbf{r}) \left(i\kappa_0 - \frac{1}{|\mathbf{r} - \mathbf{r}'|} \right) \times \right.$$
$$\left. \overline{\mu}_a(\mathbf{r}) U^{inc}(\mathbf{r}') \hat{\mathbf{n}}_d \cdot \hat{\mathbf{u}}_{\mathbf{r}-\mathbf{r}'} \mathrm{d}^3 r' \right),$$

where J^{inc} is the homogeneous contribution to the flux:

$$J^{inc}(\mathbf{r}) = \int_V S_0(\mathbf{r}') g(\mathbf{r} - \mathbf{r}') \left(i\kappa_0 - \frac{1}{|\mathbf{r} - \mathbf{r}'|} \right) \hat{\mathbf{n}}_d \cdot \hat{\mathbf{u}}_{\mathbf{r}-\mathbf{r}'} \mathrm{d}^3 r'.$$

Note that this expression is correct only for an infinite diffusive medium, since otherwise Green's function g should account for the boundaries present and its gradient would not be defined as above.

A much simpler expression appears in the more realistic case of measurement done at a diffuse/non-diffuse (D-N) interface, which is the one we will find a solution for. Since we may use the boundary condition

$U(\mathbf{r}) = \alpha J_n(\mathbf{r})$, we can use the exact expressions we obtained for the normalized average intensity both for the absorption and fluorescence cases:

$$J^{nB}(\mathbf{r}) \simeq \frac{\alpha}{U^{inc}(\mathbf{r})} \left(\frac{1}{\alpha} U^{inc}(\mathbf{r}) - \frac{1}{\alpha} \int_V G(\mathbf{r}', \mathbf{r}) \overline{\mu}_a(\mathbf{r}) U^{inc}(\mathbf{r}') d^3 r' \right),$$

and

$$J_{fl}^{nB}(\mathbf{r}, \lambda_{em}) \simeq \frac{\alpha}{U^{inc}(\mathbf{r}, \lambda_{ex})} \frac{1}{\alpha} \int_V G(\mathbf{r}, \mathbf{r}', \lambda_{em}) F(\mathbf{r}', \lambda_{em}) U^{inc}(\mathbf{r}, \lambda_{ex}) d^3 r'.$$

respectively, where now G *must account for the presence of the boundary*. We can now solve these integrals (or their U^{nB} equivalents) through discretization. Note that this could not be done when dealing with the complete solution, Eq. (10.2), since our 'unknown' (the total average intensity) was also included in the integral over the volume V. Once we have approximated the contribution of the object in terms of the incident or excitation average intensities we can therefore write:

$$J^{nB}(\mathbf{r}) = 1 - \frac{1}{U^{inc}(\mathbf{r})} \sum_{j=1}^{N} G(\mathbf{r}, \mathbf{r}_j) \overline{\mu}_a(\mathbf{r}_j) U^{inc}(\mathbf{r}_j) \Delta V,$$

where we have discretized the volume into N differential elements. We can proceed in the same way for the fluorescence expression:

$$J_{fl}^{nB}(\mathbf{r}, \lambda_{em}) = \frac{1}{U^{ex}(\mathbf{r}, \lambda_{ex})} \sum_{j=1}^{N} G(\mathbf{r}, \mathbf{r}_j, \lambda_{em}) F(\mathbf{r}_j, \lambda_{em}) U^{inc}(\mathbf{r}_j, \lambda_{ex}) \Delta V.$$

Since we will have a discretized number of N_s sources (could be positions of a laser spot, or positions of a fiber) and a discretized number of N_d detectors (pixels of a CCD or detector fibers, for example) the above equations can be written in matrix form as (note that we could easily extend this to any number of wavelengths):

$$[\mathbb{1} - \mathbb{J}]_{M \times 1} = [\mathbb{W}^a]_{M \times N} [\mathbb{A}]_{N \times 1},$$

and

$$[\mathbb{J}^{fl}]_{M \times 1} = [\mathbb{W}^{fl}]_{M \times N} [\mathbb{F}]_{N \times 1},$$

for absorption and fluorescence respectively, and where $M = N_s N_d$ represents the total number of source-detector pairs. In the above the matrices $[\mathbb{J}]$ and $[\mathbb{J}^{fl}]$ represent the normalized absorption and fluorescence measurements, respectively, and the matrices $[\mathbb{W}^a]$, $[\mathbb{W}^{fl}]$ are given by:

$$[\mathbb{W}^a]_{M \times N} = \frac{1}{U^{inc}(\mathbf{r}_s, \mathbf{r}_d)} U^{inc}(\mathbf{r}_s, \mathbf{r}_j) G(\mathbf{r}_j, \mathbf{r}_d) \Delta V,$$

and

$$[\mathbb{W}^{fl}]_{M \times N} = \frac{1}{U^{inc}(\mathbf{r}_s, \mathbf{r}_d, \lambda_{ex})} U^{inc}(\mathbf{r}_s, \mathbf{r}_j, \lambda_{ex}) G(\mathbf{r}_j, \mathbf{r}_d, \lambda_{em}) \Delta V,$$

where each $(\mathbf{r}_s, \mathbf{r}_d)$ represents one source-detector pair. The above functions are usually termed the sensitivity matrix or weight matrix, where it is important to remember that defined in this manner they are valid within the first-order Born approximation. The general sensitivity matrix would be defined as above but with the Green function, G, taking into account the presence of the inhomogeneities. Also, don't forget that G accounts for the presence of the boundary: if we would like to express the above sensitivity matrix in terms of something we can actually measure *inside* a diffusive medium, we would need to use Fick's law to obtain the flux as we did in the previous paragraphs.

We now have a set of linear equations, but since the problem is ill-posed this results in many of these equations being correlated. This means that we have to take care when inverting the matrix to find a solution for the spatial distribution of absorption and fluorescence. Typical approaches to invert them are iterative methods such as the algebraic reconstruction technique (ART) or singular value decomposition, to name the most popular. How the above matrix equations can be inverted to obtain an expression for $[\mathbb{A}]$ or $[\mathbb{F}]$ is out of the scope of this book, but there is extensive literature on the subject (see [Kak and Slaney (1988)] as a good starting point and [Arridge and Schotland (2009)] for the specific case of diffuse light).

10.5   Direct Inversion Formulas

In 2001 Markel and Schotland published an inversion approach that could handle very large data sets[Markel and Schotland (2001)] and termed it 'direct Inversion' in the sense that they found an analytical expression that related directly the reconstructed image with the data measured. In their approach, which we will term from now on the direct inversion method, both source and detectors are taken to Fourier's space. This is in contrast to what we have dealt with so far, i.e. we usually represent the measured (or predicted) measurement in the Fourier's space and make use of the angular spectrum representation for the spatial variable $\mathbf{r_d}$. However, if we also scan the source we are capable of building another function for $\mathbf{r_s}$, which we can also represent in Fourier-space. This approach enables the use of very large source positions (10^3 or higher) while retaining low computation times and it is therefore very useful when dealing with very large source position scanning measurements (or using structured illumination), together with a large number of detectors (as would happen when measuring with a CCD). The derivation we will present here is the equivalent to the original one presented by Markel and Schotland in [Markel and Schotland (2001)] but represented in the notation we have used throughout this book. The only difference is that we will derive the direct inversion formula for the *normalized* fluorescence intensity in the Born approximation, Eq. (10.27), whereas their original derivation was for imaging absorption and scattering.

Even though the goal of this book is not the study of inversion methods, the direct inversion formula is particularly suitable since it makes use of the same formulation we developed in the previous chapters, being fully developed in the angular spectrum representation. In order to obtain the direct inversion formulas let us assume we have a collection of fluorophores that are between $z > 0$ and $z < L$ in an otherwise infinite and homogeneous medium of optical properties μ_a, D and n_0. Let us also assume that we have an equally spaced grid of sources at $\mathbf{r_s} \equiv (\mathbf{R_s}, z = 0)$ and a collection of equally spaced detectors at a single plane $\mathbf{r_d} \equiv (\mathbf{R_d}, z = L)$, such as a CCD image[4]. In this case, the fluorescence measured at $\mathbf{r_d}$ due to the sources at $\mathbf{r_s}$ is given by:

$$U^{fl}(\mathbf{r}_s, \mathbf{r}_d) = \int_V U^{ex}(\mathbf{r_s}, \mathbf{r}) F(\mathbf{r}) g(\mathbf{r}, \mathbf{r_d}) \mathrm{d}^3 r \,, \qquad (10.28)$$

where g is the infinite homogeneous Green function, U^{ex} is the excitation average intensity due to source $\mathbf{r_s}$ at $\mathbf{r}$ and F accounts for the fluorophore concentration (see Eq. (10.9)). From now on to simplify the notation we will assume no wavelength dependence on the optical properties of the medium at the emission and excitation wavelengths.

Making use of the Born approximation to the fluorescence intensity we recently presented in Sec. 10.2, we can approximate Eq. (10.28) to:

$$U^{fl}(\mathbf{r}_s, \mathbf{r}_d) = S_0 \int_V g(\mathbf{r_s}, \mathbf{r}) g(\mathbf{r}, \mathbf{r_d}) F(\mathbf{r}) d^3 \mathbf{r} \,, \qquad (10.29)$$

where we have written the excitation average intensity as a point source:

$$U^{ex}(\mathbf{r}) = S_0 g(\mathbf{r_s}, \mathbf{r}), \qquad (10.30)$$

---

[4]Note that the derivation presented here does not account for boundary conditions and assumes an infinite homogeneous diffusive medium. If boundary conditions are imposed the detector plane at $\mathbf{r_d} \equiv (\mathbf{R_d}, z = L)$ could be seen as the CCD pixels projected as virtual detectors onto the plane interface by the use of focused measurements (see Chap. 8, Sec. 8.6)

with S_0 representing the energy density of the source, measured in *Watts/cm³*. Note that the complete solution without any approximation cannot be written with U^{ex} in terms of the infinite homogeneous function since the presence of the fluorophore affects the average intensity distribution. As a reminder, the infinite green function is defined as:

$$g(\mathbf{r}, \mathbf{r}') = \frac{1}{4\pi D} \frac{\exp\left(i\kappa_0 |\mathbf{r} - \mathbf{r}'|\right)}{|\mathbf{r} - \mathbf{r}'|} , \qquad (10.31)$$

where κ_0 has the usual expression for the diffuse wavenumber. For simplicity we will consider the *cw* regime in which case κ_0 is given by:

$$\kappa_0 = \sqrt{\frac{-\mu_a}{D}} , \qquad (10.32)$$

which we assume independent of the excitation and emission wavelengths.

Our core equation is Eq. (10.29), which we now need to express in a way which will enable us to extract the expression for F from the integral over the volume. In order to do so, we will first Fourier transform for $\mathbf{R_d}$ which is equivalent to the angular spectrum representation we have been dealing with throughout the second part of this book:

$$U^{fl}(\mathbf{r}_s, \mathbf{r}_d) = \frac{S_0}{4\pi^2} \int_V g(\mathbf{r_s}, \mathbf{r}) \left[\int_{-\infty}^{\infty} \widetilde{g}(\mathbf{r}, \mathbf{K_d}; z_d) F(\mathbf{r}) \mathrm{d}^3 r \right] \exp(i\mathbf{K}_d \cdot \mathbf{R_d}) \mathrm{d}^2 K_d .$$

We can now introduce a second Fourier transform, this time on the spatial distribution of the sources, $\mathbf{R_s}$:

$$U^{fl}(\mathbf{r}_s, \mathbf{r}_d) = \frac{S_0}{16\pi^4} \int_{-\infty}^{\infty} \int_{-\infty}^{\infty} \int_V \widetilde{g}(\mathbf{K_s}, z_s; \mathbf{R}, z) \widetilde{g}(\mathbf{K_d}, z_d; \mathbf{R}, z) \times$$
$$F(\mathbf{R}, z) \mathrm{d}^2 R \mathrm{d}z \exp(i\mathbf{K_d} \cdot \mathbf{R_d}) \exp(i\mathbf{K_s} \cdot \mathbf{R_s}) \mathrm{d}^2 K_d \mathrm{d}^2 K_s .$$

We can now identify in the above equation the Fourier transform of $U^{fl}(\mathbf{r}_s, \mathbf{r}_d)$ both for $\mathbf{R_s}$ and for $\mathbf{R_d}$:

$$\widetilde{U}^{fl}(\mathbf{K_s}, \mathbf{K_d}) = S_0 \int_V \widetilde{W}(\mathbf{K_s}, \mathbf{K_d}; z_s, z_d; \mathbf{R}, z) F(\mathbf{R}, z) \mathrm{d}^2 R \mathrm{d}z , \qquad (10.33)$$

where we have defined $\widetilde{W}$ as:

$$\widetilde{W}(\mathbf{K_s}, \mathbf{K_d}; z_s, z_d; \mathbf{R}, z) = \widetilde{g}(\mathbf{K_s}, z_s; \mathbf{R}, z) \widetilde{g}(\mathbf{K_d}, z_d; \mathbf{R}, z) . \qquad (10.34)$$

In the above equations we have represented the Fourier transform of the Green functions in $\mathbf{R_s}$ and $\mathbf{R_d}$ respectively as[5]:

$$\widetilde{g}(\mathbf{K_s}, z_s; \mathbf{R}, z) = \frac{1}{4\pi D} \frac{i}{2\pi q_s(\mathbf{K_s})} \exp(iq_s(\mathbf{K_s})(z - z_s)) \times$$
$$\exp(i\mathbf{K_s} \cdot \mathbf{R}) , \qquad (10.35)$$

$$\widetilde{g}(\mathbf{K_d}, z_d; \mathbf{R}, z) = \frac{1}{4\pi D} \frac{i}{2\pi q_d(\mathbf{K_d})} \exp(iq_d(\mathbf{K_d})(z - z_s)) \times$$
$$\exp(i\mathbf{K_d} \cdot \mathbf{R}) , \qquad (10.36)$$

being q_s and q_d the projection of the z-component of the wave-vector for spatial frequency component $\mathbf{K_s}$ and $\mathbf{K_d}$ respectively, represented as:

$$q_s(\mathbf{K_s}) = \sqrt{\kappa_0^2 - \mathbf{K_s}^2} , \qquad (10.37)$$

$$q_d(\mathbf{K_d}) = \sqrt{\kappa_0^2 - \mathbf{K_d}^2} . \qquad (10.38)$$

Since in the *cw* case q_s and q_d are pure imaginary numbers we can rewrite them as:

$$q_s(\mathbf{K_s}) = i\sqrt{\mu_a/D + \mathbf{K_s}^2} , \mathbf{K_s} \in [-\infty, +\infty] , \qquad (10.39)$$

$$q_d(\mathbf{K_d}) = i\sqrt{\mu_a/D + \mathbf{K_d}^2} , \mathbf{K_d} \in [-\infty, +\infty] . \qquad (10.40)$$

Introducing the expressions for the Green function in Eq. (10.34) gives us:

$$\widetilde{W}(\mathbf{K_s}, \mathbf{K_d}; z_s, z_d; \mathbf{R}, z) =$$
$$\left(\frac{i}{8\pi^2 D}\right)^2 \frac{1}{q_s q_d} \times$$
$$\exp\left(i(q_s(z - z_s) + q_d(z_d - z))\right) \exp\left(i(\mathbf{K_s} - \mathbf{K_d}) \cdot \mathbf{R}\right) . \quad (10.41)$$

Let us now introduce the fact that $z_s = 0$ and $z_d = L$ into $\widetilde{W}$. This will give us the following expression for the fluorescence average intensity, where we can now split the volume integral into the z and $\mathbf{R}$ contributions:

$$\widetilde{U}^{fl}(\mathbf{K_s}, \mathbf{K_d}) = S_0 \left(\frac{i}{8\pi^2 D}\right)^2 \frac{1}{q_s q_d}$$
$$\int_z \exp\left(i\left((q_s - q_d)z + q_d L\right)\right) \mathrm{d}z \int_{\mathbf{R}} F(\mathbf{R}, z) \exp\left(i(\mathbf{K_s} - \mathbf{K_d}) \cdot \mathbf{R}\right) \mathrm{d}^2 R .$$
$$(10.42)$$

---

[5]Note that we always assume $z_s \leq z$ so that the transfer function, $\exp(iq(z - z_s))$ always decays for increasing z.

We can now identify the Fourier transform of F in $\mathbf{K_s} - \mathbf{K_d}$:

$$\tilde{F}(\mathbf{K_s} - \mathbf{K_d}, z) = \int_{\mathbf{R}} F(\mathbf{R}, z) \exp\left(i(\mathbf{K_s} - \mathbf{K_d})\mathbf{R}\right) d^2 R , \qquad (10.43)$$

and using this expression, we can rewrite Eq. (10.42) as:

$$\tilde{U}^{fl}(\mathbf{K_s}, \mathbf{K_d}) = S_0 \left(\frac{i}{8\pi^2 D}\right)^2 \frac{1}{q_s q_d} \times$$

$$\int_z \exp\left(i\left[(q_s - q_d)z + q_d L\right]\right) \tilde{F}(\mathbf{K_s} - \mathbf{K_d}, z) dz. \qquad (10.44)$$

So far we have an expression for the fluorescence in the frequency space of both sources and detectors which depends on $\tilde{F}(\mathbf{K_s} - \mathbf{K_d}, z)$. This expression is not normalized, with the consequent problems with regards not only to the ill-posed nature of the inverse problem but also to unknown calibration parameters present in an experiment (see Chap. 9, Sec. 9.4). In order to normalize our fluorescent measurement we can make use of Eq. (10.30), which we may rewrite at the detector plane, $z = L$, in Fourier space as (see Eq. (10.35)):

$$\tilde{U}_L^{ex}(\mathbf{K_s}, \mathbf{K_d}) = \frac{S_0}{4\pi D} \frac{i}{2\pi q_s} \exp(i q_s L) . \qquad (10.45)$$

In this case, dividing Eq. (10.42) by $\tilde{U}_L^{ex}$ we obtain the normalized expression in Fourier space for both sources and detectors as:

$$\tilde{U}^{nB}(\mathbf{K_s}, \mathbf{K_d}) = \frac{\tilde{U}^{fl}(\mathbf{K_s}, \mathbf{K_d})}{\tilde{U}_L^{ex}(\mathbf{K_s}, \mathbf{K_d})} =$$

$$\frac{1}{4\pi D} \frac{i}{2\pi q_d} \int_z \exp\left(i\left[(q_s - q_d)(z - L)\right]\right) \tilde{F}(\mathbf{K_s} - \mathbf{K_d}, z) dz, \qquad (10.46)$$

which can be rewritten as:

$$\tilde{U}^{nB}(\mathbf{K_s}, \mathbf{K_d}) = \frac{1}{4\pi D} \frac{i}{2\pi q_d} \exp(-i(q_s - q_d)L) \times$$

$$\int_z \exp(i(q_s - q_d)z) \tilde{F}(\mathbf{K_s} - \mathbf{K_d}, z) dz. \qquad (10.47)$$

We can now take into account that q_s and q_d are pure imaginary numbers, and introduce the change of variable $\eta = -i(q_s - q_d), q_s \geq q_d$ and $\eta = -i(q_d - q_s), q_d > q_s$ into the previous equation resulting in:

$$\tilde{U}^{nB}(\mathbf{K_s}, \mathbf{K_d}) = \frac{1}{8\pi^2 D} \frac{i}{q_d(\mathbf{K_d})} \exp(\eta L) \int_z \exp(-\eta z) \tilde{F}(\mathbf{K_s} - \mathbf{K_d}, z) dz ,$$

$$(10.48)$$

where defined in this manner η is now real and positive. Since the integral over z can span to ∞ due to the fluorophore contribution F being zero at $z < 0$ and $z > L$, we can write the above equation in terms of the Laplace transform of $\tilde{F}$ (see Appendix A):

$$\tilde{F}(\mathbf{K_s} - \mathbf{K_d}, \eta) = \int_0^\infty \exp{(-\eta z)}\tilde{F}(\mathbf{K_s} - \mathbf{K_d}, z)dz , \qquad (10.49)$$

which requires $\eta \leq 0$ which is ensured by $\eta = |-i(q_s - q_d)|$. The expression for the normalized measurement now becomes:

$$\tilde{U}^{nB}(\mathbf{K_s}, \mathbf{K_d}) = \frac{1}{8\pi^2 D}\frac{i}{q_d(\mathbf{K_d})}\exp(\eta L)\tilde{F}(\mathbf{K_s} - \mathbf{K_d}, \eta) . \qquad (10.50)$$

Since each $[\mathbf{K_s}, \mathbf{K_d}]$ pair defines a $\mathbf{K_s} - \mathbf{K_d}$ value and an $\eta = |-i(q_s - q_d)|$, we now have U^{nB} completely defined in terms of these two coefficients. This means that we can now extract the information on $\tilde{F}$ from the above equation as:

$$\tilde{F}(\mathbf{K_s} - \mathbf{K_d}, \eta) = 8\pi^2 D\sqrt{\mu_a/D + \mathbf{K_d}^2} \times$$
$$\exp(-\eta L)\tilde{U}^{nB}(\mathbf{K_s} - \mathbf{K_d}, \eta) , \qquad (10.51)$$

which is our final result and can be regarded as a direct inversion formula for F. There are several important things we have to point out: first of all, the K-space we can sample depends on the values $\mathbf{K_s} - \mathbf{K_d}$ we can generate and not on the values themselves. This is analogous to the Ewald sphere and gives us the limit of information we can obtain from the object. Secondly, our z information is encoded in $q_s - q_d$, i.e. the difference between the z-component of the wavevectors for the spatial frequencies defined by the arrangement of sources and detectors. Finally, we see we have an exponential term $\exp(-\eta L)$ which we have to be careful with, together with a division of the Fourier transforms for the fluorescence and excitation measurements. In order to deal with these we need to regularize to maintain the reconstruction bounded, at the cost of lower resolution (lower K-value accepted). This regularization can be introduced by multiplying Eq. (10.51) by a Filter, and/or including a regularization term when dividing $\tilde{U}^{fl}$ by $\tilde{U}^{(inc)}$ or finding the values of $1/\exp(\eta L)$.

Key points

From this chapter the following important points should be remembered:

- **Born Approximation:** We derived the first order Born approximation by assuming that the average intensity inside an absorbing object can be approximated to the background average intensity (i.e. the average intensity in the absence of this object).

- **Rytov Approximation:** We derived the Rytov series and the first order Rytov approximation by assuming that the average intensity in the presence of the object can be expressed as a small change in the background average intensity that can be represented as a phase. We also saw that the Rytov and Born approximations were equivalent.

- **Normalized Born Approximation:** The use of normalized measurements was already considered in Chap. 9 and was presented here in the context of the Born approximation. It should not be forgotten, however, that the use of normalized measurements does not imply the use of the Born approximation.

- **Direct Inversion Methods:** We derived an expression for a direct inversion method based on the normalized Born approximation following the derivation from Markel and Schotland. The direct inversion methods can be applied for large datasets comprising large source (or structured illumination) and detector arrays and further enhance the importance of the angular spectrum representation.

Further Reading

Inverse problems and their application in diffuse light imaging is a topic so extensive that a complete book devoted to it would probably not suffice. I have opted for including some of the most commonly used approximations such as the Born and Rytov approximations, but have not dealt at all with how one can solve the resulting system of equations to recover the absorption or fluorescence distribution. Additionally, in this chapter the Born approximation for the scattering case was not considered. All this information can be found in great detail in the PhD dissertations of Maureen O'Leary [O'Leary (1996)], David Boas[Boas (1996)], and Vasilis Ntziachristos [Ntziachristos (2000)] and I strongly recommend them as reference material for the practical aspects of imaging with diffuse light. In particular, the limits of the Born and Rytov approximations are thoroughly studied in these Theses.

With regards to the use of the normalized Born approximation for fluorescence, this was first published by V. Ntzichristos in 2001, [Ntziachristos and Weissleder (2001)], and since then it has become one of the most commonly used approximation for fluorescence imaging (other than numerical approaches such as Finite Element Methods), and a large amount of literature can now be found on the subject. Greater insight on its validity with small animal imaging applications in mind can be found in [Soubret *et al.* (2005)] and [Ntziachristos *et al.* (2002)], the latter being the first publication in the context of molecular imaging.

With regards to the broader topic of inverse problems in diffuse light imaging, I strongly recommend reading the works of both Simon Arridge and John Schotland. In particular I recommend the reviews [Arridge (1999)], [Arridge and Schotland (2009)], and [Arridge (2011)]. A very good book for different approaches to solve the inverse problem, and one which explains in detail the implementation of iterative methods such as the Algebraic Reconstruction Technique (ART), is Kak and Slaney (1988). Even though this book is on computerized tomographic imaging it is a great source of reference for image reconstruction in general.

Finally, we presented in this chapter a direct inversion approach based on the derivation presented by Markel and Schotland [Markel and Schotland (2001)]. I strongly recommend reading their original papers, in particular the series on the inverse problem in optical diffusion tomography [Markel and Schotland (2001, 2002); Markel *et al.* (2003a,b)]. Additionally, I recommend [Schotland and Markel (2001)] and [Markel and Schotland (2004)]. With regards to experimental evidence, these direct methods have been used with experimental data for imaging complex structures in [Konecky *et al.* (2008)], showcasing the potential of this approach. Even though direct inversion methods are not being currently extensively employed, I am quite certain that they will prove extremely useful in the near future and that their use will extend quite fast.

Appendix A

Useful Formulas

One typical source of error when dealing with integral transforms is due to the fact that there is a certain freedom on how we define them, specifically in terms of the factor that should appear in front of each transformation and the sign of the exponent. To avoid confusion, I here include all transforms used in our derivations and the main formulas related to them.

A.1 The Fourier Transform

We will define the Fourier transform as:

- 1-D Fourier Transform:

$$\mathcal{F}\{g(x)\} = \widetilde{g}(k) = \frac{1}{2\pi} \int_{-\infty}^{\infty} g(x) \exp(-ikx) \mathrm{d}x, \qquad (A.1)$$

- 1-D Inverse Fourier Transform:

$$\mathcal{F}^{-1}\{\widetilde{g}(k)\} = g(x) = \int_{-\infty}^{\infty} \widetilde{g}(k) \exp(ikx) \mathrm{d}k, \qquad (A.2)$$

- 2-D Fourier Transform:

$$\mathcal{F}\{g(\mathbf{R})\} = \widetilde{g}(\mathbf{K}) = \frac{1}{(2\pi)^2} \int_{-\infty}^{\infty} g(\mathbf{R}) \exp(-i\mathbf{K} \cdot \mathbf{R}) \mathrm{d}^2 R, \qquad (A.3)$$

- 2-D Inverse Fourier Transform:

$$\mathcal{F}^{-1}\{\widetilde{g}(\mathbf{K})\} = g(\mathbf{R}) = \int_{-\infty}^{\infty} \widetilde{g}(\mathbf{K}) \exp(i\mathbf{K} \cdot \mathbf{R}) \mathrm{d}^2 K, \qquad (A.4)$$

- 3-D Fourier Transform:

$$\mathcal{F}\{g(\mathbf{r})\} = \widetilde{g}(\mathbf{k}) = \frac{1}{(2\pi)^3} \int_{-\infty}^{\infty} g(\mathbf{r})\exp(-i\mathbf{k}\cdot\mathbf{r})\mathrm{d}^3 r, \qquad (A.5)$$

- 3-D Inverse Fourier Transform:

$$\mathcal{F}^{-1}\{\widetilde{g}(\mathbf{k})\} = g(\mathbf{r}) = \int_{-\infty}^{\infty} \widetilde{g}(\mathbf{k})\exp(i\mathbf{k}\cdot\mathbf{r})\mathrm{d}^3 k, \qquad (A.6)$$

Note that the different choice of factor appears so that the following relationships holds:

$$\mathcal{F}\{\mathcal{F}^{-1}\{g(\mathbf{r})\}\} = g(\mathbf{r}), \qquad (A.7)$$

which is directly related to how we define the *delta function*.

A.2 The Hankel Transform

In the special case where we have cylindrical symmetry, we may account for this symmetry in the 2-D Fourier transform:

$$\widetilde{g}(\mathbf{K}) = \frac{1}{(2\pi)^2} \int_{-\infty}^{\infty} g(\mathbf{R})\exp(-iKR\cos\theta)\mathrm{d}^2 R,$$

where we have written $\mathbf{R}\cdot\mathbf{K} = RK\cos\theta$. Expressing $\mathrm{d}^2 R = R\mathrm{d}\theta\mathrm{d}R$ we obtain:

$$\widetilde{g}(\mathbf{K}) = \frac{1}{(2\pi)^2} \int_{0}^{\infty} g(R)\left[\int_{0}^{2\pi}\exp(-iKR\cos\theta)\mathrm{d}\theta\right]R\mathrm{d}R.$$

In the above equation we can identify the expression for the Bessel function of the first kind and zero order:

$$J_0(KR) = \frac{1}{2\pi} \int_{0}^{2\pi}\exp(-iKR\cos\theta)\mathrm{d}\theta, \qquad (A.8)$$

in which case:

$$\widetilde{g}(\mathbf{K}) = \frac{1}{2\pi} \int_{0}^{\infty} g(R)J_0(KR)R\mathrm{d}R.$$

It is common to define the Hankel Transform pair as:

$$\mathcal{H}\{g(R)\} = \int_{0}^{\infty} g(R)J_0(KR)R dR, \qquad (A.9)$$

$$\mathcal{H}^{-1}\{\hat{g}(K)\} = \int_{0}^{\infty} \hat{g}(K)J_0(KR)K dK, \qquad (A.10)$$

in which case, following our notation, we obtain that the Fourier transform holds the following relation with the Hankel transform:

$$\tilde{g}(\mathbf{K}) = \frac{1}{2\pi}\mathcal{H}\{g(R)\}. \tag{A.11}$$

Note that, once again, the 2π factors are a matter of choice and it is also common to find the Hankel transform as defined above, but with an additional 2π factor for both direct and inverse transforms.

A.3 The Laplace Transform

The Laplace transform of $f(s)$ of a function $F(t)$ is defined as:

$$f(s) = \mathcal{L}\{F(t)\} = \int_0^\infty \exp(-st)F(t)\mathrm{d}t, \tag{A.12}$$

where s can be complex as long as $\Re\{s\} > 0$. When the above expression extends its limits of integration from $(-\infty, +\infty)$ it is usually termed the two-sided Laplace transform. The inverse Laplace transform is obtained as:

$$F(t) = \mathcal{L}^{-1}\{f(t)\} = \frac{1}{2\pi i}\int_{\gamma-i\infty}^{\gamma+i\infty} \exp(st)f(s)\mathrm{d}s, \tag{A.13}$$

equation which is usually known as the Bromwich integral or the Fourier-Mellin integral. The constant γ is chosen so that all the singularities of $f(s)$ are included in the complex contour.

A.4 The Delta Function

The most commonly used definition of the normalized delta function is given by:

- 1-D Delta Function:

$$\delta(x - x_0) = \frac{1}{2\pi}\int_{-\infty}^\infty \exp\left(ik(x - x_0)\right)\mathrm{d}k. \tag{A.14}$$

- 2-D Delta Function:

$$\delta(\mathbf{R} - \mathbf{R}_0) = \frac{1}{(2\pi)^2}\int_{-\infty}^\infty \exp\left(i\mathbf{K}\cdot(\mathbf{R} - \mathbf{R}_0)\right)\mathrm{d}^2K. \tag{A.15}$$

- 3-D Delta Function:

$$\delta(\mathbf{r} - \mathbf{r}_0) = \frac{1}{(2\pi)^3}\int_{-\infty}^\infty \exp\left(i\mathbf{k}\cdot(\mathbf{r} - \mathbf{r}_0)\right)\mathrm{d}^3k. \tag{A.16}$$

Once the delta function has been defined, the proof of Eq. (A.7) comes through:

$$g(\mathbf{r}) = \int_{-\infty}^{\infty} \tilde{g}(\mathbf{k}) \exp(i\mathbf{k} \cdot \mathbf{r}) d^3 k =$$

$$\int_{-\infty}^{\infty} \left[\frac{1}{(2\pi)^3} \int_{-\infty}^{\infty} g(\mathbf{r}') \exp(-i\mathbf{k} \cdot \mathbf{r}') d^3 r' \right] \exp(i\mathbf{k} \cdot \mathbf{r}) d^3 k =$$

$$\int_{-\infty}^{\infty} g(\mathbf{r}') \left[\frac{1}{(2\pi)^3} \int_{-\infty}^{\infty} \exp(i\mathbf{k} \cdot (\mathbf{r} - \mathbf{r}')) d^3 k \right] d^3 r,$$

where now we can easily identify the quantity within brackets as the delta function, and therefore:

$$\int_{-\infty}^{\infty} g(\mathbf{r}') \delta(\mathbf{r} - \mathbf{r}') d^3 r' = g(\mathbf{r}),$$

which proves Eq. (A.7), and establishes the nature of the $1/2\pi$ factor for the Fourier transform, where it is clear that any combination is valid as long as their product equals $1/(2\pi)^n$, being n the dimensionality. Examples of factors for the Fourier transform pairs are $[1/(\sqrt{2\pi})^n, 1/(\sqrt{2\pi})^n]$, $[1/(2\pi)^n, 1]$ (the one we use) and $[1, 1/(2\pi)^n]$.

A.5 Gaussian Function

It is convenient to define source profiles as normalized Gaussian functions of half-width w_g centered at $\mathbf{R}_s$:

$$g(\mathbf{R}) = \frac{1}{\pi w_g^2} \exp\left(-\frac{(\mathbf{R} - \mathbf{R}_s)^2}{w_g^2} \right) \tag{A.17}$$

The Fourier transform of the normalized Gaussian function is given by:

$$\tilde{g}(\mathbf{K}) = \frac{1}{\pi w_g^2} \frac{1}{(2\pi)^2} \int_{-\infty}^{\infty} \exp\left(-\frac{(\mathbf{R} - \mathbf{R}_s)^2}{w_g^2} \right) \exp(-i\mathbf{K} \cdot \mathbf{R}) d^2 R, \tag{A.18}$$

which can be solved first by the change of variable $\mathbf{R}' = \mathbf{R} - \mathbf{R}_s$:

$$\tilde{g}(\mathbf{K}) = \frac{1}{\pi w_g^2} \frac{\exp(-i\mathbf{K} \cdot \mathbf{R}_s)}{(2\pi)^2} \int_{-\infty}^{\infty} \exp\left(-\frac{\mathbf{R}'^2}{w_g^2} - i\mathbf{K} \cdot \mathbf{R}' \right) d^2 R', \tag{A.19}$$

and then by multiplying and dividing by $\exp(-|\mathbf{K}|^2 w_g^2/4)$:

$$\widetilde{g}(\mathbf{K}) = \frac{1}{\pi w_g^2} \frac{\exp(-i\mathbf{K}\cdot\mathbf{R}_s)}{(2\pi)^2} \exp\left(-\frac{|\mathbf{K}|^2 w_g^2}{4}\right) \times$$

$$\int_{-\infty}^{\infty} \exp\left(-\left(\frac{\mathbf{R}'}{w_g} + i\mathbf{K}\frac{w_g}{2}\right)^2\right) \mathrm{d}^2 R'. \quad (A.20)$$

Making once again a change in variable $\mathbf{u} = \mathbf{R}'/w_g + i\mathbf{K}w_g/2$, and using the integral formula:

$$\int_{-\infty}^{\infty} \exp(-\mathbf{u}^2)\mathrm{d}^2 u = \int_{-\infty}^{\infty}\int_{-\infty}^{\infty} \exp(-\mathbf{u_x}^2)\exp(-\mathbf{u_y}^2)\mathrm{d}u_x \mathrm{d}u_y = \sqrt{\pi}\sqrt{\pi} = \pi,$$

$$(A.21)$$

we obtain:

$$\widetilde{g}(\mathbf{K}) = \frac{1}{\pi w_g^2} \frac{\exp(-i\mathbf{K}\cdot\mathbf{R}_s)}{(2\pi)^2} \exp\left(-\frac{|\mathbf{K}|^2 w_g^2}{4}\right) w_g^2 \pi. \quad (A.22)$$

Grouping terms and identifying the half-width of the Gaussian in Fourier space as $\widetilde{w}_g = 2/w_g$ we obtain the following expression in $\mathbf{K}$-space:

$$\widetilde{g}(\mathbf{K}) = \frac{1}{(2\pi)^2} \exp(-i\mathbf{K}\cdot\mathbf{R}_s)\exp\left(-\frac{|\mathbf{K}|^2}{\widetilde{w}_g^2}\right). \quad (A.23)$$

A remarkably simple approach was discovered by Poisson to solve the Gaussian Integral, which is of general interest and I include here for completeness. Let us consider the following Gaussian integral to which we intend to find the result:

$$I = \int_0^{\infty} \exp(-x^2)\mathrm{d}x.$$

If we now consider the solution to I^2, we can write it as:

$$I^2 = \left(\int_0^{\infty} \exp(-x^2)\mathrm{d}x\right)^2 = \int_0^{\infty}\int_0^{\infty} \exp(-x^2-y^2)\mathrm{d}x\mathrm{d}y.$$

If we now include the change of variable $x^2 + y^2 = r^2$ we obtain:

$$I^2 = \int_0^{\pi/2}\int_0^{\infty} r\exp(-r^2)\mathrm{d}r\mathrm{d}\theta =$$

$$\frac{\pi}{2}\int_0^{\infty} r\exp(-r^2)\mathrm{d}r =$$

$$\frac{-\pi\exp(-r^2)}{4}\bigg|_0^{\infty} = \frac{\pi}{4},$$

which gives the solution to the integral we were looking for:

$$I = \int_0^\infty \exp(-x^2)\mathrm{d}x = \frac{\sqrt{\pi}}{2}. \tag{A.24}$$

The same approach can be done for the integral used to obtain an expression for $\widetilde{g}(\mathbf{K})$, Eq. (A.21):

$$I = \int_{-\infty}^\infty \exp(-x^2)\mathrm{d}x = \sqrt{\pi}.$$

A.6 Vector Identities

The following vector identities involving the rotational, divergence or gradient are very handy to simplify formulas:

- $$\mathbf{a} \cdot (\nabla \times \mathbf{b}) - \mathbf{b} \cdot (\nabla \times \mathbf{a}) = -\nabla \cdot (\mathbf{a} \times \mathbf{b}),$$

- $$\nabla \times \left(\nabla \times \mathbf{a}\right) = \nabla \cdot \left(\nabla \mathbf{a}\right) - \nabla^2 \mathbf{a},$$

Additionally, the following scalar and vector product formulas and relations are also very useful:

- The scalar triple product:
$$\mathbf{a} \cdot (\mathbf{b} \times \mathbf{c}) = \mathbf{b} \cdot (\mathbf{c} \times \mathbf{a}) = \mathbf{c} \cdot (\mathbf{a} \times \mathbf{b}),$$
which represents the volume of the parallelepiped defined by the three vectors.

- The vector triple product:
$$\mathbf{a} \times (\mathbf{b} \times \mathbf{c}) = (\mathbf{a} \cdot \mathbf{c})\mathbf{b} - (\mathbf{a} \cdot \mathbf{b})\mathbf{c},$$
or, equivalently,
$$(\mathbf{a} \times \mathbf{b}) \times \mathbf{c} = -\mathbf{c} \times (\mathbf{a} \times \mathbf{b}) = -(\mathbf{c} \cdot \mathbf{b})\mathbf{a} + (\mathbf{c} \cdot \mathbf{a})\mathbf{b},$$
which is the triple product expansion and the formula which gives the above relation for $\nabla \times \left(\nabla \times \mathbf{a}\right) = \nabla \cdot \left(\nabla \mathbf{a}\right) - \nabla^2 \mathbf{a}$, which is Lagrange's formula.

- Jacobi's identity:
$$\mathbf{a} \times (\mathbf{b} \times \mathbf{c}) + \mathbf{c} \times (\mathbf{a} \times \mathbf{b}) + \mathbf{b} \times (\mathbf{c} \times \mathbf{a}) = 0.$$

Appendix B

The Solid Angle

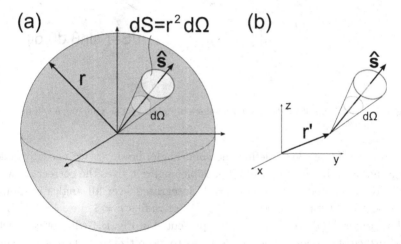

Fig. B.1 (a) Definition of solid angle for a sphere of radius r, equivalent to Fig. B.2. The solid angle is given by $d\Omega = dS/r^2$, which for the unit sphere, $r = 1$, has a value of one steradian, $d\Omega = 1sr$. In (b) it is shown how this solid angle can be used to represent the energy that flows within a solid angle $d\Omega$ into a particular direction $\hat{s}$.

The solid angle, Ω, is defined as the 2D angle subtended in 3D by an object (it's surface, really) and a point (see Fig. B.1). It is measured in *steradians*, the unit steradian being defined as the 2D angle subtended when the area considered is r^2 for a sphere of radius r (as in Fig. B.1 (a)). The use of the solid angle has become fundamental when dealing with radiosity concepts, since in radiosity (take the specific intensity, for example) the flow of energy in direction $\hat{s}$ at a point $\mathbf{r'}$ is considered within a solid angle, concept represented in Fig. B.1 (b). Note that this concept is not necessary when working with the Poynting vector since, as all vectors, it only points in a single specific direction. It is when we need to consider a volume average of these pointing vectors, such as for the specific intensity, that we must resort to using solid angles. Apart from radiosity concepts, it is useful to account for apertures seen from a distance, as for numerical apertures in geometrical optics.

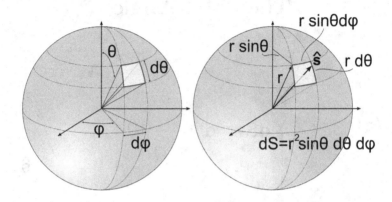

Fig. B.2 Definition of solid angle for a sphere of radius r. The solid angle will be given by $d\Omega = \sin\theta d\theta d\phi$.

In order to properly define the solid angle, however, we need to make sure that once integrated over all possible angles it gives the correct answer: defined for a sphere of radius $r = 1$, integrated over all angles it should represent the total surface of the sphere of radius $r = 1$, i.e. 4π. Clearly, taking differential areas such as that presented in Fig. B.1 will not complete the surface of a sphere. The simplest way to discretize the sphere's surface is shown in Fig. B.2, where we make use of spherical coordinates. Discretized in this manner, it is clear that the area subtended by a solid angle Ω in a sphere of radius r will be the product of the arc for θ, $rd\theta$, and the arc for ϕ, $r \sin\theta d\phi$ (see Fig. B.2):

$$dS = rd\theta r \sin\theta d\phi,$$

or equivalently:

$$dS = r^2 d(\cos\theta)d\phi. \tag{B.1}$$

Since the solid angle is defined for a sphere of radius $r = 1$, its definition in terms θ and ϕ is:

$$d\Omega = d(\cos\theta)d\phi, \tag{B.2}$$

in which case we may define the differential surface dS as:

$$dS = r^2 d\Omega. \tag{B.3}$$

If we now integrate the solid angle over all possible angles we obtain:

$$\int_{(4\pi)} d\Omega = \int_0^{2\pi} d\phi \int_{-1}^{1} d(\cos\theta) = 4\pi,$$

which considering Eq. (B.3) gives as a total surface of $S = 4\pi r^2$.

B.1 The solid angle delta function

Given the definition of the solid angle in terms of θ and ϕ one has to take great care when defining functions such as the delta function, since it is *not* a simple product of $\delta(\theta - \theta')$ and $\delta(\phi - \phi')$. The delta function for the solid angle, $\delta(\Omega - \Omega')$ must be such that:

$$\int_{(4\pi)} f(\Omega')\delta(\Omega - \Omega')d\Omega' = f(\Omega),$$

where f is a generic continuous function. Representing $d\Omega$ in terms of θ and σ, we may rewrite:

$$\int_0^{2\pi} \int_{-1}^{1} f(\theta', \phi')\delta(\Omega - \Omega')d(\cos\theta')d\phi' = f(\theta, \phi),$$

which implies that $\delta(\Omega - \Omega')$ can be defined as:

$$\delta(\Omega - \Omega') = \delta(\cos\theta - \cos\theta')\delta(\phi - \phi'). \tag{B.4}$$

As a side note, in [Ishimaru (1978)] you can find the definition of this delta function as $\delta(\theta - \theta')\delta(\phi - \phi')/\sin\theta'$, which is equivalent but it needs to be taken care of near $\sin\theta \simeq 0$ (at $\sin\theta' = 0$ it becomes $\delta(\phi - \phi')\delta(\theta)$).

B.2 The solid angle and the unit direction vector

So far, we have defined the solid angle integral over $d\Omega$ for a scalar quantity. However, the solid angle integral becomes most useful when we apply it to actual directions which we may want to consider, i.e. $\hat{s}$ in Fig. B.1 (b). Considering a specific direction of propagation $\hat{s}$, we may define this unit vector in Cartesian coordinates as:

$$\hat{s} = \cos\phi\sin\theta\mathbf{u}_x + \sin\phi\sin\theta\mathbf{u}_y + \cos\theta\mathbf{u}_z, \tag{B.5}$$

where $\mathbf{u}_x$, $\mathbf{u}_y$, and $\mathbf{u}_z$ represent the unit vectors. It can be shown that the solid integral over all angles of $\hat{\mathbf{s}}$ is:

$$\int_{(4\pi)} \hat{\mathbf{s}} d\Omega = \int_0^{2\pi} \cos\phi d\phi \int_{-1}^1 \sin^2\theta d\theta \mathbf{u}_x +$$

$$\int_0^{2\pi} \sin\phi d\phi \int_{-1}^1 \sin^2\theta d\theta \mathbf{u}_y + 2\pi \int_{-1}^1 \cos\theta d(\cos\theta) \mathbf{u}_z,$$

which is, as expected, zero:

$$\int_{(4\pi)} \hat{\mathbf{s}} d\Omega = 0,$$

since $\int_0^{2\pi} \cos\phi d\phi = \int_0^{2\pi} \sin\phi d\phi = 0$, and $\int_{-1}^1 \cos\theta d(\cos\theta) = 0$. In a similar way, the product of two directional vectors $\hat{\mathbf{s}}$ and $\hat{\mathbf{s}}'$ when integrated over all directions results in:

$$\int_{(4\pi)} \hat{\mathbf{s}} \cdot \hat{\mathbf{s}}' d\Omega = \int_{(4\pi)} \cos\theta d\Omega = \int_0^{2\pi} d\phi \int_{-1}^1 \cos\theta d(\cos\theta) = 0,$$

which when considered only a single hemisphere, take the upper half as an example, gives:

$$\int_{(2\pi)+} \hat{\mathbf{s}} \cdot \hat{\mathbf{s}}' d\Omega = \int_{(2\pi)+} \cos\theta d\Omega = \int_0^1 2\pi \cos\theta d\cos\theta = \pi.$$

The general formula for any power n of the product of two directional vectors can be defined as:

$$\int_{(4\pi)} (\hat{\mathbf{s}} \cdot \hat{\mathbf{s}}')^n d\Omega = 2\pi \int_{-1}^1 (\cos\theta)^n d(\cos\theta) = \frac{2\pi}{n+1} \left(1 - (-1)^{n+1}\right),$$

which is of course zero for odd values and $4\pi/(n+1)$ for even values of n.

We can also define a delta function, exactly as in the previous section, but accounting for the different directional vectors that define each solid angle, Ω and Ω' as:

$$\int_{(4\pi)} \delta(\hat{\mathbf{s}} - \hat{\mathbf{s}}') d\Omega' = 1,$$

or equivalently:

$$\int_{(4\pi)} f(\hat{\mathbf{s}}')\delta(\hat{\mathbf{s}} - \hat{\mathbf{s}}') d\Omega' = f(\hat{\mathbf{s}}). \tag{B.6}$$

Note that the generic function f does not need to be a vector, but simply depends on the unit vector $\hat{\mathbf{s}}$. As an example, the specific intensity, $I(\mathbf{r}, \hat{\mathbf{s}}')$, is not a vector but depends on $\hat{\mathbf{s}}$ through θ and ϕ. We could identically define a specific intensity as $I(\mathbf{r}, \theta, \phi)$ which would represent the same quantity. Expressing radiosity quantities in terms of $\hat{\mathbf{s}}$, however, simplifies the notation.

Some other useful integrals that involve unit directional vectors and the solid angle are:

$$\int_{(4\pi)} \hat{\mathbf{s}} \cdot (\hat{\mathbf{s}} \cdot \mathbf{A}) d\Omega = \frac{4\pi}{3} \mathbf{A},$$

$$\int_{(4\pi)} \hat{\mathbf{s}} \cdot [\hat{\mathbf{s}} \cdot \nabla(\mathbf{A} \cdot \hat{\mathbf{s}})] d\Omega = \int_{(4\pi)} \hat{\mathbf{s}} \cdot [\nabla \cdot \mathbf{A}] d\Omega = 0.$$

where $\mathbf{A}$ represents a generic vector. Clearly, any function which has no angular dependence (i.e that is not dependent on θ or ϕ) can be placed outside of the solid angle integral:

$$\int_{(4\pi)} \frac{\partial U(\mathbf{r}, t)}{\partial t} \hat{\mathbf{s}} d\Omega = \frac{\partial U(\mathbf{r}, t)}{\partial t} \int_{(4\pi)} \hat{\mathbf{s}} d\Omega = 0,$$

whereas functions, such as the specific intensity, which have an angular dependence must be integrated carefully:

$$U(\mathbf{r}) = \int_{(4\pi)} I(\mathbf{r}, \hat{\mathbf{s}}) d\Omega = \int_{-1}^{1} \left[\int_0^{2\pi} I(\mathbf{r}, \theta, \phi) d\phi \right] d(\cos\theta),$$

where we have used the average intensity $U(\mathbf{r})$ as an example.

An integral which dealt with most of the above-mentioned properties was encountered in Sec. 4.3 of Chap. 4:

$$\int_{(4\pi)} p(\hat{\mathbf{s}} \cdot \hat{\mathbf{s}}') \hat{\mathbf{s}} d\Omega,$$

which solution must be a vector. By using the triple vector product rule, $\mathbf{a} \times (\mathbf{b} \times \mathbf{c}) = \mathbf{b}(\mathbf{a} \cdot \mathbf{c}) - \mathbf{c}(\mathbf{a} \cdot \mathbf{c})$ identifying $\mathbf{a} = \mathbf{c} = \hat{\mathbf{s}}'$ and $\mathbf{b} = \hat{\mathbf{s}}$ we find the relation:

$$\hat{\mathbf{s}} = \hat{\mathbf{s}}' \times (\hat{\mathbf{s}} \times \hat{\mathbf{s}}') + \hat{\mathbf{s}}'(\hat{\mathbf{s}}' \cdot \hat{\mathbf{s}}),$$

and therefore the integral becomes:

$$\int_{(4\pi)} p(\hat{\mathbf{s}} \cdot \hat{\mathbf{s}}') \hat{\mathbf{s}} d\Omega = \frac{g\mu_s}{\mu_t} \hat{\mathbf{s}}' + \int_{(4\pi)} p(\hat{\mathbf{s}} \cdot \hat{\mathbf{s}}') (\hat{\mathbf{s}}' \times (\hat{\mathbf{s}} \times \hat{\mathbf{s}}')) d\Omega,$$

where we have made use of the definition for the phase function $p(\hat{\mathbf{s}} \cdot \hat{\mathbf{s}}')$:

$$\int_{(4\pi)} p(\hat{\mathbf{s}} \cdot \hat{\mathbf{s}}')(\hat{\mathbf{s}}' \cdot \hat{\mathbf{s}}) \hat{\mathbf{s}}' d\Omega = \hat{\mathbf{s}}' \int_{(4\pi)} p(\hat{\mathbf{s}} \cdot \hat{\mathbf{s}}')(\hat{\mathbf{s}}' \cdot \hat{\mathbf{s}}) d\Omega = \hat{\mathbf{s}}' \frac{g\mu_s}{\mu_t}.$$

We have left to solve the integral of the triple vector product:

$$\int_{(4\pi)} p(\hat{\mathbf{s}} \cdot \hat{\mathbf{s}}') \, (\hat{\mathbf{s}}' \times (\hat{\mathbf{s}} \times \hat{\mathbf{s}}')) \, d\Omega.$$

It is clear from geometrical considerations that the resulting vector of $\hat{\mathbf{s}}' \times (\hat{\mathbf{s}} \times \hat{\mathbf{s}}')$ will be contained within the plane defined by $\hat{\mathbf{s}} \cdot \hat{\mathbf{s}}'$ and will be perpendicular to $\hat{\mathbf{s}}'$ whatever the value of $\hat{\mathbf{s}}$, in which case the resulting integral is zero if we integrate over 2π. We can prove this directly by decomposing in Cartesian coordinates assuming $\hat{\mathbf{s}}'$ points towards z (since the integral is over all angles and p depends on the cosine of the angle, this does not reduce the generality of the solution). In this case:

$$\hat{\mathbf{s}} \times \hat{\mathbf{s}}' = s_y \mathbf{u}_x - s_x \mathbf{u}_y,$$

which when applying the vector product of $\hat{\mathbf{s}}'$ clearly gives:

$$\hat{\mathbf{s}}' \times (\hat{\mathbf{s}} \times \hat{\mathbf{s}}') = s_x \mathbf{u}_x + s_y \mathbf{u}_y,$$

i.e. vector $\hat{\mathbf{s}}$ projected onto the xy plane (as expected, perpendicular to $\hat{\mathbf{s}}' = s_z' \mathbf{u}_z$). We have already encountered the expression of $\hat{\mathbf{s}}$ in terms of θ and ϕ (see Eq. (B.5)) in which case the integral we're interested in becomes:

$$\int_{(4\pi)} p(\hat{\mathbf{s}} \cdot \hat{\mathbf{s}}') \, (\hat{\mathbf{s}}' \times (\hat{\mathbf{s}} \times \hat{\mathbf{s}}')) \, d\Omega =$$

$$\int_0^{2\pi} \cos\phi d\phi \int_{-1}^1 p(\cos\theta) \sin\theta d(\cos\theta) \mathbf{u}_x +$$

$$\int_0^{2\pi} \sin\phi d\phi \int_{-1}^1 p(\cos\theta) \sin\theta d(\cos\theta) \mathbf{u}_y.$$

Both integrals over $d\phi$ are zero, yielding the result we expected:

$$\int_{(4\pi)} p(\hat{\mathbf{s}} \cdot \hat{\mathbf{s}}') \, (\hat{\mathbf{s}}' \times (\hat{\mathbf{s}} \times \hat{\mathbf{s}}')) \, d\Omega = 0,$$

which is the result we used in Chap. 4, Sec. 4.3.

Appendix C

An Alternative Derivation of the Radiative Transfer Equation

C.1   Derivation of the Radiative Transfer Equation

In this appendix I will go over one possible derivation of the Radiative Transport Equation (RTE) using Poynting's Theorem for energy conservation as a starting point. Note that this is not the current phenomenological derivation found in classical RTE text-books, but I believe it provides significant insight on the approximations inherent to the RTE. In order to derive the RTE, let us first go back to Chap. 2, to the equation of energy conservation, and assume for generality that we have a time-varying dependence (we modulate the intensity of our source, for example) which is much slower than the frequency of the electromagnetic oscillation ω. In this case, we may use the time-averaged expressions which for a generic direction $\hat{\mathbf{s}}_J$ were (see Eq. (2.15)):

$$\frac{1}{c_0}\frac{\partial \langle \mathbf{S}(\mathbf{r}) \rangle \cdot \hat{\mathbf{s}}_J}{\partial t} + \left\langle \frac{\mathrm{d}P_{abs}}{\mathrm{d}V}(\mathbf{r}) \right\rangle (\hat{\mathbf{s}} \cdot \hat{\mathbf{s}}_J) + \hat{\mathbf{s}}_J \cdot \nabla(\langle \mathbf{S}(\mathbf{r}) \rangle \cdot \hat{\mathbf{s}}_J) = 0, \qquad (C.1)$$

which basically ensures invariance under rotation, or equivalently, energy conservation in any particular direction $\hat{\mathbf{s}}_J$. Always remember that this expression is *only valid in the far-field of the particles* where we may assume that the electric and magnetic fields are mutually orthogonal. This is correct when we can describe the field far away from the scatterers as an outgoing spherical wave.

In order to arrive to the RTE we first need to consider the equation for energy conservation shown above in a small differential volume δV, which we shall assume contains N particles such as those defined in Chapter 2

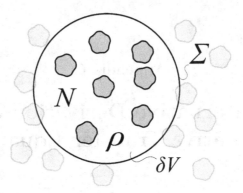

Fig. C.1 Representation of the surface Σ which encloses the differential volume δV with a total of N particles distributed with density ρ.

(see Fig. C.1). Integrating Eq. (C.1) over δV we obtain:

$$\frac{1}{c_0}\frac{\partial}{\partial t}\int_{\delta V}(\hat{\mathbf{s}}\cdot\hat{\mathbf{s}}_J)S(\mathbf{r}-\mathbf{r}')\mathrm{d}^3r'+$$

$$\int_{\delta V}(\hat{\mathbf{s}}\cdot\hat{\mathbf{s}}_J)\Big\langle\frac{\mathrm{d}P_{abs}}{\mathrm{d}V}(\mathbf{r}-\mathbf{r}')\Big\rangle\mathrm{d}^3r'+$$

$$\int_{\delta V}(\hat{\mathbf{s}}\cdot\hat{\mathbf{s}}_J)\hat{\mathbf{s}}_J\cdot\nabla_{\mathbf{r}'}S(\mathbf{r}-\mathbf{r}')\mathrm{d}^3r' = 0, \quad (C.2)$$

where we have introduced $S(\mathbf{r}) = |\langle\mathbf{S}(\mathbf{r})\rangle|$ as the magnitude of the time-averaged Poynting vector at $\mathbf{r}$. We shall now proceed to analyze each of the terms in Eq. (C.2) one by one, making use of the *volume averaged* expressions for the flow of energy defined in Chap. 3.

C.1.1 *Volume Averaged Change in Energy Density*

We will now rewrite the first term of Eq. (C.2), which defines the change in the volume averaged energy density, in terms of the volume averaged quantities. We will now write the energy flux as a sum of two contributions: the incident energy flux $\langle\mathbf{S}^{(inc)}\rangle$ (note that now this incident flux *is not* a plane-wave but rather a general function which accounts for propagation *outside* of δV) and the *scattered* contribution, $\langle\mathbf{S}^{(sc)}\rangle$:

$$\langle\mathbf{S}\rangle = \langle\mathbf{S}^{(inc)}\rangle + \langle\mathbf{S}^{(sc)}\rangle. \quad (C.3)$$

Introducing Eq. (C.3) into the first term of our starting equation, Eq. (C.2), and making use of the expression for the volume averaged flux of energy defined in Chap. 3, Eq. (3.5), we obtain:

$$\frac{1}{c_0}\frac{\partial}{\partial t}\int_{\delta V}\langle \mathbf{S}\rangle \cdot \hat{\mathbf{s}}_J dV \simeq \frac{1}{c_0}\frac{\partial}{\partial t}\|\mathbf{S}^{(inc)}\|_v w_\mathbf{r}(\hat{\mathbf{s}}_J)\delta V, \tag{C.4}$$

where we have made the assumption that:

$$\int_{\delta V}\langle \mathbf{S}^{(inc)}\rangle \cdot \hat{\mathbf{s}}_J dV \gg \int_{\delta V}\langle \mathbf{S}^{(sc)}\rangle \cdot \hat{\mathbf{s}}_J dV, \tag{C.5}$$

which basically states that the total *averaged* flux over volume δV can be seen as the averaged contribution of the flux within δV without taking the contribution of the particles inside δV into consideration. This approximation is similar to the *dipolar approximation* for solving multiple scattering of small particles, and considers that each particle does not have a self-induction term, but rather that the incident field at each particle is simply the sum of all scattered fields excluding its own. In the case represented here the dipoles would be the differential volumes δV. This approximation should hold true if δV is small enough (see Sec. 3.6.3 of Chap. 3) and the volume occupied by the medium surrounding δV that generated $\langle \mathbf{S}^{(inc)}\rangle$ is much larger than δV (i.e. $V - \delta V \gg \delta V$).

C.1.2   Volume Averaged Absorbed Power

Following a similar approach, we can solve for the second term of Eq. (C.2) which gives us the amount of power *absorbed* within this differential volume δV due to the flow energy propagating in direction $\hat{\mathbf{s}}_J$. Considering we have N particles of identical absorption cross-section σ_a, making use of the approximation of Eq. (C.5) we obtain:

$$\int_{\delta V}\left\langle \frac{dP_{abs}}{dV}\right\rangle(\hat{\mathbf{s}}\cdot \hat{\mathbf{s}}_J)dV = w_\mathbf{r}(\hat{\mathbf{s}}_J)\int_{\delta V}\left\langle \frac{dP_{abs}}{dV}\right\rangle dV \simeq$$
$$N\sigma_a\|\mathbf{S}^{(inc)}\|_v w_\mathbf{r}(\hat{\mathbf{s}}_J), \quad (C.6)$$

where $w_\mathbf{r}(\hat{\mathbf{s}}_J)$ represents the contribution of the energy flux to each angle introduced in Eq. (3.1) of Chap. 3 and we have used the volume averaged incident flow of energy, $\|\mathbf{S}^{(inc)}\|_v$, defined for the absorption cross-section (see Eq. (2.36)):

$$\sigma_a = \frac{1}{\|\mathbf{S}^{(inc)}\|_v}\int_{\delta V}\left\langle \frac{dP_{abs}}{dV}\right\rangle dV. \tag{C.7}$$

It is important to note that in order to arrive to this expression we need to assume that the incident flow of energy at any point within δV (on the absorbing particles in this case) can be approximated to the volume averaged incident flow of energy, $\|\mathbf{S}^{(inc)}\|_v$. We therefore used the following approximation:

$$\langle \mathbf{S}^{(inc)}(\mathbf{r}) \rangle \cdot \hat{\mathbf{s}} \simeq \|\mathbf{S}^{(inc)}(\mathbf{r})\|_v, \tag{C.8}$$

where you will remember that $\hat{\mathbf{s}}$ is the direction of the flow of energy at $\mathbf{r}$, i.e. $|\langle \mathbf{S}^{(inc)} \rangle| = \langle \mathbf{S}^{(inc)} \rangle \cdot \hat{\mathbf{s}}$.

C.1.3 *Volume Averaged Change in Energy Flow*

We now arrive to the third and last term $\int_{\delta V} (\hat{\mathbf{s}} \cdot \hat{\mathbf{s}}_J)\hat{\mathbf{s}}_J \cdot \nabla S dV$ of Eq. (C.2), which we may rewrite making use of $w_{\mathbf{r}}$ as:

$$\int_{\delta V} (\hat{\mathbf{s}}' \cdot \hat{\mathbf{s}}_J)\hat{\mathbf{s}}_J \cdot \nabla_{\mathbf{r}'} S(\mathbf{r} - \mathbf{r}') d^3 r' = w_{\mathbf{r}}(\hat{\mathbf{s}}_J) \int_{\delta V} \hat{\mathbf{s}}_J \cdot \nabla_{\mathbf{r}'} S(\mathbf{r} - \mathbf{r}') d^3 r'. \tag{C.9}$$

Introducing Eq. (C.3) we obtain:

$$\int_{\delta V} (\hat{\mathbf{s}}' \cdot \hat{\mathbf{s}}_J)\hat{\mathbf{s}}_J \cdot \nabla_{\mathbf{r}'} S(\mathbf{r} - \mathbf{r}') d^3 r' =$$

$$w_{\mathbf{r}}(\hat{\mathbf{s}}_J)\|\nabla \mathbf{S}^{(inc)} \cdot \hat{\mathbf{s}}_J\| + w_{\mathbf{r}}(\hat{\mathbf{s}}_J) \int_{\delta V} \hat{\mathbf{s}}_J \cdot \nabla_{\mathbf{r}'} S^{(sc)}(\mathbf{r} - \mathbf{r}') d^3 r', \tag{C.10}$$

where we have introduced the volume averaged change in $\langle \mathbf{S} \rangle$ as:

$$\|\nabla \mathbf{S}^{(inc)} \cdot \hat{\mathbf{s}}_J\| = \frac{1}{\delta V} \int_{\delta V} \hat{\mathbf{s}}_J \cdot \nabla_{\mathbf{r}'} S^{(inc)}(\mathbf{r} - \mathbf{r}') d^3 r'. \tag{C.11}$$

At this point, we will assume that within δV we are always located at the far field of the particles and therefore if δV is small $\mathbf{r} \simeq \mathbf{r}'$. Since Eq. (C.11) should be a smooth function of $\mathbf{r}$ (which it should, given that $\langle \mathbf{S} \rangle$ is continuous and differentiable), we will introduce the following approximation:

$$\|\nabla \mathbf{S}^{(inc)} \cdot \hat{\mathbf{s}}_J\| \simeq \hat{\mathbf{s}}_J \cdot \nabla \|\mathbf{S}^{(inc)}\|, \tag{C.12}$$

which implies that we may interchange $\nabla_{\mathbf{r}} \leftrightarrow \nabla_{\mathbf{r}'}$. Making use of this approximation, Eq. (C.10) becomes:

$$\int_{\delta V} (\hat{\mathbf{s}}' \cdot \hat{\mathbf{s}}_J)\hat{\mathbf{s}}_J \cdot \nabla_{\mathbf{r}'} S(\mathbf{r} - \mathbf{r}') d^3 r' = w_{\mathbf{r}}(\hat{\mathbf{s}}_J)\hat{\mathbf{s}}_J \cdot \nabla \|\mathbf{S}^{(inc)}(\mathbf{r})\|_v \delta V +$$

$$w_{\mathbf{r}}(\hat{\mathbf{s}}_J) \int_{\delta V} \hat{\mathbf{s}}_J \cdot \nabla_{\mathbf{r}'} S^{(sc)}(\mathbf{r} - \mathbf{r}') d^3 r'. \tag{C.13}$$

We will now concentrate on the second right-hand side of Eq. (C.13) which accounts for the scattered electromagnetic field. We will make use of Gauss' Theorem, but with great care: in this specific case, the surface that encloses our volume δV does not only have outward flux, since we have multiple particles *outside* of δV that will generate flux crossing the surface *inward*. Considering Σ is the surface that encloses δV, this is expressed as:

$$\int_{\delta V} \hat{s}_J \cdot \nabla S^{(sc)}(\mathbf{r} - \mathbf{r}')d^3r' = \Sigma^+(\mathbf{r}) - \Sigma^-(\mathbf{r}), \qquad (C.14)$$

where we have defined the outward and inward fluxes respectively as (see Fig. C.2):

$$\Sigma^+(\mathbf{r}) = \int_\Sigma S_{out}^{(sc)}(\mathbf{r} - \mathbf{r}')\hat{s}_J \cdot \hat{s}'dS' \qquad (C.15)$$

$$\Sigma^-(\mathbf{r}) = \int_\Sigma S_{in}^{(sc)}(\mathbf{r} - \mathbf{r}')\hat{s}_J \cdot \hat{s}'dS'. \qquad (C.16)$$

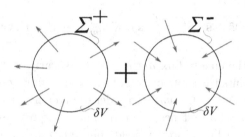

Fig. C.2   Representation of the outward and inward flux, Σ^+ and Σ^- respectively, from surface Σ which encloses the differential volume δV.

Following the same approach used to obtain the absorption energy within δV and using the equivalent formula of Eq. (C.7) for the scattering cross-section, we may rewrite the scattering coefficient in terms of the volume averaged flow of energy and approximation Eq. (C.8) as (see Eq. (2.38) in Chap. 2) :

$$\sigma_{sc} = \frac{1}{\|\mathbf{S}^{(inc)}(\mathbf{r})\|_v} \int_\Sigma S_{out}^{(sc)}(\mathbf{r} - \mathbf{r}')\hat{s}_J \cdot \hat{s}'dS', \qquad (C.17)$$

in which case, considering we have N scattering particles, Eq. (C.14) becomes:

$$\int_{\delta V} \hat{s}_J \cdot \nabla S^{(sc)}(\mathbf{r} - \mathbf{r}')d^3r' =$$

$$N\sigma_{sc}w_{\mathbf{r}}(\hat{s}_J)\|\mathbf{S}^{(inc)}(\mathbf{r})\|_v - \Sigma^-(\mathbf{r}). \qquad (C.18)$$

At this point, we have only one term left to deal with: the contribution from *outside* of volume δV represented by $\int_{\delta V} S_{in}^{(sc)} \hat{\mathbf{s}} \mathrm{d}S$. This term is an extremely important one since it will represent the contribution of scattering from outside of δV and we will dedicate the next subsection to it.

C.1.4   *The Scattering Contribution*

If you have not given up in following the derivations presented up to now, we are now left with the contribution of *inward* flow through the surface Σ which encloses our differential volume δV, which we have represented by $\Sigma^-(\mathbf{r})$. First of all, let us try to understand the meaning of this term. If V is the total volume occupied by the particles, $V' = V - \delta V$ will represent the total volume *excluding* δV. Making use again of Gauss' Theorem on $\Sigma^-(\mathbf{r})$ we may express it as:

$$\int_\Sigma S_{in}^{(sc)}(\mathbf{r} - \mathbf{r}')\hat{\mathbf{s}}_J \cdot \hat{\mathbf{s}}' \mathrm{d}S' = \int_{V'} \hat{\mathbf{s}}_J \cdot \nabla S_{in}^{(sc)}(\mathbf{r} - \mathbf{r}')\mathrm{d}^3 r', \forall \mathbf{r}' \notin \delta V \quad \text{(C.19)}$$

where $S_{in}^{(sc)}$ is the energy flux *outside* of δV. Clearly, it can be seen that this term represents the contribution of the whole outer volume V', i.e. *excluding* the contribution from δV. In order to find a manageable expression for Eq. (C.19) we will first need then to find an expression for the energy flux $S_{in}^{(sc)}$ at the surface Σ for the field going *inward*. To do so we have to remember that we have a medium with *average* optical properties which are homogeneous, i.e. the statistical distribution of our particles is equivalent in all points in space. We may then consider the situation depicted in Fig. C.3, where the total inward flux may be considered equivalent to the total outward flux due to the particles inside δV.

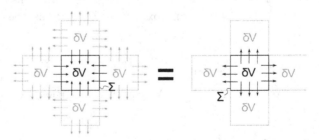

Fig. C.3   A schematic representation of how the inward flux might be considered equivalent to the outward flux through surface Σ which encloses δV. This has been represented in 2D for a square volume; however, a similar approach can be used for any geometry of Σ.

Since in δV we have N particles, the total flux contribution would be:

$$\int_{\Sigma} S_{in}^{(sc)}(\mathbf{r}-\mathbf{r}')\hat{\mathbf{s}}_J \cdot \hat{\mathbf{s}}' dS' \simeq \sum_{i=1}^{N} \int_{\Sigma} \left(\langle \mathbf{S}^{(sc)}(\mathbf{r}) \rangle_i \cdot \hat{\mathbf{s}}_i \right) \hat{\mathbf{s}}_J \cdot \hat{\mathbf{s}}' dS' \; \forall \mathbf{r} \in \Sigma, \quad (C.20)$$

where $\langle \mathbf{S}^{(sc)}(\mathbf{r}) \rangle_i$ is the energy flux scattered by particle i within δV defined by Eq. (2.58). Since we have assumed all particles to be the same, they will all contribute equally to the total flux through surface Σ. Making use of the approximation of Eq. (C.8), in this case we may write Eq. (C.20) in terms of the volume averaged flow of energy as:

$$\int_{\Sigma} S_{in}^{(sc)}(\mathbf{r} - \mathbf{r}')\hat{\mathbf{s}}_J \cdot \hat{\mathbf{s}}' dS' \simeq$$

$$\sigma_{tot} \sum_{i=1}^{N} \|\mathbf{S}^{(inc)}(\mathbf{r})\|_v \int_{\Sigma} w_{\mathbf{r}}(\hat{\mathbf{s}}_i) \frac{p(\hat{\mathbf{s}}_i, \hat{\mathbf{s}}_J)}{|\mathbf{r}' - \mathbf{r}_i|^2} dS', \quad (C.21)$$

Since the total flux traversing Σ for particle i will be the same irrespective of where it is within δV, we may identify $dS/|\mathbf{r} - \mathbf{r}_i|^2$ as the solid angle and finally obtain:

$$\int_{\Sigma} S_{in}^{(sc)}(\mathbf{r} - \mathbf{r}')\hat{\mathbf{s}}_J \cdot \hat{\mathbf{s}}' dS' \simeq$$

$$N\sigma_{tot} \|\mathbf{S}^{(inc)}(\mathbf{r})\|_v \int_{(4\pi)} w_{\mathbf{r}}(\hat{\mathbf{s}}') p(\hat{\mathbf{s}}', \hat{\mathbf{s}}_J) d\Omega'. \quad (C.22)$$

C.1.5   *The Radiative Transfer Equation*

Finally, after quite some approximations and derivations, we have found an expression for each of the three terms in Eq. (C.2) that we started out with. If we put all these expressions together we obtain:

$$\frac{1}{c_0}\frac{\partial}{\partial t}\|\mathbf{S}^{(inc)}\|_v w_{\mathbf{r}}(\hat{\mathbf{s}}_J)\delta V + N\sigma_a \|\mathbf{S}^{(inc)}\|_v w_{\mathbf{r}}(\hat{\mathbf{s}}_J) +$$

$$\hat{\mathbf{s}}_J \cdot \nabla \|\mathbf{S}^{(inc)}\|_v w_{\mathbf{r}}(\hat{\mathbf{s}}_J)\delta V + N\sigma_{sc} \|\mathbf{S}^{(inc)}\|_v w_{\mathbf{r}}(\hat{\mathbf{s}}_J) -$$

$$N\sigma_{tot} \int_{(4\pi)} \|\mathbf{S}^{(inc)}\|_v w_{\mathbf{r}}(\hat{\mathbf{s}}') p(\hat{\mathbf{s}}_J, \hat{\mathbf{s}}') d\Omega' = 0, \quad (C.23)$$

which normalizing by δV becomes:

$$\frac{1}{c_0}\frac{\partial}{\partial t}\|\mathbf{S}^{(inc)}\|_v w_{\mathbf{r}}(\hat{\mathbf{s}}_J) + \mu_a \|\mathbf{S}^{(inc)}\|_v w_{\mathbf{r}}(\hat{\mathbf{s}}_J) +$$

$$\hat{\mathbf{s}}_J \cdot \nabla (\|\mathbf{S}^{(inc)}\|_v w_{\mathbf{r}}(\hat{\mathbf{s}}_J)) + \mu_s \|\mathbf{S}^{(inc)}\|_v w_{\mathbf{r}}(\hat{\mathbf{s}}_J) -$$

$$\mu_t \int_{(4\pi)} \|\mathbf{S}^{(inc)}\|_v w_{\mathbf{r}}(\hat{\mathbf{s}}') p(\hat{\mathbf{s}}_J, \hat{\mathbf{s}}') d\Omega' = 0. \quad (C.24)$$

Eq. (C.24) is the *radiative transport equation* for the time-averaged energy flux[6] $\langle \mathbf{S}^{(inc)} \rangle$ flowing in direction $\hat{\mathbf{s}}_J$ *averaged* over a differential volume δV. I believe that at this point it is useful to summarize what each of the terms in Eq. (C.24) represents and where they came from. Starting from the left, we may remember that the first term accounts for the overall change in time of the energy density at $\mathbf{r}$; the second accounts for the average absorption present at $\mathbf{r}$ (within a differential volume δV); the third accounts for the local net flow of energy within δV; and, finally, the third *and* fourth terms represent the overall change in flow due to scattering (the third term subtracts energy flow in direction $\hat{\mathbf{s}}_J$ while the fourth adds energy flow from directions $\hat{\mathbf{s}}'$ into $\hat{\mathbf{s}}_J$). It is very important to remember that the μ_t factor from the fifth term appears *through our definition of the phase function*, Eq. (2.46). If we decide to express the phase function as $p(\hat{\mathbf{s}}, \hat{\mathbf{s}}_0) = |f(\hat{\mathbf{s}}, \hat{\mathbf{s}}_0)|^2$, as it is also commonly done, the μ_t factor will be absent. It is the $\mu_t p(\hat{\mathbf{s}}, \hat{\mathbf{s}}_0)$ product which yields the contribution which is due solely to scattering, i.e. the absorption term that appears through $\mu_t = \mu_a + \mu_s$ has no direct physical meaning.

If we now define the specific intensity as (see Chap. 3):

$$I(\mathbf{r}, \hat{\mathbf{s}}) = \frac{1}{4\pi} \|\mathbf{S}^{(inc)}\|_v w_{\mathbf{r}}(\hat{\mathbf{s}}),$$

the equation for radiative transfer, Eq. (C.24), recovers its well-known appearance

$$\frac{1}{c_0} \frac{\partial}{\partial t} I(\mathbf{r}, \hat{\mathbf{s}}) + \hat{\mathbf{s}} \cdot \nabla I(\mathbf{r}, \hat{\mathbf{s}}) + \mu_t I(\mathbf{r}, \hat{\mathbf{s}}) - \mu_t \int_{(4\pi)} I(\mathbf{r}, \hat{\mathbf{s}}') p(\hat{\mathbf{s}}, \hat{\mathbf{s}}') d\Omega' = 0.$$

C.1.6 *Summary of Approximations*

It is now convenient analyze the main approximations that took place in order to reach the equation for the radiative transfer for $\|\mathbf{S}^{(inc)}\| w_{\mathbf{r}}(\hat{\mathbf{s}}_J)$, Eq. (C.24). The approximations taken can be summarized as follows:

(1) Average of the flow of energy:

$$\langle \mathbf{S}^{(inc)}(\mathbf{r}) \rangle \cdot \hat{\mathbf{s}} \simeq \|\mathbf{S}^{(inc)}(\mathbf{r})\|_v = \frac{1}{\delta V} \int_{\delta V} S(\mathbf{r} - \mathbf{r}') d^3 r'. \tag{C.25}$$

(2) Average of the gradient of flow of energy:

$$\|\nabla \mathbf{S}^{(inc)} \cdot \hat{\mathbf{s}}_J\| \simeq \hat{\mathbf{s}}_J \cdot \nabla \|\mathbf{S}^{(inc)}\|_v,$$

$$\int_{\delta V} \nabla_{\mathbf{r}'} S(\mathbf{r} - \mathbf{r}') d^3 r' \simeq \nabla_{\mathbf{r}} \int_{\delta V} S(\mathbf{r} - \mathbf{r}') d^3 r'. \tag{C.26}$$

---

[6]Remember that $\langle \mathbf{S}^{(inc)} \rangle$ is the total flow of energy incident onto δV and *not* the incident field in a homogeneous medium

which implies that the far-field approximation is always applicable and that δV is much smaller than $\mathbf{r} - \mathbf{r}'$.

(3) Average incident flow of energy much greater than local average scattered flow of energy:

$$\frac{1}{\delta V} \int_{\delta V} \langle \mathbf{S}^{(inc)} \rangle \cdot \hat{\mathbf{s}} dV \gg \frac{1}{\delta V} \int_{\delta V} \langle \mathbf{S}^{(sc)} \rangle \cdot \hat{\mathbf{s}} dV. \tag{C.27}$$

(4) Incoherent scattering:

$$\langle \mathbf{S}^{(sc)} \rangle \simeq \sum_{i=1}^{N} \langle \mathbf{S}^{(sc)} \rangle_i. \tag{C.28}$$

where $\langle \mathbf{S}^{(sc)} \rangle_i$ is the scattered flow of energy from particle i.

(5) Statistically equivalent optical properties throughout the medium:

$$\mu_a = \frac{1}{\delta V} \sum_{i=1}^{N} \sigma_a^i \simeq \frac{1}{V} \sum_{i=1}^{N_v} \sigma_a^i,$$

$$\mu_s = \frac{1}{\delta V} \sum_{i=1}^{N} \sigma_{sc}^i \simeq \frac{1}{V} \sum_{i=1}^{N_v} \sigma_{sc}^i, \tag{C.29}$$

where N_v represents the total number of particles in the total volume V that contains them.

(6) Far-field approximation:

$$\langle \mathbf{S}^{(sc)} \rangle_i \simeq \langle \mathbf{S}^{(inc)}(\mathbf{r}_i) \rangle \frac{p(\hat{\mathbf{s}}_i, \hat{\mathbf{s}})}{|\mathbf{r} - \mathbf{r}_i|^2} dS. \tag{C.30}$$

If you remember, in order to use this expression we had to assume that we are modeling the electromagnetic field in the far-field (see Eq. (2.31)).

(7) De-polarization neglected:

$$\nabla(\nabla \cdot \mathbf{E}) \simeq 0. \tag{C.31}$$

In order to use the scalar wave equation we neglected this term which accounts for depolarization on interaction with the particles.

Once we have established the main approximations that have been taken, we can then consider: how small must δV be in order to be 'small enough'? This question has no direct answer, but a good approximation to δV should be a volume which is much smaller than the total volume but still holds the same statistical optical properties. Following this definition, the concentration of particles must not be too low otherwise equations Eq. (C.25) and Eq. (C.26) will not hold.

Bibliography

Albaladejo, S., Marqués, M., Laroche, M. and Sáenz, J. (2009). Scattering Forces from the Curl of the Spin Angular Momentum of a Light Field, *Physical Review Letters* **102**, 11, pp. 1–4, doi:10.1103/PhysRevLett.102.113602.

Alberts, B., Bray, D., Hopkin, K., Johnson, A., Lewis, J., Raff, M., Roberts, K. and Walter, P. (2009). *Essential Cell Biology*, 3rd edn. (Garland Publishing, New York), ISBN 978-0-8153-4130-7.

Arfken, G. B. and Weber, H. J. (1995). *Mathematical Methods for Physicists* (Academic Press, New York).

Aronson, R. (1995). Boundary conditions for diffusion of light, *J. Opt. Soc. Am. A* **12**, pp. 2532–2539.

Aronson, R. and Corngold, N. (1999). Photon diffusion coefficient in an absorbing medium, *J. Opt. Soc. Am. A* **16**, pp. 1066–1071.

Arridge, S. and Schotland, J. (2009). Optical tomography: forward and inverse problems, *Inverse Problems* **25**, p. 123010, doi:10.1088/0266-5611/25/12/123010.

Arridge, S. R. (1999). Optical tomography in medical imaging, *Inverse Problems* **15**, 2, pp. R41–R93.

Arridge, S. R. (2011). Methods in diffuse optical imaging. *Philosophical transactions. Series A, Mathematical, physical, and engineering sciences* **369**, 1955, pp. 4558–76, doi:10.1098/rsta.2011.0311.

Arridge, S. R., Cope, M. and Delpy, D. T. (1992). The theoretical basis for the determination of optical pathlengths in tissue: temporal and frequency analysis, *Phys Med Biol* **37**, 7, pp. 1531–1560.

Arridge, S. R. and Lionheart, W. R. B. (1998). Nonuniqueness in diffusion-based optical tomography, *Opt Lett* **23**, 1, pp. 882–884.

Axelsson, J., Svensson, J. and Andersson-Engels, S. (2007). Spatially varying regularization based on spectrally resolved fluorescence emission in fluorescence molecular tomography, *Opt Express* **15**, 21, pp. 13574–13584.

Barabanenkov, Y. (1991). Asymptotic solution of the Bethe-Salpeter equation and the Green-Kubo formula for the diffusion constant for wave propagation in random media, *Physics Letters A* **154**, 1-2, pp. 38–42, doi:10.1016/0375-9601(91)90425-8.

Barabanenkov, Y. (1995). Diffusion asymptotics of the Bethe-Salpeter equation for electromagnetic waves in discrete random media, *Physics Letters A* **206**, 1-2, pp. 116–122, doi:10.1016/0375-9601(95)00576-O.

Blackman, R. B. and Tukey, J. W. (1959). *The Measurement of Power Spectra, From the Point of View of Communications Engineering* (Dover Publ., New York).

Blum, G. (2008). Use of fluorescent imaging to investigate pathological protease activity, *Current Opinion in Drug Discovery and Development* **11**, pp. 708–716.

Boas, D. A. (1996). *Diffuse photon probes of structural and dynamical properties of turbid media: theory and biomedical applications*, Ph.D. thesis, University of Pennsylvania.

Boas, D. A., Oleary, M. A., Chance, B. and Yodh, A. G. (1993). Interference, Diffraction, and Imaging With Diffuse Photon Density Waves, *Biophysical Journal* **64**, 2, pp. A357–A357.

Boas, D. a., O'Leary, M. a., Chance, B. and Yodh, a. G. (1994). Scattering of diffuse photon density waves by spherical inhomogeneities within turbid media: analytic solution and applications. *Proceedings of the National Academy of Sciences of the United States of America* **91**, 11, pp. 4887–91.

Bohren, C. F. and Huffman, D. R. (1998). *Absorption and Scattering of Light by Small Particles* (Wiley-Interscience).

Bolin, F., Preuss, L., Taylor, R. and Ference, R. (1989). Refractive index of some mammalian tissues using a fiber optic cladding method, *Applied optics* **28**, 12, pp. 2297–2303.

Born, M. and Wolf, E. (1999). *Principles of Optics*, seventh edn. (University Press, Cambridge).

Brown. Jr., W. P. (1966). Validity of the Rytov Approximation in Optical Propagation Calculations, *Journal of the Optical Society of America* **56**, 8, p. 1045, doi:10.1364/JOSA.56.001045.

Busch, D. R., Guo, W., Choe, R., Durduran, T., Feldman, M. D., Mies, C., Rosen, M. a., Schnall, M. D., Czerniecki, B. J., Tchou, J., DeMichele, A., Putt, M. E. and Yodh, A. G. (2010). Computer aided automatic detection of malignant lesions in diffuse optical mammography, *Medical Physics* **37**, 4, p. 1840, doi:10.1118/1.3314075.

Cannell, D. M. (1999). George Green: An Enigmatic Mathematician, *The American Mathematical Monthly* **106**, 2, p. 136, doi:10.2307/2589050.

Case, K. M. and Zweifel, P. F. (1967). *Linear Transport Theory* (Addison-Wesley).

Chandrasekhar, S. (1960). *Radiative Transfer* (Dover, New York).

Chen, Y., Mu, C., Intes, X. and Chance, B. (2002). Adaptive calibration for object localization in turbid media with interfering diffuse photon density waves, *Appl Opt* **41**, 34, pp. 7325–7333.

Cheong, W. F., Prahl, S. and Welch, A. (1990). A review of the optical properties of biological tissues, *IEEE Journal of Quantum Electronics* **26**, 12, pp. 2166–2185, doi:10.1109/3.64354.

Cherry, S. R. (2004). In vivo molecular and genomic imaging: new challenges for imaging physics, *Physics in Medicine and Biology* **49**, 3, pp. R13–R48, doi:10.1088/0031-9155/49/3/R01.

Cohen-Tannoudji, C., Dupont-Roc, J. and Grynberg, G. (1997). *Photons and atoms: introduction to quantum electrodynamics*, Wiley professional paperback series (Wiley), ISBN 9780471184331.

Deliolanis, N. C., Kasmieh, R., Wurdinger, T., Tannous, B. a., Shah, K. and Ntziachristos, V. (2011). Performance of the red-shifted fluorescent proteins in deep-tissue molecular imaging applications. *Journal of biomedical optics* **13**, 4, p. 044008, doi:10.1117/1.2967184.

Drezek, R., Dunn, A. and Richards-kortum, R. (1999). Light scattering from cells: finite-difference time-domain simulations and goniometric measurements, *Applied Optics* **38**, 16, pp. 3651-3661.

Ducros, N., D'andrea, C., Valentini, G., Rudge, T., Arridge, S. and Bassi, A. (2010). Full-wavelet approach for fluorescence diffuse optical tomography with structured illumination. *Optics letters* **35**, 21, pp. 3676-8.

Dunn, A. and Richards-Kortum, R. (1996). Three-dimensional computation of light scattering from cells, *IEEE Journal of Selected Topics in Quantum Electronics* **2**, 4, pp. 898-905.

Dunn, A. K. (1997). *Light Scattering Properties of Cells*, Ph.D. thesis, The University of Texas at Austin.

Durduran, T., Culver, J. P., Holboke, M. J., Li, X. D., Zubkov, L., Chance, B., Pattanayak, D. N. and Yodh, A. G. (1999). Algorithms for 3d localization and imaging using near-field diffraction tomography with diffuse light, *Optics Express* **4**, pp. 247-262.

Economou, E. N. (2006). *Green's functions in quantum physics*, Springer series in solid-state sciences (Springer), ISBN 9783540288381.

Einstein, A. (1956). *Investigations on the Theory of the Brownian Movement* (Dover Publ.)

Elaloufi, R., Carminati, R. and Greffet, J.-J. (2003). Definition of the diffusion coefficient in scattering and absorbing media, *J. Opt. Soc. Am. A* **20**, 4, pp. 678-685.

Farrell, T. J., Patterson, M. S. and Wilson, B. (1992). A diffusion theory model of spatially resolved, steady-state diffuse reflectance for the noninvasive determination of tissue optical properties in vivo, *Med Phys* **19**, 4, pp. 879-888.

Feynman, R. P. (1998). *Quantum electrodynamics*, Advanced book classics (Westview Press), ISBN 9780201360752.

Fick, A. (1855). On liquid diffusion, *Philosophical Magazine and Journal Science* **10**, pp. 30-39.

Fishkin, J., Fantini, S. and Gratton, E. (1996). Gigahertz photon density waves in a turbid medium: theory and experiments, *Physical Review E* **53**, 3, p. 2307.

Fishkin, J. and Gratton, E. (1993). Propagation of photon-density waves in strongly scattering media containing an absorbing semi-infinite plane bounded by a straight edge, *JOSA A* **10**, 1, pp. 127-140.

Flock, S. T., Jacques, S. L., Wilson, B. C., Star, W. M. and van Gemert, M. J. C. (1992). Optical properties of Intralipid: A phantom medium for light propagation studies, *Lasers in Surgery and Medicine* **12**, pp. 510-519.

Fourier, J. B. J. and Freeman, A. (1878). *The analytical theory of heat* (The University Press).

Frisch, U. (1965). Wave propagation in random media, Part II, Multiple scattering by N bodies, Tech. rep., Institut d'Astrophysique, Paris.

Froufe-Pérez, L., Carminati, R. and Sáenz, J. (2007). Fluorescence decay rate statistics of a single molecule in a disordered cluster of nanoparticles, *Physical Review A* **76**, 1, pp. 1–5, doi:10.1103/PhysRevA.76.013835.

Froufe-Pérez, L. S. and Carminati, R. (2008). Lifetime fluctuations of a single emitter in a disordered nanoscopic system: The influence of the transition dipole orientation, *Physica Status Solidi (a)* **205**, 6, pp. 1258–1265, doi: 10.1002/pssa.200778176.

Furutsu, K. (1980). Diffusion equation derived from space-time transport equation, *JOSA* **70**, 4, pp. 360–366.

Graaff, R. and Rinzema, K. (2001). Practical improvements on photon diffusion theory: application to isotropic scattering, *Phys. Med. Biol.* **23**, pp. 3043–3050.

Green, G. (1828). *An essay on the application of mathematical analysis to the theories of electricity and magnetism* (Nabu Press), ISBN 1173567852.

Guo, X., Wood, M. F. G., Ghosh, N. and Vitkin, I. A. (2010). Depolarization of light in turbid media: a scattering event resolved Monte Carlo study, *Applied Optics* **49**, 2, pp. 153–162.

Hadamard, J. (2003). *Lectures on Cauchy's Problem in Linear Partial Differential Equations*, Dover phoenix editions (Dover Publications), ISBN 9780486495491.

Harvey, E. N. (1920). *The Nature of Animal Light* (Monographs on Experimental Biology, Philadephia).

Harvey, E. N. (1957). *A History of Luminescence: From the earliest times until 1900* (The American Physical Society, Philadelphia).

Haskell, R. C., Svaasand, L. O., Tsay, T., Feng, T., Mcadams, M. S. and Tromberg, B. J. (1994). Boundary conditions for the diffusion equation in radiative transfer, *J. Opt. Soc. Am. A* **11**, pp. 2727–2741.

Henyey, L. G. and Greenstein, J. L. (1941). Diffuse radiation in the galaxy, *AstroPhys. J.* **93**, pp. 70–83.

Hillman, E. M. C. and Moore, A. (2007). All-optical anatomical co-registration for molecular imaging of small animals using dynamic contrast. *Nature photonics* **1**, 9, pp. 526–530, doi:10.1038/nphoton.2007.146.

Hyde, D., de Kleine, R., MacLaurin, S. A., Miller, E., Brooks, D. H., Krucker, T. and Ntziachristos, V. (2009). Hybrid FMT-CT imaging of amyloid-beta plaques in a murine Alzheimer's disease model. *Neuroimage* **44**, 4, pp. 1304–1311, doi:10.1016/j.neuroimage.2008.10.038.

Intes, X., Chance, B., Holboke, M. and Yodh, a. (2001). Interfering diffusive photon-density waves with an absorbing-fluorescent inhomogeneity. *Optics express* **8**, 3, pp. 223–31.

Intes, X., Ntziachristos, V. and Chance, B. (2002). Analytical model for dual-interfering sources diffuse optical tomography, *Opt Express* **10**, 1, pp. 2–14.

Ishimaru, A. (1978). *Wave propagation and scattering in Random Media*, Vol. 1 (Academic, New York).

Ito, S. (1984). Comparison of diffusion theories for optical pulse waves propagated ni discrete random media, *Journ. Opt. Soc. Am. A* **1**, 5, pp. 502–505.

Itoh, T., Yamada, T., Kodera, Y., Matsushima, A., Hiroto, M. and Sakurai, K. (2001). Hemin (Fe 3+)- and Heme (Fe 2+)-Smectite Conjugates as a Model of Hemoprotein Based on Spectrophotometry, *Communications* , pp. 10–13.

Jackson, J. D. (1962). *Classical Electrodynamics* (John Wiley and Sons, New York).

Jacques, S. L. (1996). Origins of tissue optical properties in the UVA, visible and NR regions, *Advances in Optical Imaging and Photon Migration* **2**, pp. 364–370.

Jacques, S. L. (2003). Monte Carlo simulations of fluorescence in turbid media, in M.-A. Mycek and B. W. Pogue (eds.), *Handbook of Biomedical Fluorescence*, chap. 6 (Marcel-Dekker, New York).

Jacques, S. L. and Ramella-Roman, J. C. (2004). Polarized Light imaging of tissues, in G. Palumbo and R. Pratesi (eds.), *Lasers and Current Optical Techniques in Biology*, comprehens edn., chap. 19 (The Royal Society of Chemistry, Cambridge).

Kak, A. C. and Slaney, M. (1988). *Principles of Computerized tomographic imaging* (IEEE Press, New York).

Kienle, A., Lilge, L., Vitkin, I. A., Patterson, M. S., Wilson, B. C., Hibst, R. and Steiner, R. (1996). Why do veins appear blue? A new look at an old question, *Applied Optics* **35**, 7, pp. 1151–1160.

Konecky, S. D., Panasyuk, G. Y., Lee, K., Markel, V., Yodh, A. G. and Schotland, J. C. (2008). Imaging complex structures with diffuse light, *Opt Express* **16**, 7, pp. 5048–5060.

Lakowicz, J. R. (1999). *Principles of Fluorescence Spectroscopy*, 2nd edn. (Springer).

Lamb, W. E. (1995). Anti-photon, *Applied Physics B Laser and Optics* **60**, 2-3, pp. 77–84, doi:10.1007/BF01135846.

Li, X., Pattanayak, D. N., Durduran, T., Culver, J. P., Chance, B. and Yodh, A. G. (2000). Near-field diffraction tomography with diffuse photon density waves, *Phys Rev E Stat Phys Plasmas Fluids Relat Interdiscip Topics* **61**, 4 Pt B, pp. 4295–4309.

Li, X. and Yao, G. (2009). Mueller matrix decomposition of diffuse reflectance imaging in skeletal muscle, *Applied Optics* **48**, 14, pp. 2625–2631.

Lukic, V., Markel, V. A. and Schotland, J. C. (2009). Optical tomography with structured illumination, *Opt Lett* **34**, 7, pp. 983–985.

M. Lester, R. A. D. (1996). Scattering of electromagnetic waves at the corrugated interface between index-matched media, *Opt. Comm.* **132**, pp. 135–143.

MacKintosh, F. C., Zhu, J. X., Pine, D. J. and Weitz, D. A. (1989). Polarization memory of multiply scattered light, *Phys Rev B* **40**, 13, pp. 9342–9345.

Mandel, L. and Wolf, E. (1995). *Optical Coherence and Quantum Optics* (Cambridge University Press), ISBN 0521417112.

Mandelis, A. (1995). Greens functions in thermal wave physics: Cartesian coordinate representations, *Journal of applied physics* **78**, 2, pp. 647–655.

Mandelis, A. (2011). *Diffusion-Wave Fields: Mathematical Methods and Green Functions* (Springer), ISBN 144192888X.

Markel, V. A., Mital, V. and Schotland, J. C. (2003a). Inverse problem in optical diffusion tomography. III. Inversion formulas and singular-value decomposition. *Journal of the Optical Society of America. A, Optics, image science, and vision* **20**, 5, pp. 890–902.

Markel, V. A., O'Sullivan, J. A. and Schotland, J. C. (2003b). Inverse problem in optical diffusion tomography. IV. Nonlinear inversion formulas, *J Opt Soc Am A Opt Image Sci Vis* **20**, 5, pp. 903–912.

Markel, V. A. and Schotland, J. C. (2001). Inverse problem in optical diffusion tomography. I. Fourier-Laplace inversion formulas, *J Opt Soc Am A Opt Image Sci Vis* **18**, 6, pp. 1336–1347.

Markel, V. A. and Schotland, J. C. (2002). Inverse problem in optical diffusion tomography. II. Role of boundary conditions, *J Opt Soc Am A Opt Image Sci Vis* **19**, 3, pp. 558–566.

Markel, V. A. and Schotland, J. C. (2004). Symmetries, inversion formulas, and image reconstruction for optical tomography, *Phys Rev E Stat Nonlin Soft Matter Phys* **70**, 5 Pt 2, p. 56616.

Martelli, F., Del Bianco, S., Ismaelli, A. and Zaccanti, G. (2010). *Light propagation through biological tissue and other diffusive media: theory, solutions, and software*, Proceedings of the Society of Photo-optical Instrumentation Engineers (SPIE Press, Washington), ISBN 9780819476586.

Matson, C. (1997). A diffraction tomographic model of the forward problem using diffuse photon density waves. *Optics express* **1**, 1, pp. 6–11.

Matson, C. and Liu, H. (2000). Resolved object imaging and localization with the use of a backpropagation algorithm. *Optics express* **6**, 9, pp. 168–74.

Matson, C. L., Clark, N., McMackin, L. and Fender, J. S. (1997). Three-dimensional tumor localization in thick tissue with the use of diffuse photon-density waves, *Applied Optics* **36**, pp. 214–220.

Milne, A. E. (1930). Thermodynamics of the Stars, *Handbuch der Astrophysik* **3**, p. 65.

Mishchenko, M. (2006). Maxwell's equations, radiative transfer, and coherent backscattering: A general perspective, *Journal of Quantitative Spectroscopy and Radiative Transfer* **101**, 3, pp. 540–555, doi:10.1016/j.jqsrt.2006.02.065.

Mishchenko, M. I. (2002). Vector radiative transfer equation for arbitrarily shaped and arbitrarily oriented particles: a microphysical derivation from statistical electromagnetics. *Applied optics* **41**, 33, pp. 7114–34.

Mishchenko, M. I. (2010). Poynting-Stokes tensor and radiative transfer in discrete random media: the microphysical paradigm. *Optics express* **18**, 19, pp. 19770–91.

Mishchenko, M. I., Liu, L., Mackowski, D. W., Cairns, B. and Videen, G. (2007). Multiple scattering by random particulate media: exact 3D results, *Optics Express* **15**, 6.

Morse, P. and Feshbach, H. (1953). *Methods of theoretical Physics, Part I* (McGraw-Hill), ISBN 007043316X.

Nassau, K. (1983). *The Physics and Chemistry of Color: The Fifteen Causes of Color* (Wiley-Interscience, New York).

Nieto-Vesperinas, M. (2006). *Scattering and diffraction in physical optics*, second edi edn. (Pergamon, New York).

Ntziachristos, V. (2000). *Concurrent diffuse optical tomography, spectroscopy and magnetic resonance imaging of breast cancer*, Ph.D. thesis, University of Pennsylvania.

Ntziachristos, V. (2006). Fluorescence molecular imaging, *Annu Rev Biomed Eng* **8**, pp. 1–33.

Ntziachristos, V., Leroy-Willig, A. and Tavitian, B. (eds.) (2006). *Textbook of in vivo Imaging in Vertebrates* (John Wiley & Sons, Ltd, Chichester, UK), ISBN 9780470029596, doi:10.1002/9780470029596.

Ntziachristos, V., Ripoll, J., Wang, L. H. V. and Weissleder, R. (2005). Looking and listening to light: The evolution of whole-body photonic imaging, *Nat. Biotechnol* **vol**, pp. 23pp313–320.

Ntziachristos, V., Tung, C.-H., Bremer, C. and Weissleder, R. (2002). Fluorescence molecular tomography resolves protease activity in vivo. *Nat Med* **8**, 7, pp. 757–760, doi:10.1038/nm729.

Ntziachristos, V. and Weissleder, R. (2001). Experimental three-dimensional fluorescence reconstruction of diffuse media by use of a normalized Born approximation. *Optics letters* **26**, 12, pp. 893–5.

Ntziachristos, V., Yodh, A. G., Schnall, M. and Chance, B. (2000). Concurrent MRI and diffuse optical tomography of breast after indocyanine green enhancement. *Proc Natl Acad Sci U S A* **97**, 6, pp. 2767–2772, doi: 10.1073/pnas.040570597.

Ogilvy, J. A. (1991). *Theory of wave scattering from random rough surfaces* (Adam Hilger, Bristol).

Ohm, G. S., Francis, W. and Lockwood, T. D. (1891). *The galvanic circuit investigated mathematically*, Van Nostrand's science series (D. Van Nostrand company).

O'Leary, M. A. (1996). *Imaging with Diffuse Photon Density Waves*, Ph.D. thesis, University of Pennsylvania.

O'Leary, M. A., Boas, D. A., Chance, B. and Yodh, A. G. (1992). Refraction of diffuse photon density waves, *Physical Review Letters* **69**, 18, pp. 2658–2661.

Patterson, M. S., Chance, B. and Wilson, B. C. (1989). Time resolved reflectance and transmittance for the non-invasive measurement of tissue optical properties, *Appl. Opt.* **28**, pp. 2331–2336.

Pierrat, R. and Carminati, R. (2010). Spontaneous decay rate of a dipole emitter in a strongly scattering disordered environment, *Physical Review A* **81**, 6, pp. 10–13, doi:10.1103/PhysRevA.81.063802.

Planck, M. (1906). *Theorie der Warmestrahlung* (Verlag Von Johan Ambrosius Barth, Leipzig).

Planck, M. (1914). *The theory of heat radiation* (P. Blakiston's Son & Co., Philadelphia).

Ripoll, J. and Nieto-Vesperinas, M. (1999a). Index Mismatch for Diffuse Photon Density Waves both at Flat and Rough Diffuse-Diffuse Interfaces, *J. Opt. Soc. Am. A* **16**, pp. 1947–1957.

Ripoll, J. and Nieto-Vesperinas, M. (1999b). Reflection and transmission coefficients for diffuse photon density waves. *Opt Lett* **24**, 12, pp. 796–798.

Ripoll, J., Nieto-Vesperinas, M. and Carminati, R. (1999). Spatial resolution of diffuse photon density waves, *J. Opt. Soc. Am. A* **16**, pp. 1466–1476.

Ripoll, J. and Ntziachristos, V. (2003). Iterative boundary method for diffuse optical tomography. *J Opt Soc Am A Opt Image Sci Vis* **20**, 6, pp. 1103–1110.

Ripoll, J. and Ntziachristos, V. (2004). Imaging Scattering media from a distance: theory and applications of non-contact optical tomography, *Modern Physics Letters B* **18**, 25, pp. 1403–1431.

Ripoll, J. and Ntziachristos, V. (2006). From finite to infinite volumes: removal of boundaries in diffuse wave imaging, *Phys Rev Lett* **96**, 17, p. 173903.

Ripoll, J., Ntziachristos, V., Carminati, R. and Nieto-Vesperinas, M. (2001a). Kirchhoff approximation for diffusive waves, *Phys Rev E Stat Nonlin Soft Matter Phys* **64**, 5 Pt 1, p. 51917.

Ripoll, J., Ntziachristos, V., Culver, J. P., Pattanayak, D. N., Yodh, A. G. and Nieto-Vesperinas, M. (2001b). Recovery of optical parameters in multiple-layered diffusive media: theory and experiments. *J Opt Soc Am A Opt Image Sci Vis* **18**, 4, pp. 821–830.

Ripoll, J., Schulz, R. B. and Ntziachristos, V. (2003). Free-space propagation of diffuse light: theory and experiments, *Phys Rev Lett* **91**, 10, p. 103901.

Ripoll, J., Yessayan, D., Zacharakis, G. and Ntziachristos, V. (2005). Experimental determination of photon propagation in highly absorbing and scattering media, *J Opt Soc Am A Opt Image Sci Vis* **22**, 3, pp. 546–551.

Roux, L., Mareschal, P., Vukadinovic, N., Thibaud, J. B. and Greffet, J. J. (2001). Scattering by a slab containing randomly located cylinders: comparison between radiative transfer and electromagnetic simulation. *Journal of the Optical Society of America. A, Optics, image science, and vision* **18**, 2, pp. 374–84.

Roychoudhuri, C. and Roy, R. (2003). The nature of light: what is a photon? *Optics and Photonics news* **3**, pp. S1–S35.

Rudin, M. (2005). *Molecular Imaging: Basic Principles and Applications in Biomedical Research*, 1st edn. (Imperial College Press), ISBN 1860945287.

Schmitt, J. M., Gandjbakhche, A. H. and Bonner, R. F. (1992). Use of polarized light to discriminate short-path photons in a multiply scattering medium, *Applied Optics* **31**, 30, pp. 6535–6546.

Schotland, J. C. and Markel, V. A. (2001). Inverse scattering with diffusing waves, *J Opt Soc Am A Opt Image Sci Vis* **18**, 11, pp. 2767–2777.

Shah, N., Cerussi, A., Eker, C., Espinoza, J., Butler, J., Fishkin, J., Hornung, R. and Tromberg, B. (2001). Noninvasive functional optical spectroscopy of human breast tissue, *Proc Natl Acad Sci U S A* **98**, 8, pp. 4420–4425.

Sheng, P. (1990). *Scattering and localization of classical waves in random media*, World Scientific series on directions in condensed matter physics (World Scientific), ISBN 9789971505394.

Sheng, P. (2006). *Introduction to wave scattering, localization and mesoscopic phenomena*, Springer series in materials science (Springer), ISBN 9783540291558.

Soubret, A., Ripoll, J. and Ntziachristos, V. (2005). Accuracy of fluorescent tomography in the presence of heterogeneities: study of the normalized Born ratio. *IEEE Trans Med Imaging* **24**, 10, pp. 1377–1386.

Spott, T. and Svaasand, L. O. (2000). Collimated light sources in the diffusion approximation. *Applied optics* **39**, 34, pp. 6453–65.

Tsang, L., Kong, J. A. and Ding, K.-H. (2000). *Scattering of Electromagnetic Waves: Theories and Applications* (John Wiley & Sons, Ltd).

Tuchin, V. V. (2007). *Tissue Optics: Light Scattering Methods and Instruments for Medical Diagnosis*, 2nd edn. (SPIE Publications), ISBN 978-0819464330.

Uhlenbeck, G. and Ornstein, L. (1930). On the Theory of the Brownian Motion, *Physical Review* **36**, 5, pp. 823–841, doi:10.1103/PhysRev.36.823.

Valeur, B. (2002). *Molecular Fluorescence: principles and applications* (Wiley-Interscience, New York), ISBN 3-527-29919-X.

van de Hulst, H. C. (1981). *Light scattering by small particles* (Dover Publ., New York).

van Veen, R. L. P., Sterenborg, H. J. C. M., Pifferi, A., Torricelli, A., Chikoidze, E. and Cubeddu, R. (2005). Determination of visible near-IR absorption coefficients of mammalian fat using time- and spatially resolved diffuse reflectance and transmission spectroscopy. *Journal of biomedical optics* **10**, 5, p. 054004, doi:10.1117/1.2085149.

Walker, S. A., Boas, D. A. and Gratton, E. (1998). Photon density waves scattered from cylindrical inhomogeneities: theory and experiments, *Applied Optics* **37**, 10, pp. 1935–1944.

Wang, L., Jacques, S. and Zheng, L. (1995). MCMLMonte Carlo modeling of light transport in multi-layered tissues, *Computer methods and programs in biomedicine* **47**, 2, pp. 131–146.

Wang, L. V. and Wu, H.-I. (2007). *Biomedical Optics: Principles and Imaging* (Wiley-Interscience, New Jersey), ISBN 978-0-471-74304-0.

Watson, D., Hagen, N., Diver, J., Marchand, P. and Chachisvilis, M. (2004). Elastic light scattering from single cells: orientational dynamics in optical trap. *Biophysical journal* **87**, 2, pp. 1298–306, doi:10.1529/biophysj.104.042135.

Watson, J. D. and Berry, A. (2003). *DNA: The secret of Life*, 1st edn. (Knopf).

Weissleder, R. and Mahmood, U. (2001). Molecular Imaging, *Radiology* , May, pp. 316–333.

Weissleder, R. and Ntziachristos, V. (2003). Shedding light onto live molecular targets. *Nat Med* **9**, 1, pp. 123–128, doi:10.1038/nm0103-123.

Wu, C., Mino, K., Akimoto, H., Kawabata, M., Nakamura, K., Ozaki, M. and Ohmiya, Y. (2009). In vivo far-red luminescence imaging of a biomarker based on BRET from Cypridina bioluminescence to an organic dye. *Proceedings of the National Academy of Sciences of the United States of America* **106**, 37, pp. 15599–603, doi:10.1073/pnas.0908594106.

Wyman, D., Patterson, M. and Wilson, B. (1989). Similarity relations for the interaction parameters in radiation transport, *Applied optics* **28**, 24, pp. 5243–5249.

Yao, G. and Wang, L. (2000). Propagation of polarized light in turbid media: simulated animation sequences. *Optics express* **7**, 5, pp. 198–203.

Index